THE AMERICAN WAY OF BOMBING

THE AMERICAN WAY OF BOMBING

CHANGING ETHICAL AND LEGAL NORMS, FROM FLYING FORTRESSES TO DRONES

Edited by Matthew Evangelista and Henry Shue

CORNELL UNIVERSITY PRESS
Ithaca and London

Copyright © 2014 by Cornell University

All rights reserved. Except for brief quotations in a review, this book, or parts thereof, must not be reproduced in any form without permission in writing from the publisher. For information, address Cornell University Press, Sage House, 512 East State Street, Ithaca, New York 14850.

First published 2014 by Cornell University Press
First printing, Cornell Paperbacks, 2014

Printed in the United States of America

Library of Congress Cataloging-in-Publication Data

The American way of bombing : changing ethical and legal norms, from flying fortresses to drones / edited by Matthew Evangelista and Henry Shue.
     pages cm
  Includes bibliographical references and index.
  ISBN 978-0-8014-5280-2 (cloth : alk. paper)
  ISBN 978-0-8014-7934-2 (pbk. : alk. paper)
  1. Bombing, Aerial—United States. 2. Bombing, Aerial—Moral and ethical aspects. 3. Drone aircraft—Government policy—United States. I. Evangelista, Matthew, 1958– editor of compilation. II. Shue, Henry, editor of compilation. III. Biddle, Tami Davis, 1959– Strategic bombardment Contains (work):
  UG703.A74   2014
  358.4'24—dc23   2014006282

Cornell University Press strives to use environmentally responsible suppliers and materials to the fullest extent possible in the publishing of its books. Such materials include vegetable-based, low-VOC inks and acid-free papers that are recycled, totally chlorine-free, or partly composed of nonwood fibers. For further information, visit our website at www.cornellpress.cornell.edu.

Cloth printing   10 9 8 7 6 5 4 3 2 1
Paperback printing 10 9 8 7 6 5 4 3 2 1

Contents

Acknowledgments vii

Introduction: The American Way
of Bombing
MATTHEW EVANGELISTA 1

Part I. Historical and Theoretical Perspectives

1. Strategic Bombardment: Expectation,
   Theory, and Practice in the Early
   Twentieth Century
   TAMI DAVIS BIDDLE 27

2. Bombing Civilians after World War II:
   The Persistence of Norms against
   Targeting Civilians in the Korean War
   SAHR CONWAY-LANZ 47

3. Targeting Civilians and U.S. Strategic
   Bombing Norms: Plus ça change, plus
   c'est la même chose?
   NETA C. CRAWFORD 64

4. The Law Applies, But Which Law?
   A Consumer Guide to the Laws of War
   CHARLES GARRAWAY 87

Part II. Interpreting, Criticizing, and Creating Legal Restrictions

5. Clever or Clueless? Observations
   about Bombing Norm Debates
   CHARLES J. DUNLAP JR. 109

6. The American Way of Bombing and International Law: Two Logics of Warfare in Tension
   JANINA DILL   131

7. Force Protection, Military Advantage, and "Constant Care" for Civilians: The 1991 Bombing of Iraq
   HENRY SHUE   145

8. Civilian Deaths and American Power: Three Lessons from Iraq and Afghanistan
   RICHARD W. MILLER   158

PART III. CONSTRUCTING NEW NORMS

9. Proportionality and Restraint on the Use of Force: The Role of Nongovernmental Organizations
   MARGARITA H. PETROVA   175

10. Toward an Anthropology of Drones: Remaking Space, Time, and Valor in Combat
    HUGH GUSTERSON   191

11. What's Wrong with Drones? The Battlefield in International Humanitarian Law
    KLEM RYAN   207

12. Banning Autonomous Killing: The Legal and Ethical Requirement That Humans Make Near-Time Lethal Decisions
    MARY ELLEN O'CONNELL   224

Notes 237

List of Contributors 301

Index 303

Acknowledgments

The papers commissioned for this project were first intensively discussed at a workshop at Cornell University in April 2011. We are grateful to two Cornell institutions for providing support—the Mario Einaudi Center for International Studies for a seed grant and the Judith Reppy Institute for Peace and Conflict Studies for workshop funding and a publishing subvention from its MacArthur Foundation grant—and their directors and staff, including Fred Logevall, Heike Michelsen, Jonathan Kirshner, Judith Reppy, Elaine Scott, and Sandra Kisner. Sandy deserves a special note of thanks for her meticulous preparation of the manuscript for publication.

Owing to limitations of space, we were unable to include all of the papers presented at the workshop, but we are grateful for the contributions, published elsewhere, of Martin Cook and of Sarah Kreps and John Kaag. Other workshop participants who provided valuable comments include Charli Carpenter, Brian Cuddy, Fred Logevall, Muna Ndulo, Jens David Ohlin, and Judith Reppy. We are particularly indebted to Karl Mueller and an anonymous reader for Cornell University Press for their incisive and thorough reviews of all of the chapters. Michael McGandy, our editor at the Press, was an enthusiastic supporter of the project from the outset and we are grateful to him and his colleagues—including Karen Laun, Sarah Grossman, and copyeditor Kate Babbitt—for shepherding it through to completion.

THE AMERICAN WAY OF BOMBING

Introduction

The American Way of Bombing

MATTHEW EVANGELISTA

Aerial bombardment as a form of warfare is just about 100 years old and is showing no signs of decline. In the twenty-first century the United States has deployed airpower in military missions ranging from the wars in Afghanistan and Iraq to the "humanitarian intervention" in Libya and the surveillance and attacks on suspected terrorists by remotely piloted drones in Pakistan, Yemen, Somalia, and elsewhere. The norms governing bombing—and particularly the harm it imposes on civilians—have evolved considerably over a century: from deliberate attacks against rebellious villagers by Italian and British colonial forces in the Middle East to institutionalized practices seeking to avoid civilian casualties in the U.S. counterinsurgency and antiterrorist wars of today. In between, the strategic bombing campaigns of World War II caused great civilian destruction through fire-bombing of cities and, ultimately, the atomic attacks against Hiroshima and Nagasaki. What accounts for the dramatic changes in ethical and legal norms governing air warfare over time?

This volume brings together prominent military historians, practitioners, legal experts (civilian and military), political scientists, philosophers, and anthropologists to explore the sources of the evolution of bombing norms. The authors focus mainly on the United States—the world's preeminent military power and the one most frequently engaged in air warfare—because we expect that its behavior has influenced normative change in this domain and will continue to do so. The United States sets the standard for bombing practices and remains the focus of efforts to change those practices. Even as observers note an overall decline in interstate war since the end of the Cold War, the role of air power has not diminished. Moreover, with the proliferation of remotely piloted aerial vehicles—known colloquially as "drones"—air power has taken on new forms. Several of our authors address the challenges that drones, as well as more thoroughly automated killing machines, pose to the existing legal regime.

On 15 August 2010, the possibility that NATO jets accidentally killed five Afghan civilians merited prominent coverage in the *New York Times*.[1] Such media attention to civilian harm had become a commonplace during the U.S.-led wars of the early twenty-first century. The vastly more destructive air campaigns of World War II also prompted press comment on damage to civilians, especially in the early days, and protestations from political and military leaders on all sides that civilians were not the intended targets of bombing raids. As Sahr Conway-Lanz points out in his contribution to this volume, officials "continued to talk as if they were trying to uphold the prohibition against targeting civilians, even though the reality of civilian deaths strained the credibility of their claims."[2] Moreover, statements by Allied leaders to the effect that bombing hastened the end of the war by undermining civilian morale cast doubt on any commitment to protect civilians. Even in more recent wars—during the NATO air campaign against Serbia in 1999, for example—one could hear in the press conferences and congressional testimony of high-level political and military leaders claims for the beneficial "morale effect" of bombing. Yet such claims are fewer and less explicit than during World War II.

At first glance, the increasing public attention to "collateral damage" (as the unintentional harm to civilians is known) seems paradoxical. The advances in accuracy of modern weapons and the apparent concern—at least within the U.S. military command—for scrupulous adherence to the laws of war in targeting have done little to mute criticism of U.S. bombing practices. In the wars led by the United States in Iraq and Afghanistan, no one can say with any certainty how many civilians have died. The fact that estimates range from the tens of thousands to more than a million—and that the U.S. government evidently does not keep close track of the figures—makes one wonder whether enough is being done to prevent harm to civilians in modern wars.[3] Other countries' bombing practices also raise questions about protecting civilians.[4]

Superficially at least, the use of U.S. air power in Afghanistan or Iraq looks very different from its use in World War II. Much has indeed changed: military technology has advanced (munitions themselves, information technology and sensors, aircraft); the body of relevant international law has grown (the Geneva Conventions of 1949, the Geneva Protocols of 1977, and specific treaties banning land mines and cluster munitions); domestic implementation of international law, at least in some domains, has increased, with more emphasis on the role of military lawyers—judge advocates general (JAGs)—in target review, for example; and nongovernmental organizations and civil society groups have sought to codify and expand the restraints on warfare.[5]

The contributors to this volume, despite their diverse backgrounds and attitudes toward air power, all agree that government and popular attention to the legal requirements for protecting civilians has increased over the past decades. Yet far more interesting are their disagreements. Some authors, particularly those with military affiliations, find that the laws of armed conflict do not in fact pose many specific constraints on states' ability to use air power but that the United States has adhered to what constraints exist, subject to the limitations of technology. As the technical capability has evolved to identify and attack targets with more discrimination and precision, the United States has improved its compliance with its legal obligations. It may be, these authors suggest, that improvements in technology have encouraged critics of U.S. military power to raise the standards of compliance beyond what is reasonable or militarily feasible.

Other authors disagree about both what the law requires and how well the United States has met its legal obligations since the advent of air power for military purposes. To the extent that they perceive a trend toward greater protection for civilians, they attribute that trend to popular pressure and the strengthening of international legal norms. In their view, technology has done more than generate the means to make bombing more discriminating. In some cases—cluster munitions, for example—the technological advance has posed greater risks to civilians and has necessitated political action to bring it under control. Several of the contributors focus on the threats posed by new technologies—remotely piloted drones and fully automated weapons, for example—and the popular campaigns that have arisen to address them. Some authors make specific proposals for interpreting and improving the existing body of law, highlighting areas where official U.S. interpretation of its obligations has fallen short of what is required for adequate protection of civilians.

U.S. political and military leaders claim—and many analysts agree—that the United States takes unprecedented caution to prevent civilian harm in its air campaigns and overall military strategy, an evident departure from the vast civilian destruction of World War II.[6] What accounts for the change?

Why did the United States conduct the bombing of Iraq in 1991, for example, so differently from the bombing of North Korea in the 1950s? Have developments in international law—the First Geneva Protocol of 1977, for example—made a difference for U.S. practice? Or should one argue the United States at the turn of the millennium has simply been fighting different wars rather than fighting wars differently?

One purpose of this book is to try to identify some pattern for the changes that have occurred and evaluate plausible explanations for those changes. We consider such factors as the evolution of international humanitarian law (laws of war); changes in technology; the emergence of new types of wars, which include different adversaries and goals; changes in domestic public opinion; and the influence of "global civil society."

If changing international law matters, how does it matter? The practical legal and military experience of many of our participants allows focus on questions that have received little attention in the public or specialist literature. For example, how much difference does the routine review of targets by JAGs actually make? More fundamentally, how does the very definition of "military objective" influence targeting? The First Geneva Protocol states that a military objective should "make an effective contribution to military action," a phrase that the United States tends to give a broader reading as "effectively contribute to the enemy's war-fighting or war-sustaining capability." How much of the civilian economy is included in "war-sustaining capability"?

In part as a reaction to what they consider excessively permissive definitions of military objective, scholars and practitioners have made efforts to revise the relevant guidelines, as in the Harvard-based project on International Humanitarian Law in Air and Missile Warfare, discussed by Charles Garraway in his chapter. Within the U.S. Department of Defense, sensitivity to civilian casualties in the context of the war in Afghanistan and wars of "counterinsurgency" in general led to changes in policy associated with Generals Stanley McChrystal and David Petraeus. Finally, the use of drone aircraft offers the promise of greater precision in attacks but still comes in for criticism for unintentional civilian casualties. Moreover, the operation of the drones by the Central Intelligence Agency and private contractors raises basic concerns about compliance with the laws of war. The widespread use of drones has fundamentally affected the spatial and temporal dimensions of modern warfare.

These are among the topics our authors take up in the chapters that follow. This introduction puts their contributions into broader historical and legal context and highlights the points of engagement and disagreement among the contributors. The disagreements are substantial enough that we should acknowledge in advance that we do not fulfill the promise of our title by identifying *the* American way of bombing. Most of our authors might agree that the extent to which law infuses U.S. bombing norms and practices is distinctively American, but beyond that, disagreements on interpretation and implications abound.

What We Mean By Norms

The authors of this volume share a common interest in how the norms governing aerial bombardment have changed, do change, and should change—with some authors focused more on *explaining* change and others more on *advocating* particular changes. What do we mean by *norms*? In an earlier work, Neta Crawford, one of our authors, offered a distinction that we find useful between norms, normative beliefs, and prescriptive norms. For Crawford, norms are descriptions of dominant practices or behaviors—what is considered "normal" within a given community. "Principles, rules, and laws are prescriptive normative statements that rest on *normative beliefs.*" Thus, normative beliefs serve to underpin or justify norms. Prescriptive norms are claims "about what dominant practices ought to be."[7] In this volume, the authors try to distinguish between norms as practices and norms as prescriptive standards of behavior. In doing so, however, we must recognize that conventional language often uses *norm* to mean *prescriptive norm*. For example, the concept, increasingly popular in the study of international politics, of a "norm entrepreneur" only makes sense if we intend *prescriptive norm*. Norm entrepreneurs are typically not promoting existing practices or behavior (the status quo) but are advocating new standards or the strengthening or expanding of existing standards.

There is inevitably some overlap in our language, and we see it reflected in the broader discussion of law as well. Consider the comment in a standard textbook on international law: "International Law does not necessarily consist of what a number of States might actually do. Rather, it is a blend of their respective expectations and their actual practice."[8] This observation applies in particular to *customary* law, which depends on the concept of *opinio iuris*—that states believe that certain practices are obligated by law, even if they sometimes stray from those practices. In Crawford's terms, customary law consists of norms that states follow out of a sense of legal obligation. Because *opinio iuris* is to some degree subjective it can be influenced by such factors as predominant legal opinion of specialists—including those who do not represent governments.[9] In 2005, for example, the nongovernmental

International Committee of the Red Cross (ICRC) compiled a 5,000-page study identifying 161 rules governing warfare (including bombing) that, in its view, had attained the status of customary law: each state—according to the ICRC's interpretation—was obliged to obey them, even if it had not signed specific treaties agreeing to do so.[10] The ICRC exercise, even if described as a codification of existing law, can also be understood as a form of advocacy—trying to influence states' beliefs about their legal obligations as *opinio iuris*.

The authors of this volume also seek to influence beliefs about the past evolution of norms governing aerial bombardment, and some explicitly seek to shape those norms for the future.

Two ethical-legal concepts are most relevant to our discussion: the principle of *distinction* (or discrimination) holds that only combatants should be deliberately targeted. Noncombatants—civilians and wounded or captured soldiers—should be protected from direct attack. Much of the controversy surrounding this principle—the topic of considerable discussion by our authors—concerns just who counts as a civilian or a combatant. The principle of *proportionality* holds that the military benefit expected from an attack should outweigh the damage to noncombatants. Several of our authors parse the legal treaty language relevant to the question of what constitutes a legitimate military objective—an issue highly consequential to the fate of civilians caught in war. The principles of distinction and proportionality lie at the heart of just war theory's guidelines for conduct in war (*ius in bello*). Just war theory, in turn, provides much of the basis for what specialists variously call international humanitarian law or the law of armed conflict (LOAC), commonly known as the laws of war.

Why Bombing, and Why the United States?

Readers may wonder why our focus on *bombing* and why we devote so much attention to the historical and contemporary practices of the *United States*. After all, civilians risk harm from warfare in many forms—as inadvertent or deliberate casualties of attacks from ground or naval forces; indirectly through damage to or deprivation of supplies of food and water, sanitary and electrical systems, and medical care; or as victims of sanctions, sieges, and forced migrations. And the United States is hardly the only country whose behavior in war merits scrutiny. Yet, aside from the obvious inability to cover everything in a single volume, we justify our choice on a number of grounds.

First, large numbers of civilians killed directly by military action during the previous century were victims of aerial bombardment (even if many more civilians died from dislocation, starvation, and illness indirectly attributed to war).[11] Thus, bombing lends itself especially well as a subject for legal and ethical analysis related to protection of civilians.

Second, the laws governing targeting for air warfare are relatively underdeveloped compared to other areas, such as treatment of prisoners of war, and there are many controversial issues to debate.[12] Confronting the topic of harm to civilians, our authors reflect major disagreements in the broader community of scholars and practitioners about both history and contemporary practices. In particular, we disagree about the extent to which civilians have been, are, or should be targets of bombing. To be sure, few observers argue that states should deliberately bomb civilians. Yet the tolerance that some observers express for allowing civilians—especially the subjects of particularly odious and belligerent regimes—to suffer the consequences of their governments' engagement in war looks to other observers like insufficient concern for the principles of distinction and proportionality.

Third, given our interest in what shapes the norms of aerial bombardment, it makes sense to focus on the United States—the world's preeminent military power and the one most frequently engaged in air warfare—because we expect that its practice has influenced normative change in this domain and will continue to do so. Other countries, such as Russia and Israel, have engaged in air campaigns in recent years, and brutal dictators in Libya and Syria have set the normative clock back a century with the punitive bombing of (their own) civilians. But we expect the United States to continue to set the standard for bombing practices and to remain the focus of efforts to change those practices. As a leading scholar of customary international law explains, "It is generally easier for more powerful states to engage in behavior which will significantly affect the maintenance, development, or change of customary rules than it is for less powerful states to do so."[13] The editor of the ICRC study of customary international humanitarian law makes a similar point: "It is not simply a question of how many States participate in the practice, but also which States"—with those "whose interests are specially affected" most relevant to the establishment of customary rules.[14]

Fourth, we find that even as observers note an overall decline in interstate war since the end of the Cold War, the role of air power has not diminished. High-profile examples of so-called humanitarian intervention—the use of military force ostensibly to protect civilians, as in Kosovo in 1999 or Libya in 2011—depend heavily on bombing. The paradox of humanitarian intervention—that states will be reluctant to risk their own citizen-soldiers (and pilots) for the altruistic purpose of "saving strangers"—leads to a preference for bombing over deploying boots on the ground.[15] Given that the

United States has led the way in organizing the air campaigns at the heart of recent interventions, we find our emphasis on bombing and on the United States doubly justified.

Historical and Theoretical Perspectives

The early theorists of air power, such as Giulio Douhet, Billy Mitchell, and Hugh Trenchard, expected that civilians would be—and should be—the primary targets of air attack. Both Douhet and Mitchell rejected the principle of distinction between combatants and civilians. In a memorandum from 1919, Mitchell, one of the pioneers of U.S. air power, argued that "the entire nation is, or should be, considered a combatant force."[16] Douhet, an Italian general, argued in his 1921 treatise *The Command of the Air* that owing to the vulnerability of civilians to air attack, "there will be no distinction any longer between soldier and civilian."[17] With the experience of World War I fresh in mind, Douhet and the others expected that directly targeting civilians would shorten a war and that future wars themselves would consequently become more humane than the mutual slaughter of trench warfare.[18] World War II disproved the air theorists' predictions, as both sides fought on for years despite the devastation that aerial bombardment wrought on its cities. Tami Davis Biddle's chapter in this volume provides an expert overview of the theories of air power associated with Douhet, Mitchell, Trenchard, and others.

The earliest use of air power during the first decades of the twentieth century presaged some of the most vexing issues that would continue to confront states for the next hundred years. The armed forces of European states first employed aerial bombardment not to fight among themselves but to maintain domination over colonies in Africa and the Middle East. The Italians were the first to use air power in this fashion, dropping grenades on villagers, some of whom were armed, in Libya in 1911 in the course of war with Ottoman Turkey.[19] If it had ever been made, the distinction between rebels and the populations that supported them soon began to blur, as the British example demonstrates.[20] During the period 1919–1922 Winston Churchill served as Britain's secretary of state for war, secretary of state for air, and secretary of state for the colonies, in which role he was tasked with enforcing order among people who resisted British rule. One of the tools he advocated was aerial bombardment of tribal areas by poison gas, particularly in Iraq but also in India and Afghanistan—even when his advisors warned him that it could "kill children and sickly persons." "I am strongly in favour of using poison gas against uncivilised tribes," wrote Churchill to Trenchard, chief of the Air Staff. In the event, the British used aerial bombardment against many villages in Kurdistan and gas against Iraqi rebels (although not delivered by air) with, in Churchill's words, "excellent moral effect."[21]

The laws of war relevant to aerial bombardment were rather undeveloped at the outbreak of World War II. The Hague Conventions, which were adopted in 1899 and revised in 1907, prohibited "attack or bombardment of towns, villages, habitations or buildings which are not defended" (Article 25) and required the attackers to warn the relevant authorities on the other side in advance (Article 26) and to take all necessary steps "to spare as far as possible edifices devoted to religion, art, science, and charity, hospitals, and places where the sick and wounded are collected, provided they are not used at the same time for military purposes" (Article 27). These prohibitions applied to the signatories of the Conventions only and only in their interactions with each other, not in their colonial territories.[22]

The criterion that a population center be undefended in order to be spared bombardment left a rather large loophole, as the presence of any troops or military facilities might disqualify it. Nevertheless the onus against intentional killing of civilians was evident in the reactions of British and U.S. leaders to the German air campaign of 1939 and the Japanese attacks in Manchuria. Churchill, as Britain's prime minister, condemned Hitler's bombing of Warsaw and Rotterdam as "a new and odious form of attack"—although it was not in principle much different from Britain's bombing of Kurdish villages under Churchill's command a couple of decades earlier—and vowed that his government would not "bomb nonmilitary objectives, no matter what the policy of the German Government may be." The U.S. government issued a statement in response to Japan's bombing campaign, reinforcing its view that "any general bombing of an extensive area wherein there resides a large populace engaged in peaceful pursuits is unwarranted and contrary to the principles of law and humanity." At the outbreak of World War II, President Franklin Roosevelt invoked both the legal prohibition of the Hague Conventions and the broader moral principle of civilian immunity when he addressed "an urgent appeal to every government which may be engaged in hostilities publicly to affirm its determination that its armed forces shall in no event and in no circumstances, undertake the bombardment from the air of civilian populations or of unfortified cities."[23]

The British and other colonial powers appeared to hold two standards regarding killing civilians by air—one for "civilized" peoples and another for everybody else. At the 1932 World Disarmament Conference, the British delegation was willing to restrict aerial bombardment but not "the use of such machines as are necessary for police purposes in outlying places," that is, in the colonies. Employing a common racial epithet, Prime Minister David Lloyd George "insisted on the right" to bomb the native inhabitants.[24]

By the early 1940s, restraints even among the "civilized" belligerents had weakened as the Allies came to depend on air power as a key element of their strategy. From initially forswearing deliberate attacks on civilians, both Britain and the United States came increasingly close to justifying intentional targeting of the enemy population to destroy its morale. A 1943 joint U.S.-British operational plan, for example, envisioned "the progressive destruction and dislocation of the German military, industrial, and economic system, and the undermining of the morale of the German people to a point where their capacity for armed resistance is fatally weakened."[25]

The fire-bombing of cities such as Hamburg, Dresden, and Tokyo, which resulted in tens of thousands of victims for each, made no distinctions between civilians and combatants. Nor did the atomic bombings of Hiroshima and Nagasaki—which were unusual not for the number of victims but for the fact that they were killed by dropping only a single bomb over each city.[26] Allied leaders appeared to disregard completely the distinction between combatants and civilians. Some evidence conveys the impression that they sought to punish the latter for supporting aggressive dictatorships. This impression is strongest in Churchill's pronouncements, as when in 1943 he told the U.S. Congress to "begin the process so necessary and desirable of laying the cities and the other military centres of Japan in ashes, for in ashes they must surely lie before peace comes to the world."[27] Franklin Roosevelt was more circumspect and has left less direct evidence. He was apparently not consulted, for example, on the fire-bombing of Dresden in February 1945.[28] Earlier, however, the president had requested of Secretary of War Henry Stimson "some indication of the psychological and morale effect" on the population of a city burdened by an influx of evacuees from bombing campaigns. Ronald Schaffer argues, in part on the basis of this request, that "anyone conversant with the morale bombing controversy could understand" that "attacks aimed at terrorizing enemy civilians were acceptable to the commander in chief."[29]

Even when the military strategies rendered it impossible to discriminate between civilian and military objects in practice, Allied leaders appeared to want to do so in their intentions. Consider the remarks of President Harry Truman in the wake of the obliteration of Hiroshima: "The world will note that the first atomic bomb was dropped on Hiroshima, a military base. That was because we wished in this first attack to avoid, insofar as possible, the killing of civilians."[30] Even in the wake of the indiscriminate destruction of Hiroshima and Nagasaki, where many tens of thousands of civilians perished, these same leaders acknowledged the continuing hold of *normative beliefs* about avoiding civilian harm. Yet the *norm*—the dominant practice—of strategic bombing of cities rendered that goal unachievable.

If the strategic bombing of World War II represented a nadir in protection of civilians from deliberate harm, what of the wars that followed? In this volume Neta Crawford takes up this question in her overview of changes over time in U.S. bombing strategy. She finds considerable continuity through the U.S. wars in Korea and Vietnam in attitudes toward killing civilians, although she identifies the Vietnam bombing campaign as a kind of turning point. Political and military leaders became increasingly sensitive to public concern about civilian casualties, a point that Richard Miller also makes in his comparison in this volume of the U.S. wars in Afghanistan and Iraq to the experience of Vietnam.

Much of our understanding of the development of U.S. bombing practices in the wake of World War II comes from the painstaking archival research of Sahr Conway-Lanz. In this volume, he contributes a chapter that provides both a detailed reconstruction of bombing during the Korean War and a clue to the paradox that characterizes much of the U.S. military and political thinking about bombing. He writes of the Korean conflict that "American leaders continued to claim throughout the war that U.S. air power was being used in a discriminate manner and was avoiding harm to civilians as they had asserted even during the height of the bombing in World War II." He describes how the expanding capabilities of U.S. bombers brought new categories of plausible military targets within range—"transportation networks, arms factories, and to their workers" located in cities deep behind enemy lines. He also calls attention the "dynamics of escalation," as U.S.-led forces became bogged down and increased the level of destruction as a way to force a decision—a phenomenon observed by other scholars in a range of cases.[31] For U.S. leaders and the public, according to Conway-Lanz, "the crucial dividing line between justifiable and unjustifiable violence increasingly became whether their armed forces intentionally harmed civilians."

The prosecution of the Korean War coincided with the buildup of a U.S. arsenal of atomic bombs and the development of the vastly more destructive thermonuclear weapons. The emerging strategy of nuclear deterrence—seeking to forestall an act by threat of nuclear retaliation—arguably works only by invoking the fear of mass civilian casualties. Yet political and military authorities claimed to uphold the principle of noncombatant immunity by insisting that their *intention* was not to kill civilians. Although early U.S. plans for war against the Soviet Union, for example, entailed attacking the single city of Moscow with seventy atomic bombs, Defense Department officials insisted that the targets—military, economic, and political—were strictly related to the Soviet ability to prosecute the war. The United States did not target civilians per se, even though millions of them would die—and the expectation of those deaths was presumably what was supposed to deter Soviet aggression. As Conway-Lanz points out in this volume and elsewhere, focusing on intentions rather than behavior serves as a way of reconciling putative American values with actions that clearly violate them.[32] It is possible, however, that the cognitive dissonance was just too great between the claim not to intend harm to civilians and nuclear strategies that would lead to the death of millions. This predicament, some have argued, helps explain the fact that nuclear weapons have not been used in war since the destruction of Hiroshima and Nagasaki.[33] Unfortunately, space limitations preclude us from discussing what U.S. nuclear weapons policy implies for arguments about how norms governing bombing evolve. At a minimum, the continued U.S. reliance on the threat of mass-casualty nuclear retaliation for a range of purposes beyond deterrence of a nuclear attack against the United States should call into question any notion of an ironclad commitment to protect civilians in war.[34]

The question of intention looms large in the chapters by both Conway-Lanz and Crawford. Although they agree on some basic points, their arguments differ in subtle ways. Crawford's distinction between professed normative beliefs and dominant practices helps show how. Both authors argue that bombing practices since the end of the Vietnam War diverge considerably from the vast civilian destruction wrought in World War II and Korea. During those wars American military forces killed many civilians through bombing, and this number declined substantially in U.S. wars after Vietnam. They disagree on the question of how professed normative beliefs changed. Crawford suggests that U.S. leaders considered targeting civilians morally justified before Vietnam and changed their minds after Vietnam. Conway-Lanz remains unconvinced that currently available evidence supports such a "progress narrative," in part because we lack access to internal government and military deliberations about bombing policy for the more recent period. His chapter suggests an alternative explanation for why Americans have killed fewer civilians in recent wars: such wars have been highly asymmetrical, posing little threat to U.S. forces and none to the U.S. homeland. The elements that scholars have identified as those that drive states to target civilians—desperation to reduce military casualties or avert defeat—are both absent in recent wars. Thus, to use the distinction raised at the beginning of this chapter, the view of Conway-Lanz is closer to the position that the United States is "fighting different wars" in the late twentieth and early twenty-first centuries, whereas Crawford's is closer to the position that the United States is "fighting differently."[35] Crawford's focus on the role of *institutionalization* of practices to protect civilians during bombing implies that such practices should carry across types of war. The true test of her claim in a "total war" of desperation we hope never to see.

Interpreting, Criticizing, and Creating Legal Restrictions

Much of the controversy surrounding the question of intentional harm to civilians stems from a long-standing belief among air power enthusiasts—dating back to Douhet and Mitchell—that undermining popular morale contributes to military victory. Yet unlike the early theorists, their modern-day counterparts—aware of the international legal prohibitions—rarely argue for deliberately targeting civilians. To be sure, they have complained about undue concern for civilian casualties. In March 1968, during the Vietnam War, Secretary of the Air Force Harold Brown argued for permitting "bombing of military targets without the present scrupulous concern for collateral damage and casualties."[36] In a retrospective account of the Rolling Thunder air campaign in Vietnam, W. Hays Parks refers to "a paranoiac fixation with regard to any civilian casualties" among the U.S. civilian leadership.[37] In our volume, Charles Dunlap describes current NATO policy regarding civilian casualties as "misguided." He cites claims by its spokespeople that NATO would not fire on positions if it "knew there were civilians nearby" or "if there is the likelihood of even one civilian casualty." Dunlap observes that "the law certainly does not require such extraordinary measures to avoid civilian casualties—and for good reason." These are debates mainly about what the principle of proportionality requires.

Other critics focus on the related question of what constitutes a legitimate military target. They wonder whether promotion of legal definitions of *military objectives*, intended to limit damage to civilians, go too far in prohibiting military practices crucial for success in war. Much attention, including in our volume, focuses on Article 52(2) of Protocol I (1977) to the 1949 Geneva Conventions:

> Attacks shall be limited strictly to military objectives. In so far as objects are concerned, military objectives are limited to those objects which by their nature, location, purpose or use make an effective contribution to military action and whose total or partial destruction, capture or neutralization, in the circumstances ruling at the time, offers a definite military advantage.[38]

The United States never ratified the 1977 Geneva Protocols, but it considers the restraints on targeting articulated in Protocol I to constitute customary law, by which U.S. policy is bound.[39]

Nevertheless, numerous advocates of U.S. air power chafe at the restrictions of Article 52(2). Parks, for example, likens the Article to the failed attempt to provide Rules for air warfare at The Hague in 1923. According to Parks, states' "condemnation of the rules was virtually unanimous," and they "drifted into obscurity, adopted by no nation, and completely ignored by most aviation historians"—mainly because the Hague Rules were inconsistent with state practice, and "the law of war is based upon the practice of nations."[40] He prefers British air marshal Trenchard's definition of military objectives as "any objectives which will contribute effectively towards the destruction of the enemy's means of resistance and the lowering of his determination to fight."[41] Parks summarizes favorably a 1930 legal assessment of the customary norms of air warfare based on the experience of World War I. It claimed that during that war, "demoralization of the enemy by means of widespread bombardment was accepted by the military services as part of the functions of the aviation bombardment groups, as it was for artillery."[42] In our volume, Charles Garraway, in an overview of the development of the laws of war, likewise takes up the question of the role of the 1923 Hague Rules of Aerial Warfare. For him, "the question that has to be asked is whether these Rules were not only ahead of their time in legal terms but also ahead of the technology available at that time." In an account more nuanced than Parks's, Garraway writes that although "a number of states were fundamentally opposed to the Rules, a number of others stated that they would seek to comply with them." In any event, World War II demonstrated the "weakness of reliance on unenforceable standards in the crucible of conflict."

The focus on civilian morale as the weak point in a country's ability to wage war served to justify the Allied strategic bombing campaigns of World War II.[43] It was clearly not the only or primary focus of the bombing campaigns, whose contributions to defeating the German war effort are summarized in Dunlap's chapter. The findings of the U.S. Strategic Bombing Survey, conducted during the waning days of the war, do, however, cast doubt on the effectiveness of seeking to undermine morale, as opposed to direct targeting of, for example, oil production.[44] Nevertheless, as Conway-Lanz's chapter demonstrates, leading U.S. military figures continued to be drawn to the psychological effects of bombing in the wake of World War II.

Conway-Lanz concludes his chapter by wondering whether "American commanders in recent wars have resisted the temptations" to which their predecessors "succumbed of justifying bombing attacks for their political and psychological effects instead of for their directly military impact." Crawford's chapter provides evidence that the link between civilian morale and a state's ability to wage war continued to attract U.S. military planners throughout the postwar period. She quotes, for example, from a study of the possibilities that air power could dislodge Iraqi forces from Kuwait in August 1991 by inflicting damage not only on the troops in the field but also on the population at home in Iraq: "The psychological impact on the Iraqi populace of being open to unremitting air attacks will be a powerful reminder of the bankruptcy and impotence of the Saddam Hussein regime."[45] Affecting civilian morale was an important objective of the military leaders of NATO's air war against Serbia over Kosovo in March 1999, even if few civilians were killed compared to the casualties of the bombing practices of an earlier era. Charles Dunlap, among many others, considers the Kosovo war a successful example of the coercive use of air power for political purposes—a model for such subsequent actions as NATO's bombing campaign of twelve years later to impose a no-fly-zone over Libya.[46]

Dunlap's discussion of what he calls the "legal cacophony" governing aerial bombardment evinces some sympathy for the position that civilian morale is a legitimate focus in war: "Experience shows that the erosion of the 'will' of an adversary through the *indirect* effects of aerial bombardment on civilians is a key element of victory in modern war." In his view, "the law, properly applied, recognizes that incidental civilian deaths in connection with an otherwise bona fide attack on a military target are acceptable and often *expected*." He devotes considerable attention to questioning the concept of civilian innocence. "It is typically impossible," he writes, "in the midst of military operations to determine which civilians are authentically 'innocent.'" He describes as "deeply flawed" the "notion that combatants are legally or morally obliged to take more risk than those holding civilian status." In his view, "the life of the civilian—'innocent' or not—is not intrinsically more worthy than that of the combatant."[47]

In his chapter, Henry Shue takes on precisely this issue. He agrees with Dunlap to a point: "Not only is there nothing disreputable about a commander's attempting to protect the lives of the members of the military force, it is a commander's duty to avoid all unjustified risks." Contrary to Dunlap, however, Shue argues that we may "reasonably expect soldiers to run greater risks than civilians" because, owing to their military training and their weapons, "soldiers are better prepared in multiple ways to survive the risks, while civilians are typically vulnerable and largely defenseless." Thus, "one does not choose between the protection of the civilians and the safety of the troops—one holds on to both, but with a finger on the scales on the side of civilians. The real challenge always is to strike the proper balance."

Shue's chapter focuses on the 1991 war against Iraq, led by the United States, in which air power played a major role. He draws upon the relevant Articles of the First Geneva Protocol, particularly Article 57(3), which he understands to have customary status in international law. He describes the U.S. aerial destruction of the entire Iraqi electrical grid as a violation of the norms of proportionality, an excessive favoring of force protection and military advantage over due care for civilians. Shue quotes the Pentagon's *Final Report* to Congress justifying the decision: "Disrupting the electricity supply to key Iraqi facilities degraded a wide variety of crucial capabilities, from the radar sites that warned of Coalition air strikes, to the refrigeration used to preserve biological weapons (BW), to nuclear weapons production facilities. To do this effectively required the disruption of virtually the entire Iraqi electric grid, to prevent the rerouting of power around damaged nodes." As Shue points out, "if one turns off the refrigeration for biological weapons and radar sites by disabling the entire grid, one also turns off the electricity for perishable medicines, perishable foods, water pumping, water purification, sewage treatment, operating rooms in hospitals, and all the rest of civilian life."

Shue's account of the 1991 Gulf War highlights the long-term consequences to civilians of U.S. bombing choices. Such an account, and the further details provided in Miller's chapter, would seem to qualify Neta Crawford's cautious optimism about the institutionalization of protections to civilians which she found amply demonstrated in that war. Shue quotes the 1997 critique of the (unrelated) Commander J. W. Crawford of the U.S. Naval War College: "Targeting considerations must extend beyond direct effects. Collateral damage, by legal definition, must include a requirement to examine the reverberative effects of military action." In his view, "The devastation suffered by Iraq as a result of systematic targeting of its national electric power system is the best evidence, in the modern era, of the enormous destructive impact that aerial attacks can have on noncombatants."[48] On the more positive side, Shue cites evidence that Commander Crawford's notion of incorporating the "reverberative effects" of attacks into the norm of proportionality "is gaining a welcome foothold in different types of official U.S. military documents." According to Shue, "This is an understanding of proportionality that ought to change permanently."

Even if we are seeing some convergence on how to adhere to the principle of proportionality, Janina Dill argues that there is still considerable disagreement on the principle of *distinction*.[49] Her chapter undertakes an analysis of the definition of *military objective* as found in Article 52(2) of the First Protocol and as understood by the U.S. Air Force. According to Dill, the wording of the definition is indeterminate along two dimensions that also mark interpretive controversies among lawyers. First, it is unclear how close the connection between an object and the competition between two militaries has to be for the object to qualify as a military objective—what she refers to as the *degree of nexus*. Second, there is disagreement over how to define progress in war. Specifically, what is the point of reference for the determination of a military advantage? Is it at a maximum what she calls "generic military victory" or are belligerents allowed to attack objects whose engagement yields an advantage with regard to the *political* goals they ultimately mean to achieve with force? Neta Crawford makes a similar distinction in her chapter's discussion of the 1991 Gulf War. The military goals were limited: "to free Kuwait, to destroy Iraq's weapons of mass destruction program, and to weaken the Iraqi military." The political goals were more ambitious: "to promote a coup or popular uprising in Iraq that would overthrow Saddam Hussein."

When Dill examines U.S. doctrine, she finds that the United States is willing to consider an object a military objective based on an extremely low degree of nexus between the objects and the military competition between the two belligerents. She suggests that the United States uses the ultimate political goals of a war as the point of reference for determining what it means to achieve a military advantage. The result is a broad interpretation of Article 52(2) and an inclusive definition of military objective. Dill refers to the vision of the conduct of hostilities that emerges from this U.S. interpretation of Article 52(2) as the *logic of efficiency*.

Dill contrasts the U.S. approach with what she understands as the correct interpretation of Article 52(2): the *logic of sufficiency*. Sufficiency relies on two assumptions of what is sufficient, which in turn determine what is permissible in war: 1) It is sufficient to attack only those objects that are no more than one causal step away from the direct military competition (degree of nexus); and 2) Generic military victory is sufficient for belligerents to achieve their legitimate political goals. They may hence strive for generic military victory only, what Dill considers the most abstract allowable point of reference for the determination of a military advantage.

Dill provides an illustrative example of what constitutes causal steps and the degree of nexus: the military advantage that would result from an attack on food suppliers. The advantage "does not directly arise from that attack. It is two rather than one causal steps away from the destruction of the object in question. The result of the attack is that the industry is in ruins and food supplies decrease; first causal step—soldiers get hungry; second causal step—military effectiveness declines." The food industry is "doubtlessly part of a society's war effort. Yet it is two causal steps removed from the enemy's *military* effort, meaning the engagement of the enemy belligerent in hostilities." Dill's understanding of the requirements of Article 52(2) would prohibit such an attack—not least because of the disproportionate harm to civilians even from what was intended to achieve a strictly military advantage.

Dill's analysis clearly puts her at odds with Dunlap and other writers who consider that bombing to create psychological effects is legal and moral. She quotes Dunlap to the effect that undermining a government's political support does offer "a very 'direct and concrete' military advantage—especially in today's world." Yet for Dill, "the genuinely military advantage from undermining civilian morale arises only a couple of causal steps down the line because the contribution of civilian morale to the military effort is indirect." Thus, for her, targeting civilian morale for its purported military advantage fails the first criterion of the logic of sufficiency—the degree-of-nexus test—and is therefore illegal.

This volume's legal-ethical analyses—especially the chapters by Shue, Dill, Ryan, and O'Connell—represent a broader trend in the law of war: the incorporation of human rights. As Charles Garraway explains in his chapter, much of the law of war was developed for the "protection of combatants" (e.g., provisions for dealing with the sick and wounded and prisoners of war), yet over time emphasis has shifted to protection of civilians. He identifies a growing influence of human rights law on international humanitarian law, where the standards of the former are higher than those of the traditional laws of war.

> Under international humanitarian law, a "combatant" can be attacked at any time, simply because of who he or she is. Under the International Covenant on Civil and Political Rights, every human being has an inherent right to life. No one shall be arbitrarily deprived of his life. . . . Whereas international humanitarian law accepts that in war lives will be lost but seeks to minimize the loss of innocent lives, human rights law starts from the premise that any loss of life must be justified to a high standard. The onus is on the attacker and the loss of life itself creates a prima facie case.[50]

Garraway's chapter provides illuminating examples from the European Court of Human Rights, which has heard numerous cases related to Russia's conflict in Chechnya. The Court has often found the Russian Federation to have violated basic human rights instruments to which it is a party. Garraway argues that international humanitarian law would constitute the more appropriate standard.[51]

Other authors echo Garraway's argument about the increasing role of human rights law as they ponder the question—central to the debate over drones and automated killing—of who is considered a civilian or noncombatant. This question has bedeviled debates about the legality of various other U.S. policies carried out in the wake of the al-Qaeda attacks of September 2001.[52] It is also relevant to our study of bombing.

As Mary Ellen O'Connell describes in her chapter, states engaged in the "global war on terror" found legal definitions of who constituted a legitimate target for a military strike to be too confining. As O'Connell points out, their position seemed to find support when, "in the mid-2000s, the ICRC sought to broaden the category of persons subject to intentional targeting because of their status as persons in a 'continuous combat function.'" The ICRC insists, however, that such persons may be targeted only when it is necessary to do so. The standard of necessity depends on the circumstances, whether the situation constitutes armed conflict or not. Thus, although the ICRC guidance broadens the category of people who may be targeted, it raises the standard of necessity for when and where they may be targeted.

In his chapter, Richard Miller has identified a very different definitional innovation. Evidently under considerable influence by U.S. officials, in 2009 the United Nations Assistance Mission in Afghanistan defined the category of protected civilian or noncombatant—when it applies to the Coalition forces and the Afghan government it supports—quite broadly to include "members of the military who are not being used in counter insurgency operations and not taking a direct part in hostilities including when they are off-duty." In this respect, the notion of "not taking a direct part in hostilities," originally intended to protect civilians, becomes a protection for soldiers. The U.S. interpretation of the law under these conditions favors its own soldiers and those of its allies over the civilians—some of whom, no doubt, are engaged in unlawful belligerency—on the other side. If the category of protected noncombatant is in such evident flux, with considerable disagreement among specialists, one wonders how to go about programming automated killing machines—the subject of O'Connell's chapter—to make correct application of the principle of distinction.

The importance of the distinction between combatants and noncombatants looms large in wars characterized as counterinsurgencies. Most authors would probably agree that in U.S. military policy of the early twenty-first century, counterinsurgency came back into favor in a big way, but not the use of air power in support of counterinsurgency operations. Several of our authors cite the 2006 counterinsurgency field manual of the U.S. Army and Marines and its emphasis on protecting civilians from harm. Miller, for example, quotes the remark that "an operation that kills five insurgents is counterproductive if collateral damage leads to the recruitment of fifty more insurgents."[53] Crawford indicates that such "collateral damage" is associated at the highest levels of the U.S. military with air power, which, as she quotes General McChrystal, "contains the seeds of our own destruction if we do not use it responsibly."

Wars of counterinsurgency pose particular ethical and legal concerns, and here again our volume addresses them from a wide range of perspectives. Dunlap departs from the emerging conventional wisdom characterized by McChrystal's caution about air power during counterinsurgency wars. In contrast to the new (old) emphasis on winning the hearts and minds of civilians, Dunlap insists that killing insurgents is the way to win such wars and that air power is an effective means of killing them without excessive harm to civilians. Elsewhere Dunlap has argued that the 2006 counterinsurgency field manual "reflects a one-dimensional, ground-centric perspective almost exclusively"; is fundamentally biased against advanced technology; and fails to take into account air power's potential as a means of "defeating the enemy's military capability without excessive reliance upon the close fight; especially since the close fight is so costly in human terms and can generate intractable political issues for US decision-makers."[54]

Miller also calls into question the assumptions behind McChrystal's skepticism about bombing, but from a different perspective. He finds that caution in the use of air power stems not only from its potentially counterproductive nature in the theater of war but also because bombing that causes substantial immediate civilian harm can lead to popular opposition back home. He cites opposition to U.S. war crimes during the Vietnam War as an example. By contrast, not all the effects of bombing dual-use or civilian infrastructure, as in the 1991 Gulf War, are immediately evident, even if their impact on civilians can be equally devastating. Miller maintains that U.S. policymakers regarded such bombing of Iraq—in combination with imposition of sanctions—"as a possible way to deprive Saddam Hussein of support and encourage his overthrow," the same logic we saw operating in the Kosovo War and elsewhere.[55] This is what Janina Dill criticizes as the *logic of efficiency*, a logic aimed at a political goal well beyond immediate military victory.

Constructing New Norms

Traditional understandings of the evolution of the norms governing warfare have focused on state practice, which is assumed to stem from the rational pursuit of the "national interest."[56] Alternative approaches adduce other

INTRODUCTION   21

reasons for normative changes. One school of thought that is associated with rationalist accounts of international politics focuses on how internal domestic political groups might seek a state's adherence to a particular norm by having the state join an international institution or sign a treaty.[57] Another school—often known as "social constructivism"—considers the influence of a state's self-identity. If conformity with a certain norm becomes the mark of a "civilized" state, for example, certain states will be more likely to conform. If regional identities are important and most states in the region are adopting a norm, then the rest might do so as well.[58] Important work at the intersection of several of these approaches highlights the effect of "transnational advocacy networks" in making the link between domestic politics and international norms.[59]

In our volume, Margarita Petrova draws on insights from a range of such sources to address the question of the development of new norms governing warfare. She explains how nongovernmental organizations employed arguments from international humanitarian law regarding proportionality to make the case for a ban on cluster munitions. Her focus on the principle of proportionality links this chapter to the ethical-legal discussions of Shue and Dill in particular. Although not in precisely the same way as in the legal instruments, proponents of cluster munitions, and before that of land mines, argued for the military necessity of the weapons. Critics fought them on that ground, but also with reference to prevailing legal restrictions. Challenging the military benefits of cluster munitions, for example, the critics pointed to the likelihood that their "use undermined the larger political goals of today's military campaigns to win 'the hearts and minds' of local populations."[60] Thus, the activists adapted General McChrystal's implied critique of air power as counterproductive to counterinsurgency campaigns and applied it to cluster munitions, which can be delivered by both ground and air forces.

When appealing to legal restrictions, proponents of a ban on cluster munitions stressed the disproportionate harm to civilians and highlighted the temporal dimension that is mainly absent in existing law: unexploded cluster "bomblets," like land mines, could pose threats to civilians long after the conflict ended. Such an argument is related to concerns about the long-term effects of bombing civilian infrastructure, as in Shue's discussion of the destruction of Iraqi electric grid in 1991. By insisting that the long-term effects of weapons must be a part of the proportionality assessment, civil society advocates engaged in the interpretation of international humanitarian law that is usually left to governments, law professors, and the specialists of the International Committee of the Red Cross. Their engagement, according to Petrova, seems to make a difference. In the case of cluster munitions (and land mines before them), the focus shifted from limiting disproportionate harm to civilians to outright stigmatization and the banning of the weapons as such. Moreover, expanding the meaning of proportionality to include the longer-term impact on civilians served to place "more weight on humanitarian considerations compared to military ones." Thus, Petrova's empirical study reinforces the normative and theoretical contributions of our other authors who note a shift toward emphasis on the priority of human rights—particularly the right to life—even in the domain of warfare.

When civil society groups and nongovernmental organizations engage in advocacy for limiting weapons or military practices, they typically draw for their arguments upon the analyses of scholars. Our volume presents three such analyses of some of the most controversial weapons and practices associated with contemporary warfare, by Hugh Gusterson, Klem Ryan, and Mary Ellen O'Connell. Gusterson and Ryan address the implications of attacks by remotely piloted aerial vehicles on targets far from conventional battlefields. Gusterson focuses on the "ways that drones respatialize war, shift its pacing, and rework conventional military notions of honor and courage." He expresses concern for the impact of drone technology on democratic accountability for decisions on war and peace, as the U.S. public comes to perceive war as "casualty-free" to its country's armed forces and receives little information about the victims of drone attacks. He shares O'Connell's unease about the impact of increasingly automated decisions on targeting

Ryan's analysis of drones is grounded in the distinction he makes between the two competing paradigms of "just war" and "regular war." A particular characteristic of some versions of just war theory calls into question the legitimacy of belligerents who are not perceived to be engaged in a just cause. In the parlance of today's wars, such belligerents would be identified as terrorists and insurgents. Quoting Emerich de Vattel, the influential eighteenth-century theorist, Ryan writes of the just war approach that "since each Nation claims to have justice on its side it will arrogate to itself all the rights of war and claim that its enemy has none." Such an approach renders limitation of armed conflict ever more difficult: "the contest will become more cruel, more disastrous in its effects, and more difficult of termination."[61] From Vattel, Ryan adopts "the idea of a shared battlefield," which "fills two important functions for the regular war perspective." It separates civilians from the conflict and it "implies a direct relationship between belligerents which furthers the goal of placing limitations on war." The absence of "reciprocity and dialogue between belligerents" removes inhibitions to violence and leads to what Ryan calls "dissociation." It "comes in at least three distinct forms, each exacerbated as the distance between belligerents increases: dissociation of the agent from their violent acts; dissociation of the target from the source of the violence directed against them; and dissociation of the public from the violence committed on their behalf." In her chapter on automated killing, O'Connell worries, with regard to robots as well as drones, that "humanity has not yet taken into account the impact of increasing physical and temporal distance on our legal and moral principles on killing."[62] Ryan uses the notion of dissociation to explore some of this ethical territory.

Although all three authors are critical of U.S. deployment of armed drones and more fully automated technologies, they do not always agree on the particulars. Gusterson and Ryan, for example, disagree about the impact of long-distance killing on the drone operators themselves. Gusterson disputes the notion that the operators are at risk of developing "a 'PlayStation' mentality to killing," as Philip Alston, author of an influential United Nations report, put it. Gusterson, along with Dunlap in his chapter, points to evidence of high combat stress suffered by the operators. Ryan acknowledges the likelihood of such psychological responses but points out that "as experience with drone operations increases, the recruitment and training processes for drone operators will become ever more effective at identifying and developing personnel suitable for the role," as has happened in other domains, such as submarine warfare. Ultimately all three authors who take on the question of drones and more fully automated killing machines express reservations about their increasing role in U.S. military policy. O'Connell lays out a specific proposal for a multilateral treaty that would restrict automated weapons before they are fully developed. If Petrova's analysis is correct, we should expect to see proponents of limiting these weapons technologies adopting some of the legal and normative arguments that our authors have adduced as part of the process of constructing new norms.[63]

The rest of the volume follows the logic laid out in this introduction. Part I, Historical and Theoretical Perspectives, includes Tami Davis Biddle's overview of the historical development of theory and practice of air power through World War II, Sahr Conway-Lanz's discussion of the transition in U.S. bombing practices from World War II through the Korean War, and Neta Crawford's study of change from the Vietnam War to the present. It concludes with Charles Garraway's assessment of the relative importance of international humanitarian law and human rights law throughout the entire period. Part II, Interpreting, Criticizing, and Creating Legal Restrictions, includes Charles Dunlap's critique of some of the predominant legal and ethical assessments of U.S. air power. His essay is followed by chapters from three representatives of the views with which he disagrees: Janina Dill's elucidation of U.S. thinking on military objectives and the extent to which it differs from the legal requirements of the First Geneva Protocol, Henry Shue's analysis of the requirements of "constant care" for protection of civilians and how the 1991 air war against Iraq failed to meet them, and Richard Miller's evaluation of U.S. strategy in Afghanistan and Iraq in light of the experience of previous wars in Vietnam and Iraq.

The book's final part, Constructing New Norms, serves in lieu of a formal conclusion. It ties together the volume's main threads by combining an analysis of change over time with the political, legal, and ethical arguments that have contributed to that change. Margarita Petrova focuses on new actors and new issues in the evolution of international humanitarian law governing bombing, with particular emphasis on nongovernmental organizations; Hugh Gusterson and Klem Ryan bring the perspectives of an anthropologist and political philosopher, respectively, to bear on the questions of how norms governing the use of drones might evolve and affect the spread or limitation of warfare. Mary Ellen O'Connell applies legal and ethical analysis to the question of autonomous killing—outlining, in effect, how to create a new norm to influence, among other things, the American way of bombing.

Part I

Historical and Theoretical Perspectives

Chapter 1

Strategic Bombardment

Expectation, Theory, and Practice in the Early Twentieth Century

Tami Davis Biddle

Even before human flight was possible, the idea of "strategic bombing" was imagined by those who envisioned the prospect of flying beyond armies and navies—the traditional guardians of nations—and attacking an enemy's heartland directly. As this idea evolved, two subcomponents emerged. The first argued that crucial elements of an enemy's war economy could be identified and attacked directly, destroying the enemy's means to carry on. The second argued that an enemy population—unprotected and untrained for war—would cope poorly when bombs began to fall. That population's acute awareness of its vulnerability would produce societal and political chaos, eroding the popular will to fight and forcing the enemy government to sue for terms. Both of these, the latter in particular, had serious ramifications for the notion of "discrimination" inherent in the just war tradition, which called on belligerents to refrain from attacking noncombatants.[1]

Sometimes these ideas were articulated separately, and sometimes they were merged together. Early air theorists did not always make their working assumptions clear; indeed, they rarely explained the specific logic that linked aerial attacks and enemy capitulation. But many advocates, such as Billy Mitchell, were willing to write prolifically and leverage prevailing social and political fears to promote their ideas. The argument that it would be possible to identify specific targets in an enemy war economy and then attack them directly rests on two assumptions: 1) targets can be readily located;

and 2) bombers will reach their targets without suffering prohibitive losses. The argument that civilian populations will be particularly susceptible to air attack rests upon the premise that exposed citizens will be prone to disruption and panic in the face of a threat from the sky. These assumptions were highly problematic for a number of reasons, but because the interpretation of limited experience was highly colored by expectation and by cognitive and cultural bias—and because there was so little systematic analysis of First World War and interwar bombing—the foundational premises remained largely unchallenged going into the Second World War.[2]

For political and military elites charged with maintaining the security of states, these theories of long-range bombardment were alternately frightening and seductive. On the one hand, they threatened to nullify traditional means of preserving state security (and thus political legitimacy); on the other hand, they seemed to offer a potential "silver bullet" for both deterrence and war-fighting: a relatively inexpensive technology-intensive weapon system that might prove so daunting as to deter war, or to end it quickly should it arise. The unresolved tension between fear and seduction and the vexing issue of discrimination dominated the narrative of aerial bombing in its earliest years.

This essay will offer an overview of long-range bombing from the first years of the twentieth century through the end of the Second World War; it will compare pronouncements and expectations with actual war experience and will discuss related ideas about the possibility of protecting civilian populations in war. The failure to establish and uphold adequate protections for such populations—along with the nature of mid-twentieth-century aviation technology and the mounting pressures felt by combatants in the midst of total war—combined to produce unprecedented numbers of civilian deaths and casualties in the Second World War.[3] That war and the atomic era that followed it placed civilians directly in the crosshairs of the most powerful weapons systems that had ever been imagined or created. But even though the war experience of the early twentieth century ran roughshod over norms calling for discrimination in air warfare, these were never completely abandoned. They retained a fundamental, instinctive legitimacy throughout these years and served as an essential (if battered) focal point for moral and legal discussions about air warfare.

Early Speculation about Aerial Bombing

As scientific progress continued, speculative ideas about air war were infused with the hopes, concerns, and fears of the day. In 1862, Victor Hugo prophesied that aircraft would bring about the universal abolition of borders, leading to the end of wars and a great "peaceful revolution."[4] In 1893, Major J. D. Fullerton of the British Royal Engineers theorized about an aerial "revolution in the art of war." A year later, inventor Octave Chanute postulated that because no territory would be immune from the horrors of air war, "the ultimate effect will be to diminish greatly the frequency of wars and to substitute more rational methods of settling international misunderstandings."[5] In 1905, the British War Office's *Manual of Military Ballooning* argued that balloons dropping guncotton charges might have a "moral effect" on the enemy. The "moral effect" (pronounced "morale" but spelled without the "e," as in the French) reflected a particularly potent and widespread fixation in the European military, echoing the emphasis on "will" in warfare that Clausewitz and French writers including Foch, du Picq, Langlois, and de Grandmaison highlighted.[6]

This kind of speculation should not, perhaps, be surprising. After all, human flight opened up the prospect of warfare raining down from the skies, making all those below vulnerable in ways they had never been before. This prospect unsettled political leaders and military planners, who worried that those on the home front were already alienated from their governments because of the exploitation inherent in the industrial world, with its long hours of hard labor for low wages and its crowded and stressful living conditions. Frequent civil strife and workplace crises plagued industrial nations early in the twentieth century, and naturally these fueled speculation and worry about how workers would behave in wartime. In two lectures to the Royal United Services Institution in 1909, T. Miller Maguire associated what he called "the flotsam and jetsam of decaying British humanity" with the perversions of the "factory system."[7]

The emphasis on "moral effects" in war emphasized the qualities valued by upper-middle-class Victorian- and Edwardian-era societies—courage, tenacity, and willpower—but it also resonated with prejudices and darker trends therein, including social Darwinism. Speculation was widespread not only about how competing states would stack up against one another but also about how different races and classes within a state might affect its overall strength and cohesion under stress. In 1908, the British anxiously watched the flight tests of Count Ferdinand von Zeppelin's airships. Indeed, concern over Britain's ability to defend itself was at the center of a flurry of invasion literature that peaked between 1906 and 1909.[8]

At the same time, it was clear that the industrial environment was eroding the ability of belligerent states to discriminate between combatants and noncombatants. Were urban workers contributing to an enemy war economy to be considered noncombatants? Were cities with governmental and administrative centers to be considered military targets in war, open to aerial bombardment? The second Hague Peace Conference of 1907, which took place after heavier-than-air flight had become a reality, sought to adapt the Rules of land bombardment from the 1899 Hague Conference. Article 25 of the land warfare convention was amended to read: "The attack or bombardment *by whatever means* of towns, villages, dwellings or buildings which are undefended is prohibited." This language effectively replaced the five-year prohibition on the dropping of projectiles from aerial platforms that had been enacted at the 1899 Conference.[9] But the meaning of the term "undefended" was unclear, and in any event the proscription was muddied by guidance given to naval forces that allowed for the lawful bombardment of military objects, including depots of arms or war matériel, and workshops or plants that could be utilized for the needs of the hostile fleet or army. The naval convention also absolved commanders of responsibility for "any unavoidable damage which may be caused by bombardment."[10]

The language of the 1907 conference would cause subsequent confusion while failing to erect meaningful barriers in the path of air warfare. But this failure was not due solely to a lack of clarity in the wording of the law; neither the ideas nor the legal texts could exist in isolation from politics. Some states did not wish to deny themselves a potentially advantageous tool of war, and all feared the vulnerability that might arise from trusting one's fate to documents that one's adversaries might ignore in the heat of conflict.[11] The first bombs were dropped by aircraft during the Italo-Turkish war of 1911–12, when Lieutenant Giulio Gavotti hurled projectiles onto villagers in Libya on 1 November 1911.[12]

The First World War

Concerns over public robustness and resilience helped shape the context for military debate and planning on the role of aircraft in war prior to 1914. But there was no consensus on what these new and still primitive machines might accomplish or how they ought to be assessed against other military resources. When war came in 1914, most of the combatant states were still integrating aerial weapons into their force structures.[13] Interservice rivalries slowed progress in most states, including Britain, where aerial resources were divided between the Royal Naval Air Service (RNAS) and the army's Royal Flying Corps (RFC). But the value of aerial reconnaissance—and thus the value of air space (both the enemy's and one's own)—became immediately apparent. Other roles for aircraft evolved quickly, including communication, battlefield attack and assault, and battlefield interdiction. Manuals were written and rewritten every few months.[14]

Most states that possessed aircraft had shown interest in strategic bombing, but the development of bombing doctrine varied. The French began aerial attacks on German industry as a way of eroding the enemy's war-making capacity, and the Italians attacked enemy naval bases. These were pieces of integrated campaigns and were not typically seen as separate from army and navy efforts. As the French found themselves overwhelmed by the demands of the ground war in 1916 and 1917, they had fewer resources to devote to the air; increasingly, the French high command insisted that aerial efforts be concentrated on the battlefield.[15]

Beginning in 1915 the German Kaiser authorized the use of airships in an offensive designed to undermine both the war-making capacity and the will of their British enemy. German deputy chief of naval staff Paul Behnke speculated that the ensuing panic might "render it doubtful that the war can be continued." In 1916, Peter Strasser, commander of the German Naval Airship Division, urged as large an offensive as possible to provoke "a prompt and victorious ending of the war."[16] Throughout 1915 and early 1916, the British were vulnerable to the onslaughts. Over time, though, the trends shifted: better fighters, special incendiary bullets, and much more efficient signals communication made flying airships over England very risky by late 1916.[17]

Continued stalemate in 1917 provoked the Germans to make another attempt to break the will of the British. In the spring and summer, Gotha and four-engine "giant" bombers (Riesenflugzeuge) menaced British cities. Two small daylight raids on London (13 June and 7 July) managed to cause significant casualties, raising the indignation of the British, especially in response to bombs that hit a kindergarten. The public demanded better air defenses and retaliation in kind. This public outcry—indeed, the very fact of the public's demand for a voice in the prosecution of the war—unsettled British elites, especially in a year when strikes and industrial action had surged once again and when the Russian Revolution was in full swing.[18]

British field marshal Sir Douglas Haig found himself forced, against his will, to send fighters from the Western Front back to England. A commission led by South African general and statesman Jan Christian Smuts came to some radical conclusions about air warfare: "As far as can at present be foreseen there is absolutely no limit to the scale of its future independent war use. And the day may not be far off when aerial operations . . . on a vast scale may become the principal operations of war, to which the older forms of military and naval operations may become secondary and subordinate."[19] This radical text was among the catalysts for an independent air force in Britain.

The Germans bombed London again in September and then later in the autumn, under cover of night and with high hopes invested in a newly developed incendiary bomb. While it is no small thing to change one's defense structure in the midst of a major war, British decision makers did so because of their fears of an expanded German air offensive, their anxiety about the domestic front, and the perceived need to respond to public pressure.[20]

Neither Haig nor Major General Sir Hugh Trenchard, then commanding the army's air offensive on the Western Front, was interested in forming a new service. Nonetheless, by May 1918, Trenchard found himself overseeing a long-range bombing force created to wage attacks on Germany.[21] A newly formed Air Policy Committee argued for a campaign that would produce both a material and moral effect: "to attack the important German towns systematically . . . until the target is thoroughly destroyed" with the aim of "disrupting industry and undermining civilian morale." As historian Richard Overy has observed, this wording was a remarkable foreshadowing of language that would resurface during the Second World War.[22]

Though he was promised a sizable force, Trenchard never received it. He was well aware, though, that the eyes of the public were upon him. He directed attacks on German cities and industries when opportunities arose. But his rather haphazard approach caused exasperation among planners in London who had attempted to analyze German war industry and identify strategic targets of consequence. One analyst, Lord Tiverton, had authored a plan in 1917 that sought to identify "bottlenecks" in the German war economy; it was a precursor to the "industrial fabric" theory of bombing that would later develop in the United States during the interwar years.[23]

Because he could produce only limited physical results in Germany, Trenchard stressed instead the indirect results and the psychological impact of his air campaign. After the war he argued that his bombers produced strain on workers and lowered factory production due to ongoing air raids and alarms. He asserted frequently that the "moral effect" of bombing was "twenty times" the physical effect. Needing to justify his operations, Trenchard waged a rhetorical offensive designed to achieve what his actual air offensive could not.[24]

For their part, the Americans had shown a strong interest in the prospects for air warfare and long-range bombing. But, entering into the war late, they discovered that they could not create an aerial force overnight. While General Billy Mitchell oversaw tactical air offensives at Saint-Mihiel and Meuse-Argonne, the Americans engaged in no long-range bombardment. They did, however, keep a keen eye on the efforts of their allies and produced a plan—based entirely on the one drawn up by Tiverton in 1917—to wage a future aerial offensive against specific German industries.[25]

The experience of long-range bombing in the First World War was sporadic and inconclusive—a sideshow to the tremendous effort taking place on the ground. But much had been set in motion: the Americans were planning to enter the bomber war in a serious way in 1919. And Trenchard's bombing force was scheduled to more than triple its size by May of that year. Planners were collecting intelligence on German industry and preparing target folders and maps based on the results: "At the end of 1918, both forces were poised to conduct the first large-scale strategic air campaign against the economy and morale of an enemy power. . . . The end of the war rudely disrupted these preparations and left a strong sense of unfulfilled expectation among the airmen and officials involved."[26] Hostilities concluded without a full test of strategic bombardment, but the experience of air warfare had afforded just enough experience to encourage interested parties to stake out vigorous claims that would influence the future.

The Interwar Years

While the German 1917 offensive had prompted a reorganization of British defenses, it did not—ironically—impress the Germans with the overall utility of strategic bombardment. As had been the case during the earlier Zeppelin campaign, British defenses recovered quickly and forced the Germans into night operations that were perpetually hindered by mercurial British weather. By 1918, the Germans had retrenched and refocused their air operations on the battlefront. During the interwar period, Germany was initially stripped of its air force entirely. When the Luftwaffe reemerged after Hitler swept aside the Versailles restrictions, army perspectives dominated. While some German thinkers, including Walther Wever, expressed great interest in long-range bombing, the primary focus of German interwar thinking was on air-ground cooperation, and in this realm their doctrinal and force development was substantial.[27]

French air forces did not progress toward institutional autonomy until the early 1930s, and even then the French army kept them well in check. Air doctrine became enmeshed in the increasingly defensive outlook of French strategy generally and was pulled into ongoing civil-military tensions endemic in France in this period. France's proximity to a restive and disgruntled Germany ensured a dominant voice for French ground forces and kept the focus of air operations largely on the essential task of defending French territory.[28]

Despite the attempts of Billy Mitchell to make a dramatic case for air force autonomy in the United States, progress was slow. General John Pershing was not inclined to let air forces drift too far from armies, and Secretary of

War Newton Baker took issue with First World War strategic bombing on moral grounds. Following Mitchell's court-martial, American airmen realized that they would have to accept more gradual progress. The mood of isolationism that settled over the nation also shaped the strategic environment in a powerful way. While the American people followed Mitchell's exploits in the news and flocked to barnstorming air shows, they felt no urgency to restructure their national defenses.[29] Despite this, the U.S. Air Service continued to evolve: in 1926 it became the U.S. Air Corps and its Tactical School (ACTS) became a focal point for doctrinal development. In that same year, former ACTS instructor William C. Sherman's book *Air Warfare* identified the bomber as the "supreme air arm of destruction" and articulated the industrial fabric theory of bombing that echoed Tiverton and foreshadowed the body of thought that would dominate the ACTS in the 1930s.[30]

In postwar Britain, RAF officials developed ideas that echoed elements of their wartime experience. The first commandant of the RAF Staff College argued that in response to only 452 German flights over England, the British had been forced to send up 1,882 defensive sorties.[31] Trenchard, the postwar chief of Air Staff, made the case that Britain must be in a position, should war come, to wage a "relentless and incessant" air campaign designed to push the enemy onto the defensive before that enemy could do the same to Britain.[32]

Trenchard's thinking reflected his view that the creation of the RAF had been due to the "great popular outcry."[33] Prime Minister David Lloyd George had agonized about the stability of the home front throughout the war, especially in the aftermath of the Russian Revolution. The prewar assumption that industrial populations would be fragile under stress inclined British elites to interpret wartime experience through that particular cognitive lens.[34] And certain interwar events, including the British general strike of 1926, the Paris riots of 1934, and even the market crash and the Great Depression, seemed to reinforce this interpretation. But what elites had read as fear and panic in the face of First World War bombardment—the first steps toward popular uprising—was, in fact, something different. A close examination of wartime primary sources indicates that the public response to German bombing was dominated by anger and indignation rather than panic and fear. The consequences of this misreading would be important, since they would influence governmental choices before and during the Second World War.[35]

Trenchard fended off postwar challenges to RAF autonomy by making the case that the new service could perform two key functions: 1) provide inexpensive colonial policing by using bomber aircraft and aerial bombardment to threaten, intimidate, and coerce native populations in British-held colonial territories; and 2) deter war and, in the event deterrence failed, prosecute a powerful bombing offensive that would cause the enemy to promptly sue for terms. Both claims found political traction in a nation that had large imperial holdings, tight finances, and no stomach for any discussion of land warfare and/or continental commitments. But both claims had immense ethical implications that were largely sidestepped or subsumed under dominant narratives of power, nationalism, and economy.[36]

During the 1920s and 1930s, British public views on warfare frequently embodied two extremes: a determination to avoid the topic altogether and vest hopes in international agreements and institutions or a tendency to discuss warfare in deterministic and apocalyptic terms. Perhaps this should not be surprising in the aftermath of the shocking and unremittingly grim experience of the First World War, but the effect was to leave little room for rigorous analysis of recent experience. A general revulsion toward war in Britain left the army with limited influence. Antiwar sentiment reached its peak in the late 1920s and early 1930s, including an explosion of literature by authors such as Siegfried Sassoon, Robert Graves, and Wilfred Owen. At the other extreme, dark forebodings in the realm of popular culture resulted in a flurry of books fixed on bleak visions: *The Poison War, Invasion from the Air, War upon Women,* and *Air Reprisal.*[37] The impact of these was augmented not only by the futurist scenarios depicted in movie houses but also by the ominous events of the 1930s, including the Japanese attack on Manchuria, the Italian attack on Abyssinia, and the Spanish Civil War.

By this time, the ideas of Italian Giulio Douhet were becoming more widely known in English-speaking countries. Douhet's 1921 book *The Command of the Air* had painted a graphic vision of societal collapse in the face of air attack. Indeed, the futurist drama he conveyed—rather than the analytical rigor of his ideas—gave Douhet a lasting place in the canon of air warfare.[38] He brought to bear on his work "the intense modernist fascination with the latest advances in science and technology . . . prevalent in prewar Italian protofascist *avant-garde* culture."[39]

Though both British and American airmen had developed indigenous theories of air warfare, Douhet's ideas were cited widely in the 1930s. His view was one of technological determinism: "The brutal but inescapable conclusion we must draw is this: in the face of the technical developments of aviation today, in case of war the strongest army we can deploy . . . and the strongest navy we can dispose . . . will provide no effective defense against determined efforts . . . to bomb our cities."[40] He argued that in future wars, there would be no distinction between combatants and noncombatants. Articulating a silver lining in this daunting scenario, he argued that air wars would, at least, be brief: "Mercifully, the decision will be quick in this kind of war, since the decisive blows will be directed at civilians, that element of the countries at war least able to sustain them. These future wars may yet prove to be more humane than wars in the past in spite of all, because they may in the long run shed less blood."[41] Most readers would have found this to be cold comfort.

In placing an extreme emphasis on the offensive capability of air power, Douhet (and to a lesser extent Trenchard and Mitchell) turned a blind eye to the long history of warfare. As historian Michael Sherry has pointed out, Douhet's idea of the future rested on crude extrapolation, and like many other interwar prophets of air power, he failed to see how it "might evolve unpredictably, strengthening the defense as well as the offense, creating its own futile charges and bloody stalemates."[42]

Set against Douhet's technological determinism, however, was the hope that peoples and nations, using the instrument of international law, might be able to prevent or circumscribe the worst effects of aerial bombing. The Washington Naval Conference of 1922 was followed by a conference to wrestle with the problem of aircraft in war. Concluding that the limitation of military aircraft was not a viable proposition, delegates at the Hague struggled to articulate laws that would embrace long-held principles from the just war tradition and the longings of humanity at the time.[43] Article 22 argued bluntly: "Aerial bombardment for the purpose of terrorizing the civilian population, of destroying or damaging private property not of a military character, or of injuring non-combatants is prohibited." Article 24 stated that "Aerial bombardment is legitimate only when directed at a military objective, that is to say, an object of which the destruction or injury would constitute a distinct military advantage to the belligerent." Article 25 specified the targets to be avoided if at all possible, including historic monuments and buildings dedicated to worship and care of the sick.[44]

But the problem of what constituted a "military" target remained vexing. Factories filled with civilians fed the voracious appetites of industrial powers at war, and capital cities were the command posts of enterprises capable of vast destruction. During the interwar years most air forces made their personnel aware of the Hague Rules of Aerial Warfare, but with the proviso that they were not binding.[45] In a 1938 article titled "The Chaotic State of the International Law Governing Bombardment," the British authority J. M. Spaight lamented that the laws concerning bombardment were "in a state of baffling chaos and confusion which makes it almost impossible to say what in any given situation the rule really is." In a 1937 essay for *The New Republic*, essayist Jonathan Mitchell argued that the Hague Rules "still remain unratified, and many experts believe them too vague to be of practical importance."[46] But even as they were debated, the Hague Rules remained alive as a basic normative standard: even if no state adopted them formally, they would be called upon and referenced repeatedly as the clouds of war gathered ominously over Europe. Their fundamental intention was never in question; they were meant to protect noncombatants to the greatest extent possible. Building on the frameworks of 1899 and 1907, they helped to erect a moral trellis that could not be entirely ignored and would not be entirely torn down, even in the midst of brutal combat yet to come.

Attempts at constraint manifested themselves in another form: a new effort, under the League of Nations, to reduce and limit armaments through international agreement. Meeting first in the winter of 1932 at Geneva after many years of preparation, the World Disarmament Conference saw energetic and genuine efforts to grapple with the problem of aerial bombardment. Those efforts came to focus on "qualitative disarmament," the abolition of those weapons "whose character is the most specifically offensive or the most efficacious against national defence, or most threatening to civilians."[47] The British government, still squeezed by serious financial concerns and lacking the confidence of an earlier era, sought to head off an aerial arms race that would upend the status quo and further the instability that was beginning to overwhelm European politics. Each participant brought its own perspectives and incentives to the table; at one point, a genuine if clumsy intervention by U.S. president Herbert Hoover had the effect of dissolving the momentum that had gathered by the spring. And internal politics plagued national efforts.[48] Britain found itself in an embarrassing position when its Air Ministry—against Foreign Ministry opposition—insisted on retaining the right to use bombers against native peoples in the remote parts of the empire. Facing a storm of protest, the British government agreed to abandon the reservation if it would stall the conference.[49] But other issues came to block the momentum the delegates struggled to maintain.

The negotiations at Geneva were endangered by Hitler's ascent to power in March 1933, and they effectively collapsed later that year when the Germans withdrew from the discussions in October.[50] Limiting aircraft proved more technically vexing than limiting surface ships, and efforts at qualitative disarmament foundered on the riptides of competing interests and the shoals of mistrust. Even so, many governments remained interested in diplomatic efforts to emplace and enforce some form of constraint upon air warfare.[51] Ultimately, however, fears would triumph over hopes.

Toward War, Again

When Trenchard handed the RAF over to his successors in 1929, it was secure in its autonomy. It was not, however, in a position to carry out the offensive policy it had espoused. Trenchard had not stressed the questions that should have dominated the service agenda: Can the bomber always get through? Under what circumstances? How do bombers find and hit targets accurately and reliably? The rise of Hitler had been ominous, not least because Britain's financial situation had been perilous after the First World War and remained so. Rearmament, the Treasury realized, would raise the prospect of bankruptcy. Prime Minister Neville Chamberlain, having sought policies of appeasement, felt his next best hope might rest in using bombers as a deterrent. He therefore authorized monies for the RAF more liberally than for the other services in the late 1930s.[52] But as Hitler set about to rearm the Luftwaffe, the RAF found itself increasingly insecure about its own capabilities. By 1937 the head of the RAF's Bomber Command, Sir Edgar Ludlow-Hewitt, discovered to his disquiet that his force was "entirely unprepared for war, unable to operate except in fair weather, and extremely vulnerable both in the air and on the ground."[53]

The Luftwaffe, which had remained principally focused on supporting land warfare, was not such a potent bombing force as many in Britain assumed—but perception mattered more than reality in the terribly tense atmosphere of the late 1930s, when aerial bombing during the Sino-Japanese War and the Spanish Civil War raised concern to a fever pitch. During the Munich crisis of 1938, Chamberlain facilitated the handing over of a piece of Czechoslovakia to Hitler in a desperate effort to head off war—and the terrible aerial bombardment it was expected to entail. Whatever effect it had had on the enemy, the RAF's interwar rhetoric surely had been a deterrent to British statesmen as well.

In the United States, the Air Corps toed the official army line while simultaneously intensifying its thinking about strategic bombing. Sherman's *Air Warfare* had declared: "Industry consists . . . of a complex system of interlocking factories, each of which makes only its allotted part of the whole. . . . It is necessary to destroy certain elements of the industry only, in order to cripple the whole. . . . On the declaration of war, these key plants should be made the objective of a systematic bombardment."[54] The theory depended upon two important and ultimately problematic assumptions: 1) that intelligence work would be able to identify the "key plants" and that the enemy could not disperse them or find substitutes for their products; and 2) that bombers would be able to find and strike such factories in daylight without suffering prohibitive losses. Emerging American doctrine had been shaped in part by the early opprobrium heaped upon bombing by Secretary Baker, by conceptions of American exceptionalism, by the influence of the industrial efficiency movement, and by the luxury of distance from enemies that actually could strike the United States, which allowed Americans to take a more antiseptic approach than their European counterparts.[55]

As the crisis in Europe intensified, President Franklin Roosevelt came to believe that bombers might provide an effective deterrent—a means for keeping the United States out of war while keeping Germany and Japan in line. Like Chamberlain, he placed an initial focus on spending for air power.[56] This inclination—in both Britain and the United States—should not be surprising. Throughout their histories, both nations had eschewed large standing armies, which they recognized to be economically draining and politically and socially disruptive. The geopolitical situations of both nations seemed to facilitate a search for alternatives to the costly and bloody land warfare they had seen less than twenty years earlier.

Upon the outbreak of war in 1939, Roosevelt issued a plea calling on the belligerent states to confine their bombing to strictly military objectives. Germany, Britain, and France all agreed but reserved the right to rescind their commitment if their enemy reneged.[57] More than a year earlier, on 21 June 1938, Prime Minister Chamberlain had proposed three Rules to guide bombardment:

1. It is against international law to bomb civilians as such and to make deliberate attacks upon civilian populations.
2. Targets which are aimed at from the air must be legitimate military objectives and must be capable of identification.
3. Reasonable care must be taken in attacking those military objectives so that by carelessness a civilian population in the neighborhood is not bombed.[58]

Britain's chief of Air Staff commented, "I feel sure that this instruction will not last very long, but we obviously cannot be the first to 'take the gloves off.'"[59] Nonetheless, in late August 1939 the Air Ministry sent out instructions urging a prohibition on the bombing of civilians, lest the good opinion of neutral states (principally the United States) be jeopardized. Stricter than the Hague Rules, they could be abandoned if the enemy commenced indiscriminate bombing. The instructions, which remained in effect until 4 June 1940, were sent all the way down to squadron level; they defined acceptable targets as "purely military objectives in the narrowest sense of the word."[60]

Into the Crucible

By the time the Wehrmacht and Luftwaffe launched their war, they had taken the doctrinal lessons of modern combined arms and refined them into a mode of war-fighting that looked, for a time, unstoppable. The Blitzkrieg of 1939–40—which was nothing more than the application of armor and air power to the ground war innovations of 1918—was daunting to states that had not concentrated so effectively on tactical integration.

Sobered by the prospect of war following the Munich crisis, however, the British had made rapid progress in air defense. By the time the Luftwaffe launched the Battle of Britain in the summer of 1940, the British had shaped an effective communications net into which radar had been successfully inserted. A late-in-the-day push to build adequate numbers of reliable fighters enabled the British to hold out against the aerial onslaught of an overconfident enemy. But the failure of the Germans in the Battle of Britain did nothing to dissuade the British from trying their own air offensive against Germany. This paradox reflected the desperate straits the British found themselves in by 1940–41. In May of 1940, Prime Minister Winston Churchill had made bombing a main pillar of his argument that Britain should refrain from seeking terms with Hitler and should instead follow an economic and peripheral strategy while awaiting help. Standing alone, with his ground forces driven off the European continent, Churchill could not countenance the possibility that aerial bombing might not work. He believed that an offensive against the enemy was essential for British morale in an extreme crisis; it was the way forward.[61]

But the RAF's interwar lacunae in analysis and training all came painfully to the surface in the early years of the war. Bomber Command's initial missions—dropping propaganda leaflets and waging anti-shipping attacks—pointed out just how woefully unprepared the organization was for war. These initial efforts told of the difficulties of finding distant cities, of the constant battles with weather, and of the physical and emotional stresses crews encountered in such operations. German defenses pushed operations increasingly into the nighttime hours, when darkness could afford some protection. Crews were sent out with maps, astro-sextants, and directional radio. With such means they were expected to find their way about. In essence, crews were expected to navigate at night by observation, a largely impossible challenge under the weather conditions that so frequently prevailed. This problem was exacerbated by the fact that Britain's early bombers were wholly inadequate to the task.[62]

Concerned to avoid provoking the Germans into terror attacks on British citizens, the Air Staff focused its early attention on German transport and oil targets. But finding such targets consistently at night proved nearly impossible.[63] In the summer of 1941, the first thorough reconnaissance analysis of British bombing accuracy stunned the leaders of Bomber Command: only one in five bombers was getting within five miles of its target.[64] The news came at a desperate moment, since Britain had suffered one disastrous military setback after another in 1941. Chief of Air Staff Sir Charles Portal felt there was little choice but to turn to the only targets crews could find and hit reliably in darkness: cities.[65]

The Blitz campaign of German air attacks on Britain in 1940–41, along with earlier German air attacks on Warsaw and Rotterdam, had eroded British concerns about restraint. Indeed, the RAF had waged its first "area" attack, on the city of Mannheim, on 16 December 1940, following the German attack on the British town of Coventry.[66] But even as the British began to abandon discrimination, they saw evidence at home that civilians were more robust than many had expected them to be. During the interwar years there had been frequent and often dire predictions about the psychological upheaval that civilians would experience under air attack. By 1941, however, British medical and scientific journals were beginning to retract their extremist claims. In a typical example from *The Lancet*, Dr. Felix Brown explained that "the incidence of genuine psychiatric air-raid casualties has been much lower than might have been expected; the average previously healthy civilian has proved remarkably adjustable."[67]

In the absence of perceived alternatives, however, the British continued to rest their hopes on an expansion of their air war. On 14 February 1942, Bomber Command came under a new directive calling for attacks on "area" targets; the objective was to undermine "the morale of the enemy civil population and in particular, of the industrial workers."[68] The change, sanctioned at the highest governmental and military levels, was ominous indeed. Though the new directive had an economic component, it shifted the primary emphasis back to Trenchard's point of focus—the morale of the enemy—and it pulled dramatically away from the letter and spirit of the Hague Rules. While both the Russians and the Americans had been dragged into the war in 1941, their ability to aid in the anti-German cause was still in the distance, and—in the Russian case—was still in the balance in early 1942. At the same time, German wolfpack submarines ran amok in the Atlantic, threatening to strangle Britain economically.[69]

One week after the new directive was issued, Sir Arthur Harris was named to head Bomber Command. While the directive had predated him, he was in fact a proponent of city bombing; he believed that cities concentrated everything important to modern industrial nations. His theory of victory combined the moral and material elements of aerial bombing.[70] Harris set about making his crews technically proficient; throughout 1942–43 he pushed them to master both night flying and the new navigational and target-finding instruments that British science and technology rushed to place in their hands. His crews faced a grinding attritional campaign with, at times, staggeringly high loss rates due to ongoing improvements in German defenses.[71] But heavy attacks on industrial cities along the Ruhr in the spring of 1943 put a ceiling on German minister of supply Albert Speer's plan for an exponential increase in German munitions production.[72] In July, Bomber Command waged an intense attack on the city of Hamburg that produced an unprecedented firestorm. In the autumn and winter of 1943 Bomber Command turned its focus to Berlin, waging a long but largely fruitless campaign against the German capital city over the winter of 1943–44.[73]

In 1942 the American entry into the air war had been, as in 1917, frustratingly slow. While Harris waged thousand-bomber raids on German cities, the Americans flew twelve-bomber raids to the coastal edges of France. The desultory progress caused no end of grief for General Henry "Hap" Arnold, who was commanding the newly named U.S. Army Air Forces. Churchill, fearing that the American determination to employ "precision bombing" would fall victim to the same nemeses the British had faced, implored the Americans to join the nighttime area offensive. But the Americans refused: they were determined to maintain the distinction (especially in the minds of the American public) between their daylight "precision" campaign and the nighttime area raids of the British. It was not until January 1943 that the Americans first flew over Germany proper.[74]

Their heady faith in "precision" daylight bombing by self-defending groups came a cropper in the late summer and autumn of 1943, however, when raids deep into Germany proved so costly as to be unsustainable. In four raids carried out over six days in October, 148 American bombers failed to return to their bases.[75] Forced to reevaluate, the Americans embraced the long-range escort fighter equipped with droppable, self-sealing auxiliary fuel tanks. Taking bombers to targets the Germans felt compelled to defend, they set up duels between American escorts and German short-range defenders that began to erode Luftwaffe dominance. By the spring of 1944, the Americans had fought their way out of a desperate situation, paved the way for the Normandy invasion, and relieved much of the strain on Bomber Command.

But the Americans also backed away, quietly but steadily, from their focus on "precision" bombing. Though he eschewed the term "blind bombing," General Arnold issued a directive in early November 1943 that allowed crews to abandon visual sighting and aiming when weather conditions did not support them. This improved the tempo of American operations, which were often shut down for days on end by poor weather.[76] But the Americans did not adapt easily to the change; even in the winter of 1944–45, 42 percent of American bombs dropped through cloud fell more than five miles from the target. In order to increase the impact of these bad weather raids, the Americans added higher numbers of incendiary bombs to their ordnance mix. Usually aimed at railway marshaling yards, these raids did not differ much in their practical effects from British area bombing.[77]

Even though the tide of the war in Europe had been changing in 1943, the Anglo-American bomber forces felt themselves under great stress. Late in that year both the British and American bomber forces had faced a collapse of their efforts just as the amphibious assault into France was fast approaching. The steady falling away of constraint was a consequence not only of the shattering of assumptions underpinning long-range bombing but also the heightened stakes produced by the war itself. Fear of failure and the perceived obligation to pave the way for invasion and make clear progress toward victory had the effect of driving out other considerations.

In the spring and summer of 1944 the Anglo-American strategic bombers came under the control of Supreme Allied Commander General Dwight Eisenhower, who used them principally to pave the way for the Normandy invasion and support the subsequent land campaign.[78] By August 1944, that campaign had made great progress; indeed, after the liberation of Paris, many believed that the Third Reich's days were numbered. But then the tide began to turn again. German V-1 rockets, which landed in London over the summer, were joined in September by the much more ominous V-2s—the world's first ballistic missiles—that could not be stopped by British defenses. Other weapons, including the Messerschmitt 262 jet fighter, seemed to make manifest a long-standing fear that the Reich would rely on an array of heinous "secret weapons."[79]

The momentum of the ground war was halted at Arnhem in mid-September, and the bloody battle at Hürtgen Forest foreshadowed what many had imagined to be impossible: a German counteroffensive in December, the Battle of the Bulge. The dramatic change and the emotional upheaval it produced for the Anglo-Americans provoked a frenzied effort to examine ways to expedite the war's end. British decision makers worried about the staying power of their home front after six long years of war; the Americans knew they had a brutal battle ahead on the Japanese home islands.

On 25 January 1945 the Joint Intelligence Sub-Committee suggested an urgent review of the use of the strategic bomber forces, insisting that "the degree of success achieved by the present Russian offensive is likely to have a decisive effect on the length of the war." Well-timed attacks against

Berlin might assist the Russians, especially if they could be coordinated with the isolation of East Prussia and the fall of Breslau.[80] An intervention by Churchill—made on the eve of the Yalta discussions—pushed the process along; the prime minister inquired what plans the RAF had for "basting the Germans in their retreat from Breslau."[81]

Harris was told to undertake attacks "with the particular object of exploiting the confused conditions which are likely to exist in the above mentioned cities during the successful Russian advance."[82] Portal discussed the plan with his American peer, Lieutenant General Carl Spaatz, and the latter articulated the plan to the Allied Air Commanders' Conference on 1 February. Attacks on German synthetic oil supplies would be the first priority target, but the second priority would now be Berlin, Leipzig, Dresden, "and associated cities where heavy attack will cause great confusion in civilian evacuation from the East and hamper movement of reinforcements from other fronts."[83]

The Americans launched a powerful raid on Berlin on 3 February, and the British and Americans attacked Dresden on 13–14 February. Bomber Command waged two separate nighttime attacks on the city; the Americans followed up the next day with attacks on marshalling yards and then hit the city a fourth time on the 15th, when bombers, failing to reach their primary target, bombed Dresden as a secondary target. During the initial British raid, Mosquito target-marker bombers managed to mark in a particularly tight pattern, allowing follow-on bombers to discern a clear, well-defined aim point. The concentrated bomb fall produced an inferno that led to a firestorm.[84]

The vast destruction in the architecturally distinguished city and the high civilian death toll (among women, children, and elderly seeking to escape the fighting on the Eastern Front) provoked unusually close questioning of the raid in the United States and Britain in the last months of the European war.[85] These discussions, which took place when it was clear that Hitler's forces were finally in the last throes of defeat, resurrected and further catalyzed debates—and later heated engagements—over discrimination in air war.

Also in February–March 1945, the American strategic air offensive in Japan moved into a new phase. On the night of 9–10 March 1945, American B-29 bombers of the XXI Bomber Command under Major General Curtis LeMay launched what proved to be the most deadly raid of the war: sixteen square miles of Tokyo were burned out and more than 100,000 residents were killed in a single night. The raid marked a turn in the American air campaign in the Far East.[86]

The initial commander of the XXI, Major General Haywood Hansell, an author of the "industrial fabric" theory, had sought to carry out selective attacks on Japanese industry.[87] But constant cloud and the prevailing winds of the jet stream made it nearly impossible for bombers to sustain such a campaign.[88] General Arnold, impatient for progress, replaced Hansell with LeMay—making it clear to the latter that his tactical choices would not be constrained by those above him. The new commander stripped his B-29s of their defensive armament, filled their bomb bays with incendiaries, and flew them over Japanese cities at low level. Against the bloody backdrop of the battles of Iwo Jima and Okinawa, fear and escalation—once again—drove out constraint.[89]

Over the course of the following months, the Americans waged an area-bombing campaign of terrible fury, attacking sixty-six Japanese cities with incendiary weapons.[90] Like the brutal raids in Germany, these revealed just how much punishment nations could endure without collapsing. The willingness of American planners and policymakers to repeatedly prosecute mass fire raids represented a descent to a new level in the hell of total warfare. On 6 August over Hiroshima and on 9 August at Nagasaki, no moral threshold was crossed that had not been crossed much earlier in the year.

Even as the atomic attacks were launched, the conventional bombing continued, along with naval blockades, mining, and war at sea—all in a desperate effort to compel the Japanese to capitulate. During the end phases of the war in Europe and in the Pacific, spasms of unprecedented violence fell with previously unimaginable weight upon increasingly vulnerable civilian populations. This crescendo of destruction was orchestrated by democratic states that responded with fear-driven fury to the end-stage doubling-down by their enemies as those regimes sought to resist the imposition of unconditional surrender. It revealed the full extent to which peoples who considered themselves civilized—protectors of human rights and Enlightenment principles—could be brutalized by fear (of defeat, of humiliation) and by the uniquely pernicious spiral of total war.

But not even this Armageddon would eliminate or eviscerate the most fundamental notions of *jus in bello* as they pertained to aerial warfare. The questioning that followed in the wake of the Dresden raid and the heated discussion that began to develop about World War II bombardment just as it became clear that Hitler truly was defeated were testament to the fact that norms—while battered—had not ceased to exist. And while Americans celebrated victory over Japan and were quick to associate the atomic attacks with Japanese surrender, they also found themselves daunted by the ramifications inherent in the nature of the terrible weapons they had produced and used in the last days of the war.[91] As Richard Overy has argued, if the bombing of civilians had not been viewed as "legally problematic," the major powers would not have agreed, so quickly to the protective language written into the new Geneva Conventions of 1949.[92] The questions and pangs of conscience that bore down upon those who fought the Second World War would begin to shape the environment for the postwar world and its struggles with a new and more terrifying form of aerial bombardment.

CHAPTER 2

Bombing Civilians after World War II
The Persistence of Norms against Targeting Civilians in the Korean War

SAHR CONWAY-LANZ

World War II demonstrated an enormous shift in the technological capability of the United States to bring death and destruction to the civilian populations of its enemies through aerial attack. The American air forces undertook strategic bombing campaigns that pulverized and burned numerous German and Japanese cities, culminating in the nuclear devastation of Hiroshima and Nagasaki. This bombing killed hundreds of thousands of civilians. Although the massive killing of noncombatants did not provoke widespread protests or recriminations among Americans at the time, the aftermath was not a simple story of acceptance of the practice as a common and legitimate method of warfare in a new technological age of air power. The experience of the Korean War demonstrated that American moral scruples against targeting civilians did not disappear with the bombing in World War II, as some historians have argued.[1] Instead, American norms about bombing civilians followed a more complicated evolution.

Only five years later, the Korean War followed the pattern set by World War II of massive civilian destruction inflicted by bombing. Nevertheless, American leaders continued to claim throughout the war that U.S. air power was being used in a discriminate manner and was avoiding harm to civilians, as they had asserted even during the height of the bombing in World War II. The elasticity of the definition of a "military target" helped make these claims of discrimination more plausible. The new bombing capabilities contributed to stretching the definitions of military targets because they brought new portions of civilian societies, such as transportation networks, arms factories, and their workers, within reach and under consideration for targeting. However, the American experience during the Korean War suggests that a dynamic of escalation stretched the definitions of "military targets" even more. As military crises threatened and the war dragged on, American commanders vastly expanded the portion of the enemy's society deemed to be a "military target." While the loose semantics of military targets made it easier to claim publicly that prohibitions on targeting civilians remained, the prohibition found active reinforcement in the United States' prominent role in the post–World War II war crimes trials of Germans and Japanese. Having held their former enemies accountable for harming civilians, Americans worked to distance themselves from similar atrocities, and the international competition of the Cold War only increased the stakes for American identity and political interests. In short, the broadly accepted moral prohibition against targeting civilians did not disappear with the bombing in World War II and Korea.

Although the norm against targeting civilians remained robust in the face of the technological transformations surrounding air power, the new bombing capabilities did foster several related changes in thinking about war's harm to civilians and in international humanitarian law. One of the most significant was the increased importance of intention in rationalizing harm to noncombatants. For Americans, the crucial dividing line between justifiable and unjustifiable violence increasingly became whether their armed forces intentionally harmed civilians. With this reasoning, unintended harm—what later would be called "collateral damage"—became a tragic but acceptable cost of war. The difficulties of controlling the violence of air power made common and widespread unintended harm plausible. American weapons might inflict massive casualties on civilians, as they had in World War II and Korea, but only intentionally targeting civilians remained a crime. International humanitarian law lagged behind the development of public norms on bombing but did eventually formally incorporate restrictions on bombing and in particular reflected this growing emphasis on intention. While other changes in thinking about bombing civilians are more difficult to assess because of the changing nature of American wars after Korea and limited access to sources related to more recent conflicts, Americans did come to accept that certain portions of civilian society that directly supported the fighting capabilities of armed forces, such as arms factories and their workers, were justifiable targets for attack.

The World War II Background

On the eve of World War II, American leaders strongly condemned the bombing of civilians. Following Japanese air strikes in China and fascist bombing in Spain, the U.S. Senate issued its own "unqualified condemnation of the inhuman bombing of civilian populations" in 1938. When Germany invaded Poland in 1939, President Franklin D. Roosevelt urgently appealed to all sides in the hostilities to affirm publicly that their armed forces "shall in no event, and under no circumstances, undertake the bombardment from the air of civilian populations or of unfortified cities." He condemned such attacks as "inhuman barbarism."[2] As the fighting in Europe escalated, the American press contained regular discussion of the bombing of civilians by both the Germans and the British.[3] These public expressions of concern suggested that Americans supported a transnational norm against attacks on civilians, from bombing or otherwise, or that, at least, American leaders and journalists thought this norm had widespread support. World War II offered further evidence of this norm's existence.

Indeed, judged from the perspective of what American leaders said about the bombing of civilians, little changed during World War II, even at the height of the air campaigns against Germany and Japan. They continued to talk as if they were trying to uphold the prohibition against targeting civilians, even though the reality of civilian deaths strained the credibility of their claims. U.S. armed forces described their strategic bombing methods as precision bombing throughout the war.[4] When American planes joined the British Royal Air Force in burning Dresden in February 1945, Secretary of War Henry L. Stimson assured the public: "We will continue to bomb military targets and . . . there has been no change in the policy against conducting 'terror bombings' against civilian populations." When asked off the record about the burning of Tokyo at a press conference, Air Force spokesman General Lauris Norstad denied that there had been any change in the Air Force's basic policy of "pin-point" precision bombing.[5] Even President Harry S. Truman in his initial public statements described the attack on Hiroshima as a strike against "a Japanese Army base" and said that "we wished in this first attack to avoid, insofar as possible, the killing of civilians."[6]

So even in the face of these gross violations of the custom of actually sparing civilians, American leaders persisted in publicly deferring to a norm against targeting civilians, insisting that they were only bombing military targets and rarely claiming that attacking civilians directly was legitimate. There is still much work to be done to answer the question of whether these statements by American leaders reflected wider public sentiments or political calculation. A better assessment of the breadth and depth of the American public's attachment to the norm against attacking civilians during World War II is also needed. After all, American reactions to the bombing of civilians seem to have been quite muted during the war, and little protest against the bombing occurred.[7] However, several factors could help explain why this apparent quiescence was not proof of Americans abandoning the norm against targeting civilians in war. One was the relative novelty of the extensive killing of civilians through bombing and the limited information Americans had about the attacks during the war, especially when official sources were continuing to claim that air power was being used precisely. Another could have been beliefs that the violence in World War II was exceptional even for war, justified as retribution for German or Japanese aggression and atrocities or because such tactics were a lesser evil than the feared consequences of defeat by the Axis powers.

Although Americans were quiet about the harm to civilians resulting from U.S. bombing, they spoke out loudly against German and Japanese atrocities. Condemnation and prosecution of Axis atrocities after World War II provided the strongest reinforcement of the norm against attacking civilians. The Nuremberg tribunals in Germany and a similar set of war crimes trials of the Japanese focused international attention on the harm that Axis leaders and soldiers had inflicted on civilians and held them criminally accountable for it. This assertive application of international law and the leading role that the United States played in these prosecutions reinforced the impression that Americans remained committed to the norm against attacking civilians. However, conscious of the snares of hypocrisy, none of the tribunals prosecuted any of the defendants for promiscuous bombing of civilians. As U.S. relations with the Soviet Union deteriorated, Americans increasingly sought to distinguish clearly American killing of civilians in the past war and their strategies for fighting future wars in an atomic age from the crimes of Nazi Germany and Imperial Japan. In clashes with the United States, the Soviet Union enthusiastically condemned the American armed forces for relying on barbarous methods of bombing civilians to fight imperialistic wars.[8]

The Korean War

When the United States intervened in the fighting on the Korean peninsula in 1950, Americans continued to proclaim a norm against targeting civilians, even though, like World War II, the Korean War would become massively destructive of civilian lives and property. However, the devastation did not come immediately. American leaders explicitly rejected the fire-bombing of North Korean cities in the early days of the war. The Korean War would not begin as World War II had ended. The experiences of 1945 had not made the obliteration of cities and their populations the standard tactic for U.S. air power, only one of a range of options. Firebombing and the widespread harm to Korean civilians would only come after a process of escalation and dramatic setbacks for United Nations forces in the fall of 1950.

Only days after the outbreak of heavy fighting in Korea on June 25, 1950, President Truman ordered U.S. air attacks against North Korea as part of the American-led intervention by the United Nations. The instructions from Washington for UN commander General Douglas A. MacArthur specified a narrow range of "purely military targets" for attack such as air bases, depots, tank farms, and troop columns.⁹ MacArthur's bomber commander, General Emmett "Rosy" O'Donnell, had other ideas. O'Donnell recommended incinerating North Korea's five largest cities. MacArthur could announce to the world, O'Donnell proposed, that he was going to employ, against his wishes, the means which "brought Japan to its knees." The announcement could ease concerns over harming civilians by serving as a warning, as O'Donnell put it, "to get women and children and other noncombatants the hell out." According to O'Donnell, MacArthur responded, "No, Rosy, I'm not prepared to go that far yet. My instructions are very explicit; however, I want you to know that I have no compunction whatever to your bombing bona fide military objectives, with high explosives, in those five industrial centers. If you miss your target and kill people or destroy other parts of the city, I accept that as a part of war." MacArthur was not yet ready to destroy entire enemy-held cities but was willing to accept the risk of unintended harm to civilians.¹⁰

After rejecting O'Donnell's recommendation for incendiary attacks, MacArthur had his commander of the Far East Air Forces (FEAF), General George E. Stratemeyer, issue a directive on bombing. It forbade O'Donnell from attacking "urban areas" as targets but authorized strikes against "specific military targets" within urban areas. Two days earlier, Stratemeyer's director of operations had written a memorandum, approved by the FEAF commander, which said that "reasonable care" should be exercised in air operations "to avoid providing a basis for claims of 'illegal' attack against population centers."¹¹

Accompanying their measures to limit bombing damage to cities, American leaders strongly proclaimed their commitment to avoiding harm to civilians. "The problem of avoiding the killing of innocent civilians and damages to the civilian economy is continually present and given my personal attention,"

General MacArthur asserted in his public reports to the UN.[12] In response to a flood of accusations from communists,[13] Secretary Acheson denied that UN forces were "bombing and killing defenseless civilians."[14]

As the early months of the fighting demonstrated, the Korean War began, as World War II had, with efforts to distinguish between military targets and civilians and public condemnation of attacks against noncombatants. The devastating aerial campaigns of 1945 had not annihilated the norm against targeting civilians nor made indiscriminate destruction inevitable. However, the Korean War, like World War II, would demonstrate a dynamic of escalation that rendered the persisting norm against targeting civilians largely impotent to actually save civilians from harm.[15]

In early November 1950, when UN soldiers first fought with Chinese units, the UN Command adopted a policy of the purposeful destruction of cities in enemy hands. The Far East Air Force began incendiary raids against urban areas reminiscent of those of World War II, and MacArthur spoke privately of making the remaining territory held by the North Koreans a "desert."[16] Yet, as they had during World War II, American leaders persisted in describing their escalated aerial attacks as discriminating strikes against military targets. To do this, U.S. commanders stretched the definition of "military target" far beyond its usual meaning. This elasticity, tied to a dynamic of escalation, was visible from the opening of the UN fire-bombing campaign. As one of its first objectives, the UN command selected for destruction the city of Sinuiju, a provincial capital with an estimated population of over 60,000, that was across the Yalu River from the Manchurian city of Antung.

The previous month, General MacArthur had restrained his FEAF commander General Stratemeyer in his bombing of Sinuiju. Stratemeyer had asked for the authorization of an attack "over the widest area of the city, without warning, by burning and high explosive," but he was willing to settle for an attack only against "military targets in the city, with high explosive, with warning." Here Stratemeyer was still distinguishing between specific military targets within a city and attacks on the city as a whole. Stratemeyer offered no direct military justification for the attack but instead argued that Sinuiju could be used as the capital of North Korea once Pyongyang was evacuated, which would provide more legitimacy to the communist government than if it were a refugee government on foreign soil. He also believed the psychological effect of a "mass attack" would be "salutary" to the Chinese across the Yalu. The closest Stratemeyer came to a military justification for the attack was his observations that the city served as a rail exchange point between Korea and Manchuria and that the city had considerable industrial capacity that could provide "some means" of supporting a North

Korean government, but he did not tie either of these points to the fighting then occurring. MacArthur's headquarters returned a reply to Stratemeyer's suggestion the next day that read: "The general policy enunciated from Washington negates such an attack unless the military situation clearly requires it. Under present circumstances this is not the case." MacArthur was still refusing his air commanders' pleas for incendiary attacks, but this would not last long.[17]

On November 3, Stratemeyer again asked MacArthur for permission to destroy Sinuiju. That day Stratemeyer forwarded the request of General Earle E. Partridge, commander of the Fifth Air Force, for clearance to "burn Sinuiju" because of heavy anti-aircraft fire from the city and from Antung. Stratemeyer's conversation with MacArthur demonstrated the subjectivity of a "military target" for the UN commanders, especially when they desired escalation. General MacArthur said he did not want to burn Sinuiju because he planned to use the town's facilities once the 24th Division seized it. MacArthur did grant permission to send fighters to attack the anti-aircraft positions in Sinuiju with any weapon desired, including napalm. Stratemeyer then raised the subject of the marshalling yards near the bridge between Sinuiju and Antung, and MacArthur told him to bomb the yards if Stratemeyer considered them a military target.

At the meeting, Sinuiju was spared from burning, but another North Korean city was not so lucky. MacArthur desired an increase in the use of the B-29 bombers, which had run short of targets, so he was sympathetic to Stratemeyer's further recommendation to attack the town of Kanggye. The Air Force commander suggested the FEAF could burn several towns in North Korea as a "lesson" and indicated that Kanggye was a communications center for both rail and road and was occupied, he believed, by enemy troops. MacArthur answered: "Burn it if you so desire. Not only that, Strat, but burn and destroy as a lesson any other of those towns that you consider of military value to the enemy." MacArthur left the decision to his air commander. Apparently, MacArthur did not feel the towns to be so vitally important to the enemy's war effort that it was obvious to him that they had to be destroyed, but Stratemeyer's idea about teaching the communists a lesson appealed to him.[18]

MacArthur's prohibition on burning Sinuiju lasted only a few hours this time. The general may have changed his mind because of the intelligence he was receiving that more than 850,000 Chinese soldiers had gathered in Manchuria. By the evening, MacArthur's chief of staff told Stratemeyer that the burning of Sinuiju had been approved. On November 5, MacArthur conveyed his new instructions to his air commander. Stratemeyer wrote in his diary that the "gist" of these instructions was: "Every installation, facility, and village in North Korea now becomes a military and tactical target." The only exceptions were to be hydroelectric power plants, the destruction of which might provoke further Chinese intervention, and the city of Rashin, which was close to the Soviet border.

Stratemeyer demonstrated a single-mindedness in carrying out MacArthur's wishes even at the risk of unwanted destruction. Stratemeyer's staff pointed out to him how reported sites of POW camps, hospitals, and prisons would be vulnerable to incendiary attack. The Air Force commander later wrote in his diary about the danger to these sites, "Whether vulnerable or not, our target was to take out lines of communication and towns." Stratemeyer sent orders to the Fifth Air Force and Bomber Command "to destroy every means of communications and every installation, factory, city, and village." In reviewing Stratemeyer's orders, MacArthur had him add a sentence that explained the rationale for the escalation. Inserted immediately after the phrase about destroying all communications and settlements, the sentence read, "Under present circumstances all such have marked military potential and can only be regarded as military installations." Stratemeyer included a similar rationale in his cable to the Air Force chief of staff about the attack: "Entire city of Kanggye was virtual arsenal and tremendously important communications center, hence decision to employ incendiaries for first time in Korea."[19]

Several points are worth stressing about these remarkable exchanges between MacArthur and his air commander. Before MacArthur decided to escalate, the UN commander and Stratemeyer were distinguishing the targeting of specific structures defined as military targets from the targeting of urban areas as such. The anti-aircraft batteries in Sinuiju were the clear example of a "military" target, but even before the decision to escalate, some targets were more ambiguous, such as the city's marshalling yards. The commanders were also tempted to initiate area attacks because of their beliefs in the potential political and psychological effects the strikes might have on the enemy, even though those effects were at best indirectly related to the actual fighting then occurring.

However, it is crucial to note that the generals never explicitly defined noncombatants as legitimate targets, even though Stratemeyer readily risked the destruction of hospitals, POW camps, and prisons. The generals escalated the war by targeting the physical infrastructure of cities and sought political and psychological benefits from this destruction, but there is no evidence that they talked, even privately among themselves, about aiming to kill enemy civilians or about gaining benefits from those civilian deaths. It is conceivable that killing civilians could have been their underlying intention and motivation, but it is exceedingly difficult to demonstrate convincingly an individual's state of mind at a given time, and the historical evidence that has come to light does not suggest that the UN commanders were thinking specifically about killing civilians.

The episode did demonstrate the instability of the definition of a military target, which slid within hours from preventing the burning of Sinuiju to justifying it. Instead of defining anti-aircraft batteries and railroad yards as the only military targets in Sinuiju, MacArthur redefined the entire physical infrastructure of the city as a military target and showed how quickly structures usually considered civilian became open for attack. The generals also employed new (and possibly disingenuous or muddled) attempts to obscure or justify the escalation. The attack on Kanggye, which Stratemeyer had justified to MacArthur for its potential as a "lesson" and for its transportation capacity and its possible housing of enemy troops, suddenly became necessary because the city was a "virtual arsenal" and a "tremendously important communications center." While some of these points may sound like the second-guessing of difficult military decisions based on the limited information of historical hindsight, even if one agrees with every decision MacArthur and Stratemeyer made, their conversations suggested that pressures to escalate stretched the definition of military targets well beyond its common usage.

The "fire job," which General O'Donnell had advocated in July but Washington had forbidden, commenced in early November. On November 8, the FEAF showered 500 tons of incendiary bombs on more than one square mile of Sinuiju's built-up area, destroying 60 percent of the city. In O'Donnell's report on the work of his bombers, he declared that "the town was gone." Other towns were to follow. By November 28, Bomber Command reported that 95 percent of the town of Manpojin's built-up area was destroyed, for Hoeryong 90 percent, Namsi 90 percent, Chosan 85 percent, Sakchu 75 percent, Huichon 75 percent, Koindong 90 percent, and Uiju 20 percent. As UN units withdrew from the major North Korean cities, those cities too became targets. On December 30, the FEAF commander informed his subordinates that they had the authority to "destroy" Pyongyang, Wonsan, Hamhung, and Hungnam, four of North Korea's largest cities. The FEAF conducted the attacks without warning to the civilian population and purposefully avoided publicizing the strikes. By the end of the war, eighteen of twenty-two major cities in North Korea had been at least half-obliterated, according to damage assessments by the U.S. Air Force. The fire-bombing of North Korean communities that commenced in November made meaningless the earlier claims of the FEAF that their bombing operations avoided the destruction of residential areas.[20]

However, just as during World War II, Americans' depiction of their fighting as employing discriminating means changed little. Military officers and the press proceeded to discuss the violence in Korea as if its application remained discriminate and as if risks to noncombatants had not increased. The objects of attack were still "military targets," but the implicit definition of the term "military target" had grown to include virtually every human-made structure in enemy-occupied territory. The norm against targeting civilians survived within this definition, in the sense that Americans never came to the point of arguing that the civilian population itself was a "military target" and therefore a legitimate object of attack, but the expanded definition of the term and the acceptance of the destruction it entailed offered meager protection for Korean civilians.

While avoiding direct acknowledgment that UN forces were systematically burning North Korean cities, the UN Command did admit that it had escalated the air war. UN commanders offered new justifications for the expanded destruction that clung to the notion that its airplanes were attacking military targets. The justifications diverged from what had been the Air Force's primary vision of a strategic air offensive to destroy war-supporting industries in order to deprive the enemy's forces in the field of weapons, ammunition, and supplies.

Instead, the Air Force viewed its escalated bombing in Korea as part of a campaign to interdict the flow of weapons, supplies, and additional men to the communist army in Korea and explained it to the public as such. But the campaign went beyond precise attacks against transportation and communication systems in North Korea in which bridges, railroad yards, docks, and vehicles were targets. UN forces undertook the destruction of entire towns, particularly those along major transportation routes from Manchuria and the Soviet Union, in order to deprive the communists of shelter in which to conceal their supplies and soldiers from the UN airplanes. The destruction also stripped the enemy soldiers of protection from the elements during the winter campaign.

Nevertheless, the UN forces rarely acknowledged that this escalation was destroying entire communities and placing Korean civilians at risk. Public communiqués from the UN Command avoided discussing or justifying the destruction of Korean towns and villages directly. Instead, the press releases named "buildings," often identified as enemy-occupied or as structures for storing supplies, as the usual target of UN airplanes, disaggregating the communities into their constituent structures. Within the Air Force, the square footage of buildings destroyed eventually became a semi-official measure of progress in the air campaign.[21] The press releases of the UN Command also avoided directly acknowledging attacks on entire villages and towns by the use of the term "supply center" and similar phrases such as "communications center," "military area," and "built-up area," often implying that whole communities served military purposes.[22]

However, the reliance of the press releases on describing operations as attacks on "buildings" and "supply centers" was not always enough to quiet the UN Command's fears about the American image in Korea. In August 1951, the UN Command's Office of the Chief of Information wrote a memorandum for the Public Information Office of the Far East Air Force. The memo said that General Matthew B. Ridgway, MacArthur's replacement, had suggested that in news releases of targets destroyed by air attacks, the Air Force publicists might "specify more definite military targets" such as tanks, anti-aircraft guns, or armored vehicles. This would prevent anyone from pointing to the releases as evidence that American forces were "wantonly attacking mass objectives such as cities and towns" in North Korea. The UN Command, despite its expanded air attacks, continued to present the war it was waging as a discriminate use of force directed solely against military targets.[23]

These public relations efforts met with considerable success in the United States. Press coverage of the escalated air assault did not challenge the comforting picture the UN Command presented. Newspapers did note that the UN forces had initiated some of the largest air strikes of the war in November and occasionally acknowledged the burning of entire cities. Nevertheless, the reporting indicated the military usefulness of destroying the physical infrastructure and avoided discussing the impact of the destruction on civilians.[24] This picture of a discriminate use of air power in Korea has survived in many of the historical treatments of the war, including the official Air Force history[25] and a number of popular military histories and cursory scholarly accounts of the air war in Korea.[26] Only recently have Americans begun to acknowledge the full extent of the fire-bombing campaigns in histories of the Korean War.[27]

As in World War II, U.S. air power inflicted massive harm on civilians during the Korea War and diverged from the customary practice of sparing civilians from the violence of war. However, this violence came through a process of escalation during the war. Area bombing did not supplant precision bombing as the standard method of employing air power against an enemy, but it remained an option when the fighting escalated. Even with the undeniable widespread harm Korean civilians suffered from U.S. weapons,

Americans clung to the normative value of avoiding direct attacks against noncombatants, a norm buttressed by international humanitarian law and the precedents of Nuremberg. They almost never advocated, publicly or privately, within the armed forces or outside them, the purposeful targeting of civilian populations as such. The stunning contradictions between lethal consequences and proclaimed scrupulousness were eased by the elastic definitions of military targets, but other changes in thinking about harming civilians assisted in this tortured reconciliation as well.

One of the most significant changes was the emerging emphasis on intention as the crucial distinction between justifiable and unjustifiable harm to civilians in war. Americans and a broader transnational consensus, which was eventually reflected in international humanitarian law, placed less importance on whether civilians were killed than on whether they were killed intentionally. It was not that intentional killing was identified as a new wrong after World War II; the norm against attacking civilians had all along implied prohibition of intentional attacks. It was rather that the massive expansion of firepower that was difficult to control, as exemplified by American air power, created a novel cultural space for plausible unintentional destruction on a tremendous scale. When wars were fought with spears, or even with cannon or rifles, the relative ease with which these weapons could be directed against a specific target left little room for questions of intent. In face-to-face warfare, warriors attacked individuals that they could identify as combatants or as bystanders, and intention was usually manifest in action. Either warriors killed noncombatants purposefully or they spared them. With the introduction of weapons that killed over long distances and devastated great areas, intent no longer clearly followed from action. Common and widespread unintended destruction became plausible. The great acceleration of this trend toward uncontrollable firepower in the twentieth century contributed to making intention crucial to Americans' thinking about attacking civilians. Americans rationalized harm to noncombatants from violence that they could not control as a tragedy of war but not a crime.

The Korean War clearly illustrated this preoccupation with intention. Americans' public insistence throughout the war that they discriminated between military targets and civilians sought to demonstrate that Americans did not intend to kill civilians. In addition to their extensive talk about intentions, Americans pointed to their military's efforts to warn civilians of air attacks and evacuate them from combat areas. UN forces regularly broadcast warnings to civilians by radio and loudspeaker and conducted a number of operations where warning leaflets were dropped on communities.[28] These warnings, while of dubious value in actually protecting civilians, were well covered by the American media.[29] UN forces also tried to assist civilians by conducting several large operations to evacuate them out of harm's way during the winter retreat.[30] Even though these evacuations assisted only a small fraction of the Koreans who were threatened by the war's violence, the U.S. press lauded these operations and other well-intentioned deeds by American soldiers on behalf of civilians.[31]

After the war, the U.S. Army's revised field manual on the law of land warfare introduced a new statement that expressed as doctrine the growing importance of intention. The revised 1956 manual said, "It is a generally recognized rule of international law that civilians must not be made the object of attack directed exclusively against them."[32] Previous army manuals had left this rule unexpressed. As a subculture, military professionals may have placed even more emphasis on their intentions not to harm noncombatants even in the face of widespread civilian deaths. While the sources make it difficult to assess the personal sentiments of officers and soldiers about civilian casualties during the Korean War, it is not hard to believe that many in private did not want to think of themselves as waging war against defenseless civilians.[33]

This focus on intentions assisted in leaving the vital core of a norm against attacking civilians intact. Americans did not come to accept the targeting of civilians as a legitimate method in the Korean War. Nevertheless, the focus on intentions encouraged by new air-power capabilities created a tendency in American thinking that was extremely dangerous to civilians in war. Americans came to condone unintended civilian casualties as an acceptable human cost of war, what would later be called "collateral damage."[34]

International humanitarian law evolved slowly to reflect the changing norms about bombing and attacking civilians and the increased importance of intention, but the laws have lagged far behind broader attitudes. When the 1949 Geneva Conventions were revised following the experiences of World War II, they were almost completely silent on the threat to civilians from bombing. Although negotiators composed an entirely new convention for the protection of civilians in wartime, the protections concerned almost exclusively civilians in occupied territory, not civilians still behind their side's front lines, precisely the people who were most vulnerable to strategic bombing. At the 1949 Geneva conference, the Americans and the British opposed both the inclusion of restrictions on bombing and the Soviet Union's attempts to use the treaty to outlaw atomic weapons. Two of the American negotiators later wrote, "It is to be emphasized that these 'grave breaches' do not constitute restrictions upon the use of modern combat weapons. For example, modern warfare unfortunately and often may involve the killing of civilians in proximity to military objectives, as well as immense destruction of property."[35] The 1949 agreements shielded only hospitals from all forms of attack, including bombing, and otherwise proposed voluntary establishment of safety zones where noncombatants could be sheltered from the effects of war. Although the United States and the UN forces agreed to abide by the Geneva Conventions in Korea, the laws provided few impediments to the use of American air power. When the International Committee of the Red Cross (ICRC) and the United Nations raised the idea of the creation of safety zones in Korea to protect women, children, and the elderly from the ravages of war, the United States rejected the proposal out of concern that neutral observers could not be found to ensure that the safety zones in North Korea were not contributing to the war effort.[36]

Legacies

After the Korean War, the ICRC began to circulate draft rules for the protection of civilian populations from the dangers of indiscriminate warfare, but it took years for protections against targeting civilians to be written into international law. In 1968, the UN General Assembly affirmed a Red Cross resolution that banned attacks against civilian populations as such. In 1977, an international conference completed the drafting of two additional Protocols to the Geneva Conventions of 1949. The first and second Protocols, which related to the protection of victims of international and non-international armed conflicts, respectively, each included the provision: "The civilian population as such, as well as individual civilians, shall not be the object of attack. Acts or threats of violence the primary purpose of which is to spread terror among the civilian population are prohibited."[37] Only slowly did international law come to embody the increased importance of intention that the norm against targeting civilians had acquired.

Beyond the growing importance of intention in defining legitimate uses of force in war, it is much more challenging to assess the legacy of the rise of bombing after World War II on norms because of the changing nature of conflicts the United States fought after Korea and the unavailability of crucial sources. Despite these challenges, one normative belief appears to have been firmly established among American military leaders and to have become noncontroversial among a wider public: that the weapons of war and military supplies before they found their way to soldiers' hands were a worthy target. Bombing behind the front lines of battle opened up the possibility of destroying arms and supplies before they could be used by enemy forces, through attacks on factories or the transportation networks through which this matériel flowed. This disarming strategy was the favorite justification of bombing by commanders and civilian advocates of air power, as was clearly shown during the Korean War.[38] The U.S. Army's 1956 field manual on the law of land warfare also incorporated this new understanding into the revisions of the previous manual from 1940. In narrowing the Hague Convention prohibition on the bombardment of undefended places, the manual clarified that this did not preclude strikes against military supply. The new manual said, "Factories producing munitions and military supplies, military camps, warehouses storing munitions and military supplies, ports and railroads being used for the transportation of military supplies, and other places devoted to the support of military operations or the accommodation of troops may also be attacked and bombarded even though they are not defended."[39] These parts of civilian society behind the front line were deemed a vital component of a war effort, and few during the Korean War or since have challenged the legitimacy of these sources of supply as targets. The distinctions between civilian and military and defended and undefended became less important than the difference between noncombatant and combatant. Just as a civilian factory could produce supplies for the military, a soldier could become a noncombatant once wounded and incapacitated. An individual's or resource's relationship to the actual violence of war became the most important determinant of whether they were legitimate targets for attack.

Other changes in thinking about bombing civilians are much more difficult to assess. For example, the subjectivity in choosing "military" targets has not necessarily decreased in the wars since Korea. Given the elaborate expressions of official American concern over civilian casualties, it might be tempting to argue that the wars in the Persian Gulf, Iraq, and Afghanistan have encouraged more precise and rigid definitions of military targets. Nevertheless, these definitions have not been tested, as they were in the Korean War. These later wars have been severely asymmetrical conflicts, and American forces and commanders were not strained in the ways they were in Korea, let alone during World War II. Definitions of military targets may still be elastic, but recent wars may not have necessitated the type of escalation that encouraged this flexible thinking.

In other areas where changes in thinking about bombing civilians might seem apparent, a closer examination may reveal their superficiality. Indisputably, the United States has conducted less area bombing in its wars since Korea, but this could simply be because it has fought fewer evenly matched wars and has faced fewer desperate decisions to escalate. It might also be tempting to believe that American commanders in recent wars have resisted the temptations to which MacArthur and his air commanders succumbed of justifying bombing attacks for their political and psychological effects instead of for their directly military impact. However, limited current access to sources and records about these highly classified internal discussions hampers a full assessment.[40]

Finally, more active efforts to avoid civilian casualties in recent American wars, such as the expanded role of operational law and military lawyers in targeting, may be more a result of the rise of counterinsurgency thinking than evidence of a growing belief among Americans that killing civilians is wrong. Counterinsurgency doctrine has emphasized the importance of winning the support of civilian populations in civil wars as a means to military victory. From Vietnam to Afghanistan, American commanders have tried to limit civilian casualties in order to avoid alienating civilians.[41] The rise in counterinsurgency doctrine is an important change in military thought, but one tied more to the changing nature of American wars than to norms about bombing civilians.

In assessing changing norms about bombing after World War II, it is crucial to distinguish among the changes in values, ideas, laws, and behavior that the term "norm" can encompass. These distinctions make it easier to summarize how norms about bombing changed after World War II. The transnational normative value that prohibited attacks on civilians persisted. However, the actual protections it offered to civilians were undermined by the new bombing capabilities. Because of the difficulties with controlling the violence of modern weaponry, intention gained great significance in moral justification, and this focus helped rationalize unintended harm and contribute to a complacent stance toward the terrible human cost of collateral damage. On the other hand, normative behavior or customary practice did change, at least temporarily, during both World War II and Korea. As the wars escalated, U.S. armed forces conducted unprecedented fire-bombing and other area attacks against cities and towns that proved deadly to civilians, and the flexibility of the definition of "military targets" facilitated these area attacks. International humanitarian law also evolved to catch up with the growing significance of intentional attacks, but at a relatively slow rate. Finally, while normative beliefs about bombing civilians are the hardest to assess, Americans have come to accept the idea that bombing behind the front lines with the goal of disarming was an effective and acceptable method of fighting.

The decade after World War II and the experience of the Korean War laid a foundation for the sensitivity to civilian casualties that became evident in the American wars of the late twentieth and early twenty-first centuries.

This foundation was not built through a recovery of the norm against targeting civilians spurred by the trauma of the Vietnam War after a period when the norm had been abandoned. The role of the Vietnam War in changing American attitudes toward civilian casualties was not so crucial because many of these changes, such as the growing significance of intention, began earlier and because much about these attitudes has remained relatively constant from the 1930s to the 1970s and will likely remain so into the twenty-first century. Instead, the Korean War experience demonstrated the durability of the norm against targeting civilians even in the face of mass killing from bombing or otherwise. Adherence to the norm persisted even though the norm provided severely limited protections to civilians when bombing was employed and conventional wars escalated. In avoiding massive killing of civilians in their wars since Vietnam, Americans may not have become more virtuous but only more fortunate in not having to fight more evenly matched wars.

CHAPTER 3

Targeting Civilians and U.S. Strategic Bombing Norms

Plus ça change, plus c'est la même chose?

NETA C. CRAWFORD

United States leaders' normative beliefs about targeting civilians with conventional strategic bombing and the practices themselves have changed dramatically over the last nearly seventy years. Specifically, before and during World War II, and to a lesser degree in Korea, military and civilian leaders believed that targeting civilians was militarily necessary and effective. Perceptions of military necessity consistently trumped the value of civilian immunity, which itself was an emerging normative belief. It was considered acceptable to deliberately target civilians and to be relatively unconcerned when civilians were harmed incidentally as "collateral damage." Targeting civilian morale and economic infrastructure generally led to the same consequences as deliberately targeting civilian bodies, since those bodies were often located alongside economic assets.

During the long Vietnam War, which I argue constitutes a turning point in U.S. policy, it became less acceptable among military professionals and the public to deliberately target civilians or to strike in ways that could lead to foreseeable harm. Ad hoc procedures put constraints on bombing that could harm civilians. After Vietnam, declaratory policy and operational planning increasingly emphasized protecting civilians and U.S. authorities instituted methods to mitigate civilian casualties. The emphasis on civilian casualty avoidance and its institutionalization is seen in the first Gulf War and U.S. air operations in the Balkans. While there was still a degree of tolerance for collateral damage during the first years of the post-9/11 U.S. wars in Afghanistan, Iraq, and Pakistan, that tolerance was gradually diminished, leading to more stringent control of planned air strikes in Afghanistan in 2009. In May 2013, President Obama articulated zero tolerance for anticipated collateral damage in drone strikes outside war zones.[1]

Explaining Normative Change

Why have U.S. normative beliefs and practices of conventional strategic bombing changed?[2] I suggest three possible reasons. I then review U.S. bombing norms and practices from the Vietnam War through the post-9/11 wars, emphasizing the main turning points.

The first possible explanation suggests that changes in the *normative beliefs* (beliefs about what is right and wrong) of elites and the general public about targeting civilians caused the United States to alter its bombing practices. Prior to the Vietnam War, the U.S. military believed that bombing civilians was both effective and morally justified in the sense that it would bring a quick end to a war. When the U.S. civilian and military leadership began to believe that targeting civilians was morally wrong, it began to change its bombing practices.

If the normative-change explanation is correct, changes in the normative beliefs articulated by leaders should precede changes in doctrine and practice. Further, if changing normative beliefs about the morality of targeting civilians drives changes in bombing doctrine and practices, those beliefs will likely be consistently implemented across U.S. military practices. In other words, consistent with new beliefs about the importance of civilian immunity, the United States should change other practices, such as the use of ground forces in ways that harm civilians, and retire weapons that are indiscriminate in their effects. We will have greater confidence in the normative-change explanation if the U.S. leadership believes that targeting and hurting civilians is militarily effective, but chooses not to do so.

Why—if this first explanation is correct—did U.S. leaders come to believe that targeting civilians is wrong? To the extent they adopted the principle of civilian immunity, it was part of a global change in views about human rights—viz., that there is something called human rights and they belong to civilians of all sides, even in war. The carnage of the two World Wars, of course, encouraged the development of human rights norms; the Universal Declaration of Human Rights of 1949 and the Genocide Convention of that same year are in part an emotional and moral reaction to the attitudes that led to the World Wars and the Holocaust. This greater respect for human rights also, incidentally, underpins the increased emphasis on force protection; it is no longer acceptable to use soldiers as mere tools whose lives may be "wasted" in the thousands.

The second potential explanation for the shift from targeting civilians to minimizing harm to civilians stresses the elites' understanding of their military-strategic interests, specifically, U.S. *elite understanding of military necessity*. When leaders believed that targeting civilians was militarily productive—by hurting the adversary's civilian capacity to contribute to the war effort or by causing the adversary's civilians to withdraw their support of the war or prompting them to actively oppose their leaders—bombing civilians was an important military objective. When the U.S. military came to believe that targeting civilians was unnecessary, ineffective, or counterproductive for its military objectives, it changed its strategic bombing practices. In short, in this view, doctrine and practices changed because the United States' understanding of how to achieve its objectives changed.

If the military-necessity explanation applies, a change in the understanding of the military effectiveness of targeting civilians would precede a change in practices. The evidence should show that those who advocated civilian targeting on the grounds that it would cause civilian will to collapse lost the argument that strategic bombing was militarily effective. When military leaders decide that strategic bombing does not lead to the collapse of civilian morale or that hurting civilians causes the opposite effect—increased resistance—they should adopt a doctrine of avoiding civilian casualties. Or, in the case of counterinsurgency, if military officials decide that the only way to "win" entails winning the hearts and minds of a target population, they will move to protecting civilians. We will have greater confidence in this explanation if military doctrine emphases the *utility* or *military necessity* of moving from targeting civilians to avoiding civilian harm or emphasizes protecting civilian populations. On the other hand, consistent with the military-necessity argument, if targeting civilians for pain is understood, once again, to be militarily useful or essential, it will perhaps resume.

The third potential explanation combines elite concerns with military effectiveness and the public's sincere normative belief that harming civilians is wrong to stress the negative political effects of the loss of moral legitimacy that occurs when civilians are harmed through carelessness or deliberate intention. This is a *political-utility* explanation; military necessity is understood to require domestic and international public support and that support depends on adherence to the norms of noncombatant immunity. The concern with civilian casualties thus rests on a desire to prevent one's own population and allied populations from turning against the war.

The military and political leadership of the United States might believe that the costs to public opinion of harming civilians are high because domestic and international publics abhor targeting civilians. If concern for public opinion drives operational change, the military leadership may have no genuine or deep commitment to civilian immunity. Operations are changed because of the perception that internal and external audiences will be disturbed by harm to enemy civilians. Publics may have always abhorred targeting civilians, and what has changed, then, is the dependence of elites on public support. Or the public's views about targeting civilians may have changed. If so, the causes for changing normative beliefs among publics about civilian targeting must be investigated.

If this political-utility explanation is correct, we should find that military and civilian leaders became convinced that the support of their own public and the support of their allies depends on *not* targeting civilians. Biddle, in her chapter, notes that this was a concern of the British in World War II. Further, the change in policy and practice from favoring strategic bombing of civilians should occur when the political costs of the public's rejection of civilian killing becomes clear to military and political leaders. We should subsequently see an effort to reduce civilian casualties, combined with a considerable effort by U.S. and allied governments to manage *civilian perceptions* of treatment of noncombatants. This might take the form of arguing about the numbers of civilians killed—attempting to minimize reported deaths in any particular incident or by suggesting that the killing of civilians was inadvertent and that everything was done to reduce harm to civilians. One way to minimize the political impact of harm to civilians is to suggest that some amount of civilian killing in war is inevitable. Thus, American and international public abhorrence of harm to other civilians may be ameliorated in a two-step argument: the first part emphasizes that the harm is unintentional; the second part stresses that the harm to civilians is inevitable—always and forever a cost of war.

Sahr Conway-Lanz argues that Americans have traditionally valued the immunity of noncombatants and even after the atomic bombings of World War II "tenaciously clung to the optimistic assumption that violence in war could still be used in a discriminating manner despite the increased destructiveness of weapons."[3] When the Korean War and the hydrogen bomb made it difficult to ignore that U.S. military action had inflicted or would inflict massive harm on civilians, Americans resolved this dilemma by focusing on intention. In the early 1950s, a new interpretation of noncombatant immunity emerged that incorporated elastic definitions of a "military target" and made intention the dividing line between justifiable and unjustifiable action.

Any harm the United States inflicted on noncombatants was unintentional, a tragedy for which the responsibility was diffuse. Only the calculated killing of people who are not involved in the fighting of wars remained generally condemned as inhumane and indefensible.[4] I agree that "the centrality of intention has contributed to a complacent stance toward the problem of collateral damage."[5] Officials consistently appealed to military necessity to explain and excuse harm to civilians as unavoidable.

The evidence can support all three causal explanations. Indeed, it is possible for all of them to be true at various levels and at various points in the history of U.S. strategic bombing practice.

Targeting Civilians in Vietnam

The Vietnam War was a turning point in U.S. bombing norms and practices. Many scholars, including Tami Biddle and Sahr Conway-Lanz in this volume, have shown that although there were debates among air planners in the period from World War I through Korea, the policy of the United States was or became one of targeting civilians. Even if the United States initially avoided civilians, three theories of how bombing worked—partly inspired by European air-power enthusiasts—eventually led planners to target civilians. The first theory was that to win wars, civilian morale had to be broken and that bombing could do it. Second, strategists believed that because modern war depends on industrial might, economic infrastructure had to be destroyed and civilians who worked in war-related industries or who lived near them had forfeited civilian immunity. And third, some came to believe it was impossible to distinguish between the military and civilians because in some wars, civilian activities erased the combatant/noncombatant distinction.

In line with these theories, the U.S. bombing of civilians in World War II and Korea was deliberately devastating. Further, there was little domestic criticism. Nevertheless, Americans gradually came to reject deliberate targeting of civilians, even as they forgave unintentional harm to noncombatants. Outcry over bombing civilians during the Korean War, for example, was modest in the United States, although it was more intense abroad.[6]

The air war against Vietnam was from the outset more restrained than bombing in World War II and Korea. American air strikes included four elements: close air support of ground operations in North and South Vietnam, air strikes designed to interdict military supplies, sustained heavy bombing campaigns, and attacks on Cambodia and Laos to interdict supplies and the flow of troops from North to South Vietnam. During the early years of the war, strategic bombing focused on military and economic targets, but targeting political will and civilian morale—and therefore civilians—became increasingly important over time. As with both World War II and Korea, the perception of military necessity would ultimately trump concerns about civilians. Nevertheless, attitudes and practices changed during the Vietnam War as U.S. military leaders and politicians believed that the public political climate had turned against bombing civilians.[7]

The Rolling Thunder bombing campaign, initially conceived as an eight-week operation, extended from March 1965 to October 1968. Its military objective was to interdict the flow of material supplies and troops from North Vietnam to the South by targeting both the source of the supplies in North Vietnam and the supply routes themselves. The political objective was to pressure North Vietnam to make concessions at the bargaining table by steadily ratcheting up civilian pain: "From late 1966 on, they intended to make the North's civilian populace wince from the destruction of military objectives."[8]

Military and civilian leaders within the Johnson administration held intense debates over Rolling Thunder.[9] The military wanted the authority to destroy North Vietnam's military, while most U.S. civilian leaders were convinced that gradual escalation and the threat of more force would induce the North Vietnamese to negotiate.[10] Secretary of Defense McNamara told President Johnson in 1966 that "to bomb the North sufficiently to make a radical impact upon Hanoi's political, economic, and social structure, would require an effort which we could make but which would not be stomached either by our own people or by world opinion, and it would involve a serious risk of drawing us into open war with China."[11] On the other hand, the military felt that restricted bombing would increase U.S. vulnerability in the air and diminish military effectiveness. Secretary of the Air Force Harold Brown said in March 1968 that he wanted the restrictions eased "so as to permit bombing of military targets without the present scrupulous concern for collateral civilian damage and casualties."[12]

During Rolling Thunder, the military, the Office of the Secretary of Defense, and the State Department selected bombing targets and the White House approved them. White House control was tight. Thus, during the escalation of the bombing campaign, civilians vetted the targets according to political criteria and President Johnson himself regularly engaged in the planning. Each Tuesday after lunch, the president, McNamara, and military advisors met to discuss the details of the campaign. Potential bombing targets were categorized along four criteria: military advantage, risk to U.S. aircraft and pilots, estimated civilian casualties, and danger to third-country nationals.[13]

This process extended a long-standing devolution of power toward civilians during the early Cold War. As military historian H. R. McMaster wrote, "The president and McNamara shifted responsibility for real planning away from" the Joint Chiefs of Staff "to ad hoc committees composed principally of civilian analysts and attorneys, whose main goal was to obtain a consensus consistent with the president's pursuit of the middle ground between disengagement and war."[14] In McMaster's view, the civilian analysts did not understand war and hence their strategy of "graduated pressure" was fundamentally flawed. McNamara, he argues, "viewed the war as another business management problem" and "refused to consider the consequences of his recommendations and forged ahead oblivious of the human and psychological complexities of war."[15]

Indeed, civilians set the limits on the initial interdiction campaign. Specifically, there were two zones around Hanoi. In the inner "prohibited" zone of ten miles, attacks required White House authorization and there was no restrike authority. Outside that was a thirty-mile "restricted" area where specific White House approval was still required for striking targets; restrikes were permitted only with further White House authorization. The prohibited and restricted zones around Haiphong were ten and four miles, respectively. Specific White House authorization was required to strike bridges, dikes, levees, dams, and hydroelectric plants. In addition, when the North Vietnamese acquired surface-to-air missiles (SAMs) to augment their other anti-aircraft guns, the White House restricted attacks on those sites as well. SAMs could be targeted only if they were preparing to fire on U.S. aircraft or if they were located away from populated areas. The North Vietnamese capitalized on these limits by putting their anti-aircraft and SAM sites in restricted locations, including dikes.[16]

Bombing restrictions were gradually eased in the summer of 1967.[17] When it ended in October 1968, Rolling Thunder had killed an estimated 52,000 North Vietnamese—about 0.3 percent of the population of North Vietnam.[18] The Air Force denied that it deliberately targeted civilians, and General William Momyer, who commanded the Air Force in the theater, put the blame for some casualties on the enemy, saying that "many of the North Vietnamese claims of civilian damage came about because their own anti-aircraft rounds and SAMs missed their mark and impacted the ground."[19] The "lesson" the U.S. military took from Vietnam was that civilian oversight was too intense. Some blamed the significant losses of U.S. aircraft to North Vietnamese air defenses on the political restrictions on targeting air defenses and on the pauses in U.S. bombing that were meant to give the North Vietnamese time to think about whether they should negotiate.

Although there was a "pause" in the strategic bombing of Vietnam from to 1968 to 1972, there was no pause in the bombing of Laos and Cambodia, which were targeted because the North Vietnamese were moving ammunition and other equipment through those countries on the Ho Chi Minh trail. The bombing of Laos, which began in 1965 and continued on and off through 1973, was also intended to hurt the Laotian Communist movement, the Pathet Lao, by depriving it of food.

The United States dropped not only heavy conventional bombs but also the herbicide Agent Orange and perhaps eighty or ninety million cluster bomblets in Laos. As one villager described the bombing, "they came like birds and the bombs fell like rain."[20] Cluster bomblets, which had "dud rates" as high as 30 or 40 percent, continued to detonate in Laos, killing or maiming children or farmers who came upon them decades after the bombing was halted. Of eighteen provinces in Laos, fifteen remained contaminated with cluster bombs thirty years after the bombing ended.[21] An estimated one-quarter of the Laotian population became refugees, and 350,000 civilians—over one-tenth of the population—were killed in the strikes.[22]

The United States bombed Cambodia from 1965 to 1973, most intensively in secret during the Nixon administration in March 1969. The Joint Chiefs of Staff (JCS) noted that although their target was the Vietnamese troops, "some Cambodian casualties would be sustained in the operation."[23]

One estimate puts the number of Cambodians killed between 1969 and early 1973 during what was known as "carpet bombing" attacks at 600,000; another two million Cambodians were turned into refugees.[24] If the estimate is correct—and there are higher estimates of up to 750,000—nearly 8.6 percent of the Cambodian population was killed by the bombing.[25] The bombing ended when Congress became aware of it in 1973 and ordered a halt. Between 1965 and the end of the bombing, the United States made more than 230,500 sorties against Cambodia, attacking 113,716 sites and releasing 2,756,941 tons of ordnance.[26]

In March 1972 negotiations stalled and North Vietnamese forces initiated another offensive against South Vietnam. The United States then launched the most intensive campaign of bombing yet seen in the war, a series of operations designed to halt the North Vietnamese offensive and induce serious negotiations. On 5 April the United States resumed bombing logistics and petroleum supply points in North Vietnam near the border and began to bomb Haiphong Harbor. In May 1972, the United States began Operation Linebacker, a much wider campaign of bombing and the most intense assault on North Vietnam to that date.[27] Some restrictions on bombing were reduced or eliminated in this campaign. For example, there were no longer any prohibited areas in Hanoi or Haiphong and the restricted zone was reduced to ten and five nautical miles, respectively. During Linebacker, the White House no longer approved target selection and the concern about civilian casualties was relaxed so that the potential for incidental injury to civilians would not be grounds for halting an attack.

Yet civilian targeting in Linebacker was restrained. For instance, the 7th Air Force directed that only laser-guided bombs be used in areas of high population density. When Assistant Secretary of Defense for International Security Affairs John T. McNaughton proposed to systematically destroy the dams and dikes that controlled the Red River, which would have led to flooding the rice fields of North Vietnam and "widespread starvation," McNamara rejected the idea, even though the system of dikes could have been understood as a legitimate military target, since the waterways were important transportation routes.[28] Laser-guided bombs enabled the United States to avoid damage to large numbers of civilians when it attacked the Lang Chi hydroelectric facility in June 1972; the turbines and generators were destroyed but the dam was not breached. There was some damage to the dike system during Linebacker I attacks in 1972, but the military argued persuasively—by producing photographic evidence obtained in bomb damage assessment—that the damage was unintended and incidental to attacks on military targets and that there were few civilian casualties. Linebacker I officially ended in October 1972, although some bombing occurred in November. Guenter Lewy estimates that 13,000 civilians died from U.S. bombing operations from April to 23 October 1972.[29]

When progress at negotiations faltered again in late 1972, the United States began the Linebacker II campaign (also known as the Christmas bombings), eleven days around the clock, to pressure North Vietnam to accept a political settlement.[30] During Linebacker II, military targets in or near Hanoi and Haiphong were attacked. As with Linebacker I, the United States attempted to avoid collateral damage; the JCS instructed the Pacific and Strategic Commands to "minimize risk of civilian casualties" by using laser-guided bombs against "designated targets."[31] Further, B-52 target maps marked the locations of areas that included structures such as schools and hospitals, where civilians would be concentrated.[32] General Momyer felt the restrictions were too tight: "The concern for civilian casualties in many bombing raids allowed many legitimate military targets to go free." Momyer's argument was essentially that since war matériel was located in residential areas, those sites should be cleared for attack. He also suggests how the norms had changed from World War II and Korea:

All residential areas of Hanoi and Haiphong were filled with supplies stacked on each side of the street. Photos showed vehicles lined up bumper to bumper. My feeling was that since such material could be deployed by the enemy to inflict casualties on our forces, it should constitute a legitimate target. Such targets in World War II and Korea would have been cleared for attack without having to query higher headquarters. In Korea restraints were placed on attacks against Pyongyang on separate occasions, but these restraints did not apply to targets of military value within Pyongyang proper.[33]

Despite restrictions, civilians were killed. For example, in a widely publicized incident during the Linebacker II attacks on Hanoi, bombs strayed 1,000 feet from their target, the Bach Mai airfield, to hit Bach Mai Hospital, killing twenty-eight civilians.[34] North Vietnam stated that during Linebacker II, the bombing killed 1,318 in Hanoi and 305 in Haiphong.[35]

Vietnam was a turning point in two senses. First, the use of air power was admittedly constrained by public and political opinion, and these constraints certainly saved lives of noncombatants. Planners were explicit about their desire not to alarm public opinion, and the evidence supports Sebastian Kaempf's claim that "the bombing practice in Vietnam suggests that the sensitivity about civilian casualties had grown since World War II and the Korean War. . . . The norm of non-combatant immunity began to impose restraints on the killing of enemy civilians."[36]

Second, the restrictions intended to protect noncombatants, which were first devised and ordered by civilians, were gradually institutionalized by the military itself: rules of engagement were elaborated, the idea of "precision" strikes was further enabled by the development of more precise weapons, and bomb damage assessment became an important part of the Linebacker I operation.[37] Precision bombing was the rhetoric, and General Momyer was not wrong when he said that "by the end of 1972 we could strike point targets in heavily defended zones, using only a few aircraft, with very high probability of success and very low probability of collateral damage."[38] After the war, the ad hoc procedures and restrictions on bombing intended to protect noncombatants were codified in manuals and training documents.

But Vietnam was hardly an unambiguous turning point. First, many in the U.S. military and civilian leadership were ambivalent, at best, about restrictions on the use of airpower. W. Hays Parks, a lawyer working for the Office of Air Force History, called the concern for civilian casualties in Rolling Thunder a "paranoiac fixation with regard to any civilian casualties" and the restrictions that resulted "unreasonable burdens" that the North Vietnamese were quick to exploit. In Parks's view, the Linebacker restrictions were more reasonable.[39] Similarly, General Momyer said, "I deeply resented the proscription of attacks on North Vietnamese airfields, SAM and AAA sites and other targets. Airmen are bound to resent such restraints; it is an ugly and bitter thing to hold a hand voluntarily behind one's back while being beaten or while watching one's friend being beaten. But self-imposed restraint has been a fact in all U.S. conflict since World War II."[40] Whereas analysts such as Parks claim that restrictions on harming civilians contributed to the U.S. defeat, some members of the U.S. Air Force believe, despite such restrictions, that "it won the war. Ask many airmen about airpower in Vietnam and they will relate the myth of Linebacker Two: how using B-52s over Hanoi and other major cities for eleven days in December 1972 brought the North Vietnamese to their knees."[41]

Second, while the U.S. bombing of North Vietnam was constrained compared to previous wars, the bombing of Cambodia and Laos was less so. The military targets, to the extent that there were any that could be clearly identified, were located in and near civilian populations, but the bombing apparently occurred with little regard for nearby civilians. Neither the Johnson nor the Nixon administration felt political pressure to prevent civilian deaths in Laos and Cambodia. Indeed, few were paying attention to bombing that the Nixon administration went to great lengths to hide.

Thus, despite the elevation of the concern for noncombatants and its ad hoc institutionalization during the Vietnam War, the bombing was also very much more of the same in three ways. First, the perception of military necessity often trumped concerns about civilians: if the president or the Pentagon thought that ratcheting up the pain would bring results, they increased the bombing. Second, despite the use of the term "precision," any precision was limited by weather conditions and the choice of weapons. As it had done in Korea and World War II, weather limited the accuracy of aerial bombardment. Under the best conditions, 75 percent of the payload could land within 400 feet of the target. In less favorable conditions, "bombs often fell between 1,500 and 2,000 feet from the intended targets."[42] And less than favorable conditions were the norm: the long monsoon season meant that "the weather was rotten for nearly eight months a year."[43] Perhaps more devastating from the perspective of noncombatants, the United States used what it had in the arsenal, and many of its weapons—specifically, cluster bombs, napalm, and white phosphorous—were far from precise or discriminating. Third, in large areas known as "free fire zones," U.S. authorities deliberately abandoned discrimination between combatants and noncombatants.[44] In these areas of South Vietnam, which were thought to be dominated by

TARGETING CIVILIANS AND U.S. STRATEGIC BOMBING NORMS

anti-government forces, *anything that moved* could be targeted because it was assumed that anyone who remained in areas that had been cleared was a Viet Cong or a North Vietnamese.

The Operationalization of Restraint and the Gulf War

U.S. doctrine became more explicitly concerned with eliminating deliberate attacks on civilians and with minimizing inadvertent harm in the 1970s. The U.S. civilian leaderships' concern to prevent harm to civilians, or to at least give the appearance of protecting civilians, became a commitment by military leaders that was gradually expressed in military doctrine. This shift is exemplified in 1976, in the Air Force document AFP 110-31, *International Law—The Conduct of Armed Conflict and Air Operations*. It was first operationalized in the 1991 Gulf War.

While AFP 110-31 marks the institutionalization of procedures to minimize harm to civilians, it also articulates the tensions between civilian protection and military necessity, as Janina Dill describes in her chapter. Specifically, in AFP 110-31, the Air Force defined military necessity as "measures of regulated force not forbidden by international law which are indispensable for securing the prompt submission of the enemy, with the least possible expenditures of economic and human resources."[45] While the guide stresses that the laws of war prohibit "unnecessary suffering" and the deliberate targeting of civilians, it also states that the law "does not preclude unavoidable incidental casualties which may occur during the course of attacks against military objectives, and which are not excessive in relation to the concrete and direct military advantage anticipated."[46] Everything not forbidden is acceptable if it is needed for victory. The idea of a military objective and direct military advantage is understood to mean the use value of an object or location. If an object makes an effective contribution to the adversary's military capacities, it is understood as a legitimate target. There is nothing inherent that makes an object a military objective. Legitimate targets are those that make an "effective contribution" to the adversary's military effort.[47] Clearly then, whether a target was legitimate—regardless of civilian casualties—depended on the understanding of effective contribution. The idea of "balancing" necessity with civilian protection acknowledges civilian protection as a value. The Air Force recognized that a "careful balancing of interests is required between potential military advantage and the degree of incidental injury or damage in order to preclude situations raising issues of indiscriminate attacks violating general civilian protections."[48]

The effects of this institutionalization of a concern with civilian casualties are evident in the 1991 Gulf War. President George H. W. Bush ordered the military to minimize harm to civilians. Indeed, the United States took care not to target Iraqi civilians, and relatively few Iraqis (compared to other U.S. wars) were killed—an estimated 3,000 Iraqi civilians—during the short "hot" phase of the war as a consequence of U.S. airstrikes.[49] This is in part because the military objectives of the war were limited: to free Kuwait, to destroy Iraq's weapons of mass destruction program, and to weaken the Iraqi military. The political objective was more ambitious: to promote a coup or popular uprising in Iraq that would overthrow Saddam Hussein. The priorities for the air war were Iraq's political/military leadership, its military and industrial assets (including electrical power generation), oil facilities, railroads and bridges, airfields, and naval ports.[50] Many more civilians could have died during the war if the U.S. strategy had been to deliberately target civilians or civilian "morale."[51] Yet the long-term effects of bombing Iraqi infrastructure killed more Iraqi civilians after the war than during the hot phase.

The institutionalization of avoiding harm to noncombatants is most evident in the planning and conduct of air operations. The plan for air strikes, known as Instant Thunder, was conceived along the lines of USAF Colonel John Warden's theory of coercive air power.[52] Leadership was the most important target: if a state's leaders could be neutralized or immobilized, its war effort would fall apart. This is not to say that civilian morale was not a target. The Air Force believed that "the psychological impact on the Iraqi populace of being open to unremitting air attacks will be a powerful reminder of the bankruptcy and impotence of the Saddam Hussein regime."[53] The air strikes, it was hoped, would spark a rebellion or coup that would lead to the overthrow of Iraq's government and the installation of one more friendly to U.S. interests.[54] These were the same arguments that had been made about bombing Germany, Japan, and North Korea, but in the case of the Gulf War, the effects on morale were to be rendered psychologically rather than by killing civilians.

With AFP 110-31 and the laws of war as legal guidance, concern for civilian casualties affected the planning of the air war at several junctures. During the planning phase, a military lawyer assessed each target for its potential to cause collateral damage. Major Harry Heintzelman understood his job in evaluating targets as "balancing the importance of the target to the enemy against the potential collateral damage which might result from the attack."[55] The United States emphasized the precision of its air attacks. In truth, however, most of the aerial bombing used unguided bombs, and those bombs were only 25 percent accurate.[56] By contrast, precision-guided weapons hit their targets 90 percent of the time. The U.S. military was impressed by the performance of laser-guided and GPS-guided bombs against Iraq's bridges, tanks, air defenses, and command and control facilities.

TARGETING CIVILIANS AND U.S. STRATEGIC BOMBING NORMS 77

During the war, a lawyer for each air wing (a judge advocate general, JAG) would review targets from the perspective of necessity, proportionality, and discrimination. The JAGs were able to and did make recommendations that aim points or some other aspect of a strike, such as timing or height of burst, be altered to avoid harming civilians. In addition, during the operational phase of the war, the U.S. Air Force was instructed to call off strikes if they might cause collateral damage. Thus, the bulk of the unguided ordnance dropped by airplane was targeted on military assets; the weapons that were used near concentrations of civilians were precision laser-guided or cruise missiles.

A total of 330 weapons were used against targets in Baghdad, including air defense, airfields, government and intelligence sites, and presidential buildings. Although none of Iraq's top leaders were killed, air strikes damaged eighteen of the twenty-six leadership targets. These strikes included targets such as bunkers, where, according to the final plan drawn up by General John Horner, the intention was to decapitate the Iraqi leadership. When an attack on 13 February 1991 on the Al Firdos bunker, which the United States believed housed Iraqi political and military leaders, instead killed an estimated 200–400 civilians, authorization to strike targets in Baghdad became a matter for much greater scrutiny by the area commander, General Norman Schwartzkopf, and the chairman of the Joint Chiefs of Staff, Colin Powell.[57] The targeting of Baghdad was halted for five days, and only a few other targets in Baghdad were struck after the incident.[58] The United States apologized for the attack, saying the shelter would not have been attacked had U.S. intelligence shown that it housed civilians.[59]

While the care to protect civilians paid off during the hot phase of the war, the choices of weapons and targeting hurt civilians long after the war's end. Between the start of the war and 28 February 1991, the United States and its allies used a total of 61,000 air-dropped cluster munitions, accounting for the release of twenty million submunitions. Estimates are that 400 Iraqis (and 1,200 Kuwaitis) were killed by cluster bombs in the first two years after the conclusion of the war. More than a decade after the war, "dud" cluster bombs were still being found in Kuwait and Iraq.[60]

But more damaging for civilians over the long run were the effects of the bombing of Iraqi infrastructure, most importantly the disruption of electricity by air attacks on transformers and the subsequent closure of water purification and sewage treatment facilities. There is a debate over whether the damage to Iraq's water supply and its subsequent impact on civilians was intentional, or at least foreseen.[61] But intended or not, the attacks on Iraqi infrastructure caused tens of thousands of deaths—although some estimates suggest many times that—due to diseases including cholera and typhoid. Hundreds more civilians were killed and injured by air strikes from 1992 to early 2003 against Iraqi air defenses in "no-fly zones."[62]

Fine-Tuning Restraint in Bosnia and Kosovo

During the 1995 Bosnia and 1999 Kosovo campaigns, the United States further institutionalized procedures to minimize harm to civilians while attempting to keep risks for U.S. personnel to a minimum.

In August and September 1995, the U.S.-led North Atlantic Treaty Organization (NATO) intervened in the civil war in Bosnia and carried out an air campaign called Operation Deliberate Force. Its goal was to weaken the armed forces of the Republika Srpska and induce the government of Yugoslavia to negotiate a peace agreement on its behalf. While Bosnian and Croatian forces assaulted the Serbs on the ground with apparently little concern for civilian casualties, NATO aircraft attacked Serbian military targets over the course of two weeks. NATO and the United States—which flew the majority of missions—made a concerted effort to avoid harming civilians: "The pilots are operating under a zero tolerance policy for collateral damage. We have the technology to do it, but it can be difficult, particularly at night, in weather, while getting shot at."[63] To keep civilian casualties to a minimum, much of the bombing was scheduled for nighttime, when it was supposed that civilians would be less likely to be near targeted infrastructure. Weapons were also chosen to minimize civilian casualties. For instance, the United States chose not to employ cluster bombs (although some were inadvertently used).[64] About 98 percent of the weapons used by the United States were precision-guided, while 28 percent of the bombs dropped by non-U.S. NATO countries were precision-guided.[65] A third of the precision-guided weapons missed their targets.[66] NATO also avoided putting aircraft at risk, and there were relatively few incidents of combatant injury or death among NATO forces in either the air operations or the subsequent peacekeeping effort.

During the Kosovo Allied Force air campaign from 24 March 1999 to 10 June 1999 against Serbia, the priorities of the United States were to halt and reverse Serbian gains on the ground, specifically to stop Serbian forces from expelling Kosovar Albanians, to avoid harm to civilians, and to minimize casualties to NATO forces. The U.S. air operations in the Kosovo campaign were again characterized by the use of precision weapons, including the introduction for the first time in war of the United States' most advanced strategic bomber, the B-2 "stealth bomber." More than 5,200 precision-guided

TARGETING CIVILIANS AND U.S. STRATEGIC BOMBING NORMS

weapons were used by the United States—78 percent of the U.S. total. Planes flew from a relatively high altitude, no lower than 15,000 feet above sea level, to keep NATO aircraft relatively safe from Serbian surface-to-air attacks. Only two NATO planes were lost, and no NATO service members died in hostile action in over 14,000 sorties.

Some NATO commanders believed that constraints on air operations to avoid civilian casualties were too strict. The NATO commander, U.S. general Wesley Clark, wanted to focus attacks on Serbia's Third Army in Kosovo as well as the police and paramilitary forces that were attacking civilians there, while Air Component Commander Lieutenant General Michael Short thought the focus should be on Serbian leadership and civilian infrastructure throughout Yugoslavia.[67] Short argued that the key was morale—making the Serbian military and civilians suffer. "If you are getting pounded by B-1s and B-52s and A-10s are chasing you every day, and if you know that every time you move you are liable to be hit, at some point your spirit will break, particularly if you are not getting any help from Belgrade." Air power would "break" the Serbian military eventually: "I don't have a good feel for knowing how close they are to breaking, but I'll tell you that if we do this for two more months, we will either kill this army in Kosovo or send it on the run." Short also argued that Serbian civilians would turn against the leadership of Slobodan Milošević if they were made to suffer enough.[68]

In the end, both Serbian military forces in Kosovo and key sites in Yugoslavia, including Belgrade, were struck. The targets were selected to minimize noncombatant deaths, and military lawyers were involved in every aspect of the campaign. Yet as Ward Thomas notes, the circumstances were not optimal: "exceptionally bad weather, a more compact and densely populated theater of operations, less thorough preparation in developing an extensive target set, and, most problematic, mobile military targets that operated in close proximity to civilians, many of whom were refugees travelling the same roads as the troops who displaced them."[69]

High-profile cases of bombs killing civilians caused the United States some embarrassment and caused NATO to alter its tactics more than once.[70] When an attack on a rail bridge over the Nis River on 12 April led to the destruction of a passenger train and twenty deaths, tactics were changed.[71] According to Lieutenant General Short, "The guidance for attacking bridges in the future was: You will no longer attack bridges on weekends or market days or holidays. In fact, you will only attack bridges between 10 o'clock at night and 4 o'clock in the morning."[72] Clark ordered that all strikes in Belgrade be personally approved, and in May pilots were allowed to go lower than 15,000 feet to check targets.[73] Finally, when cluster bombs killed civilians in several incidents, the White House prohibited their use in May, although the British continued to use them. Human Rights Watch counted between 90 and 150 civilian deaths due to cluster bombs.[74]

General Clark reported that, overall, there were fewer than twenty incidents of collateral damage—"an incident rate of less than one-tenth of 1 percent"—of the more than 23,000 total bombs and other munitions delivered. "There's never been anything like it in the history of a military campaign, and I think it's a real tribute to the skill and proficiency of the men and women who flew and executed this campaign, to achieve that kind of precision."[75] In 1999, U.S. Secretary of Defense William Cohen and General Henry Shelton, the chair of the Joint Chiefs of Staff, said the Kosovo campaign was the United States' "most precise and lowest-collateral damage air campaign in history."[76]

Others disagreed. Midway through the campaign, Serbian officials claimed that the United States had killed more than 500 civilians in Yugoslavia. Although Human Rights Watch acknowledged that its estimates might have been incomplete, they estimated that about 500 Yugoslav civilians (Serbian and Kosovo Albanians) were killed in ninety separate incidents of collateral damage.[77] Thus, perhaps surprisingly, despite all the talk of precision, the rate of civilian death per ton of bombs dropped in Kosovo was nearly equivalent to the rate during the Vietnam War. Yet, considering that Belgrade, a major city, was the target of air attacks, it is remarkable that so few civilians were killed.

Bombing and Drone Strikes in the Post-9/11 Wars

The norms and practices of strategic bombing continued to evolve toward greater institutionalization of civilian protection in the post-9/11 U.S. wars. During the early phases of the wars in Afghanistan and Iraq, the United States used heavy bombers (B-52, B-1, B-2) and cruise missiles in its attacks, and bomber aircraft were used in Afghanistan on occasion through 2011. Civilian morale was still a target, as "shock and awe" (the phrase Bush administration officials used to describe the early bombing strategy in Iraq) suggests. Civilian assets and civilians themselves, however, were to be avoided. There was very little strategic bombing after the initial phases of the Iraq invasion, with the notable exception of the siege and occupation of Fallujah in 2004: the city was essentially leveled by a combination of airpower and ground-based bombardment.

For several years there was no consistent accounting of the civilian death toll in either Afghanistan or Iraq. Yet civilian casualties resulted from bombing, fixed-wing close air support, helicopter close-combat air operations, and remotely piloted (drone) aircraft strikes. In the early years of the wars, the inadvertent deaths or injury of civilians did not cause great alarm within the U.S. government. For example, when the United States inadvertently killed civilians very early in the war in Afghanistan, Secretary of Defense Donald Rumsfeld said, "Now in a war, that happens. There is nothing you can do about it."[78] As a Pentagon official said of Iraq in March 2003: "If it is a high enough value target, you accept a higher risk of casualties."[79] However, over the course of the wars, the military institutionalized a decreased tolerance of collateral damage in air operations.

Specifically, the chair of the JCS issued a directive in September 2002 that combatant commanders must estimate, evaluate, and mitigate potential collateral damage.[80] According to the 2002 directive, several algorithms that were either already in use or newly developed for use by the services were to be used to estimate both casualties and collateral damage. For planned air operations the analysis of targets, weapons, and other characteristics of a particular situation were used to calculate likely collateral damage and, when necessary, alter the use of a weapon or call off a strike. Unplanned operations in time-sensitive situations, such as close air support, did not get such rigorous analysis.

In the first days of the 2003 war against Iraq, the U.S. military and civilian planners set a ceiling of thirty anticipated noncombatant deaths before permission must be sought and given by the secretary of defense or the president. The ceiling was intended to apply to "high value" targets. It did not mean that the strike would not occur, only that permission must be sought and given by the highest level of commander, a civilian political figure. In the cases where significant loss of civilian life was anticipated, the targeting was usually altered to reduce the risk to civilians—for example, the strike would made at a time when fewer civilians might be harmed or the size of the weapon was reduced. Permission was sought and given more than fifty times in the early months of the Iraq war.[81] In this respect, and also when the United States modified its bombing tactics in Bosnia and Kosovo to minimize harm to civilians, the U.S. military has practiced a level of "due care."

That the defense secretary or the president was required to authorize such strikes highlights two additional features of the situation. First, algorithms that estimate noncombatant killing for preplanned strikes were employed much more widely than in previous conflicts. These algorithms are a set of decision rules, formulas, and inputs used to calculate risk and likely harm to noncombatants. And second, there was a threshold of risk to noncombatants that was considered acceptable given the understanding of military necessity and the estimated risk to U.S. forces.

Computerized algorithms had been used in the Gulf War and Kosovo to estimate civilian casualties and to assess methods for their mitigation.[82] During the early part of the war in Afghanistan, the United States used an older program but for the first time also used the Fast Assessment Strike Tool—Collateral Damage (FAST-CD), earlier known as "BugSplat," to "vet" about 400 targets.[83] Brigadier General Kelvin Coppock, director of intelligence for the Air Combat Command, told a reporter that BugSplat was "a significant advance." "It will allow us to target those facilities that we want to target with confidence that we're not going to cause collateral damage."[84] FAST-CD became the tool of choice for estimating collateral damage for "time-sensitive targets" in both war zones because it could take as little as five minutes to run the program. If the estimated circle of damage included civilians or civilian infrastructure, analysts could attempt to minimize the damage to civilians by changing weapons, the timing of the attack, or other technical measures.

Despite these efforts, reports of civilian casualties in Afghanistan and Iraq due to air strikes and other causes became a public relations problem. In response, the U.S. Air Force commissioned research by the RAND Corporation to assess reaction to civilian killing. RAND found that Americans care about civilian casualties and are most concerned if they think the killing is intentional or a result of negligence. According to the research, "The principal support for U.S. military operations and civilian casualties is the belief that the United States is making serious efforts to avoid civilian casualties: Those who hold this belief are more likely to support U.S. military operations than those who do not, and this result holds both in the American public and foreign publics."[85] This research included advice for Pentagon public affairs personnel to prepare the American public for the "eventuality" of collateral damage and to emphasize that as much as possible was being done to prevent civilian killing. The report recommended that it was important to improve communication with the media and the public. "Such improvements also would have the salutary benefits of reducing the likelihood of constantly changing (or even contradictory) explanations that can erode credibility."[86] The RAND report stressed the importance of shaping a positive message of U.S. care in incidents that lead to civilian death: "By emphasizing the efforts that are being made to reduce civilian casualties (e.g., increased precision, smaller blast effects, improved target verification, and so on), the Air Force and DoD can help ensure that the U.S. Congress and public have continued reason to trust that the U.S. military is seeking new ways to reduce the prospects of civilian deaths in future military operations."[87] These measures reinforce the insight of Conway-Lanz that U.S. public opinion is assuaged

TARGETING CIVILIANS AND U.S. STRATEGIC BOMBING NORMS

by the impression of its government's intention to avoid civilian harm, even if the reality does not fully conform.

By 2006 Pentagon officials and civilian planners were also increasingly convinced that the civilian casualties in Iraq were bolstering the counterinsurgency. The December 2006 *U.S. Army/Marine Corps Counterinsurgency Field Manual* articulated an emerging view that eliminating the enemy depends precisely on *not* killing civilians: "An operation that kills insurgents is counterproductive if collateral damage leads to the recruitment of fifty more insurgents."[88] The new *Field Manual* stressed calibrating the use of force so as not to unleash a backlash.[89] Ground operations in Iraq were changed, leading to a gradual reduction in civilian casualties.

But civilian casualties continued to mount in Afghanistan, and most civilian casualties at the hands of NATO, the International Security Assistance Force (ISAF), and U.S. forces were caused by air power. The belief that killing civilians breeds support for insurgents was, by early 2009, the consensus view among top Pentagon commanders and civilian leaders. The "mathematics of insurgency," as it was known among troops, was supported by econometric analysis.[90] In his June 2009 confirmation hearings to become the next U.S. commander in Afghanistan, General Stanley McChrystal argued: "I would emphasize that how we conduct operations is vital to success. . . . Our willingness to operate in ways that minimize casualties or damage, even when doing so makes our task more difficult, is essential to our credibility."[91]

In late June 2009 General McChrystal ordered the use of air strikes that were to be restricted to cases where Coalition troops were in danger of being overrun. McChrystal emphasized that "air power contains the seeds of our own destruction if we do not use it responsibly. We can lose this fight."[92] In late June 2009 McChrystal issued a revised "tactical directive" stating that "I expect leaders at all levels to scrutinize and limit the use of force like close air support (CAS) against residential compounds and other locations likely to produce civilian casualties in accordance with the guidance. Commanders must weigh the gain of using CAS against the cost of civilian casualties, which in the long run make mission success more difficult and turn the Afghan people against us."[93] The tactical directive led to complaints from soldiers that they were at greater risk. Indeed, the rate of U.S. casualties increased in the months after the directive was issued. But the number of civilians killed in air strikes declined in 2009.[94]

In his contribution to this volume, Charles Dunlap strongly takes issue with both the claim that air strikes were responsible for a large portion of civilian war deaths in Afghanistan and Iraq and that air power was counterproductive. In his view, "the best way of protecting civilians is to kill those who would kill them," preferably from the air. However, the main point is that U.S. policy was increasingly driven by a contrary understanding. This understanding eventually extended to the use of drones.

The United States began remotely piloted Predator and Reaper drone strikes targeting Al Qaeda in Yemen in late 2002 and in Pakistan, targeting the military leadership of Taliban and Al Qaeda as an extension of the Afghanistan war, in late 2004. At first the United States refused to officially confirm or deny the drone strike program. Bush and Obama officials generally minimized or denied civilian casualties in the strikes, even as civilian deaths were described in local and global media. But as with other forms of bombing, concern for civilian casualties due to drone strikes gradually increased, and the Obama administration developed procedures for mitigating collateral damage. Although drones are addressed in several other chapters, they merit consideration here as well because of the extent to which they fit the trajectory of growing official U.S. concern for harm to civilians.

The argument for the tactic, to the extent that it was publicly articulated before May 2013, was that precision drone strikes could minimize civilian casualties while keeping U.S. soldiers out of harm's way. In June 2011, the Obama administration's chief counterterrorism official, John Brennan, for example, argued that the drone strikes were "exceptionally precise and surgical in terms of addressing the terrorist threat. And by that I mean, if there are terrorists who are within an area where there are women and children or others, you know, we do not take such action that might put those innocent men, women and children in danger." Brennan further stated that in the last year, "there hasn't been a single collateral death because of the exceptional proficiency, precision of the capabilities that we've been able to develop."[95]

Most sources agree on the approximate number of drone strikes in Pakistan—between 350 and 380 strikes between 2004 and December 2013—and that the Obama administration increased the use of drones in both Pakistan and Yemen. Sources widely disagree about the number of people killed and wounded, and their identities, specifically whether they are Taliban, Al Qaeda, or civilian noncombatants.[96] Pakistan Body Count, for instance, estimated that between about 1,200 and 2,400 civilians were killed by U.S. drone strikes between 2004 and 1 December 2013, while the *Long War Journal* estimated that 156 civilians had been killed in Pakistan during the same period. Uncertainty about the numbers and proportion of civilians killed in the drone strikes clouded and inhibited debate over the program.

Sources do agree that the use of drones for targeted killing increased. From 2009 to late 2013, the Obama administration made more than 330 drone strikes in Pakistan, compared to the Bush administration, which made some 50 strikes. On the other hand, under the Bush administration, procedures for authorizing drone strikes were apparently ad hoc. Although it took almost four years, President Obama moved to institutionalize the drone strike program in late 2012. Bob Woodward reported that Pakistan's President Asif Ali Zardari told the director of the Central Intelligence Agency, Mike Hayden, in November 2009 that "Collateral damage worries you Americans. It does not worry me."[97] Yet gradually, critical domestic and international public scrutiny of the drone strikes in Pakistan and Yemen increased from 2011.

When, in May 2013, President Obama announced new criteria and procedures for limiting drone strikes, he acknowledged that concern for civilian casualties and accountability had prompted the development of the drone policy guidance: "Much of the criticism about drone strikes—both here at home and abroad—understandably centers on reports of civilian casualties." Obama admitted that "U.S. strikes have resulted in civilian casualties" and acknowledged their moral significance: "And for the families of those civilians, no words or legal construct can justify their loss. For me, and those in my chain of command, those deaths will haunt us as long as we live, just as we are haunted by the civilian casualties that have occurred throughout conventional fighting in Afghanistan and Iraq."[98]

President Obama then outlined the new guidance for drone strikes against terrorists, including the requirement that "before any strike is taken, there must be near-certainty that no civilian can be killed or injured—the highest standard we can set."[99] The norm was thus articulated as a goal. Implementation has proven more difficult.

Conclusion

In sum, I argue that the evidence supports all three explanations for the change in bombing norms and practices. First, U.S. military and civilian leaders have gradually internalized norms of civilian immunity to the point that avoiding harm to civilians is a major concern. But an elastic conception of military necessity is also evident. When U.S. leaders believe that harm to civilians (or civilian infrastructure) is necessary for victory, civilian protection becomes a lower priority, and indeed, during World War II and Korea, civilians were targeted for pain-inducing bombing. Finally, U.S. leaders have become increasingly sensitive to domestic and international public opinion, as Margarita Petrova's chapter on the transnational campaign against cluster munitions demonstrates.

From the period after World War I, when U.S. strategic bombing policy was first debated and then implemented, through the Vietnam War, the dominant belief was essentially that civilians were a legitimate and indeed essential target of conventional strategic bombing. In this period, military necessity was understood to require killing civilians—or at least crushing their morale—so deliberately targeting civilians and civilian infrastructure was the policy. This set of beliefs began to change during the Vietnam War, and restrictions on harm to civilians were set and implemented in an ad hoc fashion by civilian leaders. After Vietnam, the military gradually institutionalized restraint. When military necessity was not believed to require killing civilians but it had not become a military imperative to protect them, an ad hoc effort was made to avoid harm to civilians, as in Vietnam. The United States did not develop and implement techniques to mitigate the risk and harm to civilians at an institutional level until the 1990s.

But the care taken was less than scrupulous, and the threshold and acceptance of collateral damage could be relatively high depending on the perceived political and military stakes, as in the early years of the U.S. wars in Afghanistan and Iraq. Yet when protecting civilians becomes a political priority, the United States is able to reduce harm to civilians by air power (even keeping in mind that strategic bombing is a very different matter than close air support). This is seen very clearly in Afghanistan in 2009 with McChrystal's tactical directive. By contrast, when there are few observers and less critical public reaction, as was the case with the drone strikes in Pakistan from 2004 to 2011, U.S. officials pay comparatively less attention to the problem of civilian killing. The different attitude toward civilian casualties in Afghanistan and Pakistan is illustrated by the fact that to the time of this writing there have been no high-profile investigations of the killing of Pakistani civilians by U.S. drone strikes, while there were several investigations of collateral damage by air power in both Iraq and Afghanistan during the same period. When potential harm to civilians by drone strikes in Pakistan became a topic of wide concern, the United States moved to institutionalize criteria and procedures for authorizing drone strikes and essentially articulated a ceiling of "near certainty" for acceptable anticipated civilian casualties at zero. This development marks an important change in U.S. bombing norms.

Chapter 4

The Law Applies, But Which Law?
A Consumer Guide to the Laws of War

CHARLES GARRAWAY

                      The law is a blunt instrument. It has always been so. Law by its nature is cast in stone. While it is not, like the law of the Medes and Persians, unalterable, it is not easy to change; that requires a formal process. It is therefore an uneasy bedfellow with more informal structures such as ethics or even justice itself. These latter concepts can evolve as situations and circumstances change, while the law stands firm, in some ways a bastion in a sea of uncertainty but at other times becoming powerless against the forces of nature.

    Law and ethics have always been intertwined. This is particularly so in the field of international law. It was Hersch Lauterpacht who described international law as being at the vanishing point of law and the law of armed conflict, international humanitarian law, as at the vanishing point of international law.[1] The history of the law of armed conflict and how it is intertwined with ethics supports his thesis. Only by looking at current law in a historical context can one understand both how it has developed and how the current tensions between law and ethics are reflected in that development. In particular, the interplay between the law of armed conflict and human rights law that has come to dominate the current legal scene can be understood only in a historical context. The law of air warfare is no exception.

History of the Laws of War

The Boundaries of Humanity

From earliest times, there have been laws of war. Whether one studies Sun Tzu of ancient China, the Code of Manu in Hindu philosophy, the Koran, the Old Testament, or African culture, certain underlying principles shine through. The law, if that is what it should be called at this stage of development, has always been an uneasy balance between the interests of belligerent parties, military necessity, and what is now referred to as humanity. This Faustian pact has governed the laws of war for centuries.

In the earliest times, for the most part, the rules were not laid down. They were the customs accepted by fighting men (and it was usually, but not exclusively, men) through the ages. The conduct of hostilities was principally governed by the interests of the parties concerned, whether they were communities, religious groups, or, in latter days, states. And yet set against this were a series of fundamental principles. In the early years, they were based on religion or philosophy. In the Middle Ages in Europe, they were known as chivalry. In modern times, the word humanity is preferred. However, in practice all are based on the same concept: an underlying set of values that are universal. The law has developed subject to the inevitable tension between the interests of fighting men (military necessity) and these underlying values (humanity, or as some would call it, ethics). I would therefore argue that ethics have been a key factor in the laws of war for centuries, if not millennia. That does not mean that the relationship has always been easy. Indeed, it has not. At times, the two concepts have grown separate to such an extent that they have seemed contradictory and yet in principle they are joined at the hip. The law of war without ethics is an empty shell, while ethics without law is an aspiration. Each needs the other, and the task of ethicists and law-of-war experts is to work to ensure that each contributes to the essential balance without which the laws of war are meaningless. The laws of war without humanity are tyranny; humanity without the laws of war is idealism.

But how does this affect the laws relating to aerial bombardment? I would submit that this provides one of the classic illustrations of the need for this balance and the dangers of ignoring it. Aerial warfare, particularly bombardment, is just a new means of waging war. Weapons have changed through the millennia and, as with other new developments, have always been subject to attempts to restrict them.

Humanity and Law

In the middle of the nineteenth century, a trend toward codification of the laws of war began. As there had always been, there were two separate strands. The trend began, at least in Europe, with the efforts of Henry Dunant after the Battle of Solferino. This led to the first Geneva Convention of 1864[2] and the founding of what became the International Committee of the Red Cross (ICRC). This Geneva strand concentrated on the protection of victims of war. It was based on the principle of humanity, and while it began with an acceptance of the balance between humanity and military necessity, it has in recent years seemed to elevate the principle of humanity over and above that of military necessity. The 1868 St. Petersburg Declaration[3] referred to the need to "conciliate the necessities of war with the laws of humanity." This terminology reflected the Faustian pact, and yet the same Declaration also spoke of "the technical limits at which the necessities of war ought to yield to the requirements of humanity."[4] This would appear to indicate that there are limits where even military necessity, however great, cannot outweigh the interests of humanity.

At the same time, the Rules relating to the actual conduct of hostilities continued to develop and began to adopt the practice of written codification. The St. Petersburg Declaration fell within this strand. The Declaration prohibited the use of explosive projectiles of a weight under 400 grams. While it could be argued that the object of the treaty was humanitarian, the background would suggest a more pragmatic reason.[5]

Even this landmark Declaration has not fully stood the test of time. As new weapons and types of ammunition were developed, ways were found to avoid what seemed like an absolute prohibition. This was particularly so with the use of tracer and incendiary weapons, the former of particular importance in the context of air warfare, something not even considered in 1868.

The Hague Peace Conferences of 1899 and 1907 were set in this particular context. They sought primarily to deal, insofar as the *ius in bello* was concerned, with the conduct of hostilities. This Hague strand of law reflected the interests of states in maintaining the balance between military necessity and humanity. Weaponry was still important, and Declarations were adopted in relation to asphyxiating gases and expanding bullets (the infamous "dum-dum").[6]

At this time, the age of steam meant that naval warfare was beginning to gain prominence. There had, of course, always been great maritime powers and, where the interests of those powers collided, naval conflict. Indeed, the laws relating to that conflict had developed a life of their own, separate from the laws relating to land conflict. Much of the Hague Conferences of 1899 and 1907 was given over to naval warfare, and a number of important agreements were drafted, including those on naval neutrality, automatic submarine contact mines, and, perhaps most relevant for our purposes, a Convention concerning Bombardment by Naval Forces in Time of War.[7] This sought to reflect the principles then governing land warfare that were codified in Article 25 of the Regulations Respecting the Laws and Customs of War on Land, annexed to Hague Convention IV of 1907.[8]

The Advent of Flight

At this time, air warfare was just an idea in the fertile imaginations of great thinkers. Balloons had been used for generations, principally as a means of observation (there is reference to communications balloons in Article 29 of the 1907 Hague Land Regulations),[9] but developments in ballooning were opening up the possibilities of other uses, including crude forms of bombardment. When states considered this in 1899, the majority of states, which did not have access to this "new" technology, were somewhat alarmed by the prospect. As a result, a Declaration was adopted in 1899 that prohibited the launching of projectiles and explosives from balloons and other methods of a similar nature.[10] This last phrase could have meant that air warfare as we know it today would have been stillborn, or at least strangled at birth. However, this Declaration was limited to five years, and the technological tide was beginning to come in. It was replaced by a similar Declaration in 1907,[11] but by then it was too late. Many states that were later to develop military aviation programs did not become parties, and again the treaty was time limited. Even before the First World War, Italy, which was not a party to the 1907 Declaration, had used balloons to bombard enemy troops during the Turko-Italian War of 1911–12.[12] The Institute of International Law, meeting in Madrid in 1911, recommended that air warfare should not pose a greater danger to the civilian population than land or sea warfare.[13] However, this pious pronouncement could not stand in the way of technology, and during the First World War both sides vied with each other to use air warfare to their advantage. It is an incredible thought that only a few years after the Wright brothers proved that flight was possible on 17 December 1903, planes were taking off from ships and the first rudimentary aircraft carrier was in operation.[14]

Controlling the Genie during the Interwar Years

The First World War was something of an experimental crucible for the proponents of flight and aircraft in warfare. While the technology of flight itself advanced in leaps and bounds, the accuracy of delivery platforms did not. There was no doubt that aerial bombardment was effective, but it was much less accurate than even long-range land artillery. It would seem that the balance between military interests and humanity had gotten badly out of kilter. However, it was much too late to put the genie back into the bottle, so the best that could be hoped for was some form of restraint, either by law or by policy. But the political situation was not favorable. The interwar years in Europe, in hindsight, were not a time of genuine peace. The need for a comprehensive code dealing with air warfare was recognized, but getting agreement from states proved far more difficult.

The Conference on the Limitation of Armament, held in Washington in 1921–22, examined the issue but failed to reach agreement on air warfare. However, an eminent Commission of Jurists representing six states was established and asked to report back to the respective governments. This commission, supported by technical experts, met in The Hague from 11 December 1922 to 19 February 1923 and produced a detailed general report. This consisted of Rules concerning the Control of Wireless Telegraphy in Time of War and Aerial Warfare and Rules of Aerial Warfare. These latter consisted of draft Articles outlining Rules for the conduct of air warfare and an accompanying Commentary.[15] Unfortunately, these Rules were never adopted by states and thus remain in draft form. They never became legally binding, although, as we shall see, they did have some moral persuasion power.

In some ways, these Rules were ahead of their time; they offered provisions that did not become hard law until the adoption of the 1977 Additional Protocols to the 1949 Geneva Conventions. There was a complete Chapter on "Hostilities" that began, somewhat disconcertingly, by lifting the prohibition on explosive projectiles under 400 grams contained in the St. Petersburg Declaration insofar as it applied to the use of tracer projectiles by or against an aircraft.[16] Five Rules dealt specifically with "Bombardment," including the first attempt to define what a military objective was and a rudimentary requirement for "precautions in attack."[17] The question that has to be asked is whether these Rules were ahead of their time both in legal terms and in terms of the technology available at that time. If their effect was, in reality, to make bombardment from the air illegal, then they stood no chance of success. Although a number of states were fundamentally opposed to the Rules, a number of others stated that they would seek to comply with them.

As the world began again to spiral into all-out conflict, the need for some form of control over air warfare once more became only too apparent. Once more, the technology of flight was moving faster than that applied to the accuracy of delivery mechanisms. Air bombardment was a cause for serious concern as a result of the actions of the Italian air force during the invasion of Ethiopia in 1935–36, the German air force during the Spanish civil war in 1936–39, and the Japanese air force during the invasion of China in 1937–39.

On 21 June 1938, prior to the outbreak of the Second World War, Neville Chamberlain enunciated in the House of Commons what the British government understood as the principles of international law governing air warfare. As Tami Davis Biddle also points out in her chapter, these were that:

1. It is against international law to bomb civilians as such and to make deliberate attacks upon civilian populations;
2. Targets which are aimed at from the air must be legitimate military objectives and must be capable of identification; and
3. Reasonable care must be taken in attacking those military objectives so that by carelessness a civilian population in the neighborhood is not bombed.[18]

These principles were subsequently embodied in a Resolution unanimously adopted by the League of Nations on 30 September 1938.[19] However, Germany and Japan had already withdrawn from the League in 1933, and Italy withdrew in 1937.

The Second World War and Total Warfare

At the start of the Second World War virtually no hard law that applied to air warfare was in force. Although both Axis and Allied powers made statements supporting the draft of the Hague Rules of Aerial Warfare (1923) and regularly accused each other of violations, these were not legally binding, and the effect of any form of moral restraint broke down as the conflict developed. This led to the almost unrestricted use of air bombardment as witnessed in Dresden and Coventry. Any attempts made to attack military objectives foundered due to the lack of technology that would make accurate targeting possible. In order to ensure the destruction of a military objective, it was necessary to accept what would now be considered an utterly unacceptable degree of collateral damage.

Although many seek to blame the law for the excesses of the Second World War, it was the *lack* of law that was the problem. Cries for an ethical balance went unheeded. The weakness of reliance on unenforceable standards in the crucible of conflict was shown up. However, what was also shown was the need for law and ethics to work together to produce enforceable standards through the prism of law, but guided by ethics. To some extent, that was what the 1923 draft Hague Rules of Aerial Warfare had sought to do, but states were not prepared at that point to accept the restrictions.

Postwar Developments

Despite the appalling damage caused by air bombardment in the Second World War, the law was slow to react, not helped perhaps by the division of the world into ideological blocks. This hindered negotiations on limits to the conduct of hostilities. The ICRC was able to make progress in developing Geneva law so that the four Geneva Conventions of 1949 were able to be agreed upon.[20] However, the law relating to the conduct of hostilities remained frozen in the form it had adopted in 1907, strengthened by the pronouncement by the Nuremberg Tribunal that the Regulations Respecting the Laws and Customs of War on Land now reflected customary international law and were thus binding on all states.[21]

Humanity Fights Back through Human Rights Law

However, the Second World War also spawned a new system of law, human rights law. This grew from the way the Axis powers had treated sections of their own populations, and it sought to provide peoples with "rights" they could enforce against their own governments. Encouraged by the terms of the United Nations Charter[22] and the 1948 Universal Declaration of Human Rights,[23] human rights law initially developed slowly. The two International Covenants, on Economic, Social and Cultural Rights[24] and on Civil and Political Rights,[25] were not adopted until 1966. In the meantime, Europe had adopted its own Convention for the Protection of Human Rights and Fundamental Freedoms in 1950, which came into force in 1953.[26] Where this Convention was particularly significant was that it had a judicial enforcement mechanism in the form of the European Court of Human Rights, a court that has increasingly taken a proactive line in interpreting and enforcing the European Convention.

Although the Universal Declaration was seen as part of the law of peace, the wording of the European Convention provided for its continued applicability in times of war. Its derogation clause specifically referred to "war or other public emergency threatening the life of the nation."[27] It was difficult, therefore, to argue that human rights played no part in governing conduct in time of war, at least for European states. Nevertheless, it was generally accepted that in time of war—armed conflict between states—it was the laws of war, now increasingly known as international humanitarian law, that took priority.

State Sovereignty Rules

But armed conflict was not limited to wars between states. States no longer had a monopoly on the use of force—if they ever had—and after 1945 the majority of conflicts in the world have been internal, within a state, rather than external, between states. The ICRC had recognized this trend in 1949 and had initially sought to apply the full weight of Geneva law, as embodied in the four 1949 Conventions, to non-international armed conflict. It failed. States were not prepared to go that far in allowing international supervision in their internal affairs. The result was that only one Article, common to all four of the 1949 Conventions, was applied to non-international armed conflict. Significantly, the wording of Common Article 3 is greatly influenced by human rights law.[28] It is here, in non-international armed conflict, that the relationship between these two systems of law becomes most complex.

While Common Article 3 clearly applied to non-international armed conflict, the application of Hague law on the conduct of hostilities was much more problematic. The treaties almost exclusively dealt only with international armed conflict between states, and few, if any, commentators were prepared to argue that as a matter of custom, such law extended into non-international armed conflict. States still considered that sovereignty was an overriding consideration, and they were not prepared to allow international law to govern how they conducted operations against rebel forces on their own territory. But human rights law was already beginning to do just that.

In the light of changes in the characteristics of warfare since 1945, the ICRC prepared two draft Protocols for consideration by states. There were two notable features to these drafts. First, the text was clearly heavily influenced by human rights law. Second, the text dealt not only with Geneva law, the traditional area in which the ICRC had operated, but also contained substantial elements of Hague law dealing with the conduct of hostilities. The two draft Protocols dealt respectively with international armed conflict and non-international armed conflict. These drafts were considered by a Diplomatic Conference convened by the Swiss government between 1974 and 1977, and two texts were adopted in June 1977.[29] The original draft texts had again sought to bring together the law relating to the two distinct types of conflict, but at the last minute the text of Additional Protocol II, which related to non-international armed conflict, was substantially trimmed. States again were cautious about allowing too much international control over internal matters. What remained was almost entirely Geneva law that expanded the minimal provisions contained in Common Article 3. Furthermore, although Common Article 3 applied to any "armed conflict not of an international character occurring in the territory of one of the High Contracting Parties,"[30] Additional Protocol II had a much higher threshold that applied to non-international armed conflicts taking place "in the territory of a High Contracting Party between its armed forces and dissident armed forces or other organized armed groups which, under responsible command, exercise such control over a part of its territory as to enable them to carry out sustained and concerted military operations and to implement this Protocol."[31] Thus Hague law still was seen as having a minimal impact on non-international armed conflicts.

In international armed conflict, the increase in treaty provisions on the conduct of hostilities was significant. Flesh was placed on the bare bones of the 1907 Regulations Respecting the Laws and Customs of War on Land with the introduction of a definition of military objective that was very similar to the definition in the 1923 Hague Draft Rules of Aerial Warfare.[32] Also, the principle of proportionality, that key element of the Faustian pact, was codified for the first time, providing that attacks should not be "expected to cause incidental loss of civilian life, injury to civilians, damage to civilian objects, or a combination thereof, which would be excessive in relation to the concrete and direct military advantage anticipated."[33] The key words here are "expected" and "anticipated." It is what is in the mind of the attacker that decides the legality of the attack, not necessarily the end result. This precisely illustrates one of the key differences between Hague and Geneva law. Hague law looks at matters principally from the view of the belligerent; Geneva law looks at matters from the point of view of the victim. The attempt to amalgamate these two viewpoints in the Additional Protocols has caused some difficulties.

Additional Protocol I also defined the terms "combatant"[34] and, by negative definition, "civilian"[35] and required both attackers and defenders to take certain precautions to seek to protect the civilian population and civilians generally from the effects of hostilities.[36] However, again the focus was primarily on the rules of land warfare. To what extent did this Protocol impinge on the comparatively undefined rules of air warfare? The answer is to be found in Article 49(3), which provides that: "the provisions of this Section [general protection of the civilian population against the effects of hostilities] apply to any land, air or sea warfare which may affect the civilian population, individual civilians or civilian objects on land. They further apply to all attacks from the sea or from the air against objectives on land but do not otherwise affect the rules of international law applicable in armed conflict at sea or in the air."[37]

This provision would clearly apply to air bombardment of targets on land. The relevant Section of the Protocol includes all the key elements outlined above, particularly proportionality and the requirement for precautions in attack, and it includes language about defense. The latter is important because it would require a defending state to take precautions to protect its own civilian population from attacks from the air.[38] The burden is not wholly one-sided.

The reduced text of Additional Protocol II contains nothing on proportionality or precautions in attack save for a Part on the protection of the civilian population, including certain provisions to protect specific targets.[39] It followed that in relation to non-international armed conflict, little progress had been made.

The Surge of Nationalism after the Cold War

The end of the Cold War changed all this. Suddenly, forces of nationalism and ethnicity were unleashed that led to the unraveling of the old Soviet Empire. In particular, the world was faced with the collapse of the former Yugoslavia into its ethnic component parts and the genocide in Rwanda between Hutus and Tutsis. The response was to return to a part of international law that had been neglected since the end of the Second World War, international criminal justice. The United Nations Security Council established the International Criminal Tribunals for the Former Yugoslavia and Rwanda, and new efforts began to establish an International Criminal Court. These were ultimately successful.

The Yugoslav Tribunal in particular found itself in something of a quandary. There was considerable doubt about how to categorize the conflicts in the former Yugoslavia. Were they international (between the new states) or internal (between ethnic groups within the new states)? Indeed, did the categories change at various points, and if so, when? The rules on the conduct of international hostilities were comparatively clear after the adoption of Additional Protocol I. Although this treaty did not have the universal acceptance of the Geneva Conventions, its key provisions, including proportionality and precautions in attack, were accepted as custom even by those states that had not ratified it because of objections to other provisions. But what was the situation in non-international armed conflict? To what extent were the participants bound by Hague law on the conduct of hostilities?

The Yugoslav Tribunal met this challenge head on in their first case, that of Dusko Tadić.[40] The Appeals Chamber stated that "a number of rules and principles . . . have gradually been extended to apply to internal conflicts." However, the Appeals Chamber included the important caveat that "this extension has not taken place in the form of a full and mechanical transplant of those rules to internal conflict; rather, the general essence of those rules, and not the detailed regulation they may contain, has become applicable to internal conflicts."[41] While the judgment itself may have been understandably cautious, it opened Pandora's box. Within a very short period, the caveats of the Appeals Chamber seemed to have been forgotten.

In 1998, the International Criminal Court Statute followed the Tadić decision by transposing some of the war crimes applicable in international armed conflict to non-international armed conflict.[42] While most were still of the Geneva law type, some were clearly Hague law, including offenses of pillage and of directing attacks against protected persons and objects. The seminal ICRC Study into Customary International Humanitarian Law,[43] which identified 161 "rules" of customary international humanitarian law, found that no fewer than 147 applied across the board in both international and non-international armed conflicts.[44] The clear conclusion was that, subject to those areas where there were obvious distinctions (e.g., status of prisoners of war), the rules, particularly those relating to the conduct of hostilities, were the same. The unwillingness of states to accept such conclusions in 1949 or particularly in 1977 was thus overcome by a combination of judicial activism and interpretation of customary law.

The Legal Tension

But if the conduct of hostilities in non-international armed conflicts is now governed by the rules of international humanitarian law, where does that leave human rights law? Under international humanitarian law, it is recognized that in war people die and things get broken. Even a degree of innocent death is acceptable if it is not excessive in relation to the anticipated military advantage. This would seem to fly in the face of human rights law, with its more hardened attitude based on the rights of the victim. As international humanitarian law sought wider applicability in non-international armed conflict, it was inevitable that it would collide with human rights law, as that too sought to protect the victims of conflicts.

While other bodies have also played a part, the European Court of Human Rights has been at the forefront of this confrontation. Cases were referred to the Court arising out of the Troubles in Northern Ireland, and as the

United Kingdom never acknowledged that these ever reached the level of an "armed conflict," it was no surprise that the Court dealt with the cases purely on the basis of human rights law with no reference to international humanitarian law.[45] Slightly more problematic were cases arising out of the Kurdish insurgency in Eastern Turkey and the Turkish invasion of Cyprus in 1974 and the subsequent occupation. These cases too were dealt with solely on the basis of human rights law.[46]

The Court first became involved with air bombardment in the Bankovic case involving the bombing by NATO forces of the Serbian television station during the Kosovo air campaign.[47] The case was brought by some of those injured in the attack and by families of those killed. Had it gone to a decision on the merits, a number of crucial questions involving international humanitarian law would seemingly have become relevant. Was the TV station a military objective? If so, how should the anticipated military advantage be assessed and what was the expected incidental loss or damage? How is the proportionality balance to be calculated? How does all of this fit with the right to life under human rights law? There was no derogation under Article 15 of the Convention, so to what extent could the Court take into account international humanitarian law at all? Should the Court deal with the matter solely as a human rights issue without any reference to international humanitarian law? The Court decided on a preliminary issue that the victims of such an air attack did not fall within its "jurisdiction," so the merits were never considered.

However, this was not the end of the matter. This was an international armed conflict and it was clear that NATO had no control over the ground. Furthermore, the territory involved, Serbia, was not within the *espace juridique* of the Convention. The armed conflict in Chechnya provided a different scenario, a non-international armed conflict on the territory of a state party to the Convention. Here the jurisdictional arguments that had prevented the Court from adjudicating did not apply. The Court therefore had to bite the bullet. This conflict involved both land and air operations, and it was not long before a case involving air operations came before the Court.

The case involved the bombing from the air of what turned out to be a civilian convoy of vehicles fleeing Grozny.[48] It hinged therefore, in international humanitarian law terms, on the issue of precautions in attack. The Court, however, dealt with it entirely in human rights terms, although international humanitarian law had been discussed in argument. As it happens, the facts were such that the same result would probably have been reached under either system of law, and the Court used language very similar to that contained in international humanitarian law, particularly Additional Protocol I.

However, on the facts, the Court was able to evade some of the key issues, including that of proportionality. Had the convoy turned out to be a military objective, perhaps because of a number of military vehicles embedded in the convoy, would the issue of proportionality have been dealt with differently under human rights law and the right to life, rather than under humanitarian law where a certain measure of incidental loss and damage is acceptable? The tectonic plates are beginning to rub together.[49]

The European Court of Human Rights has shown that it is a powerful weapon for the enforcement of human rights with its ability to issue judgments that are binding on its member states. International humanitarian law has no such mechanism; its enforcement mechanisms depend principally on criminal courts. The European Court of Human Rights therefore offers a very attractive option to victims, particularly if there is the chance that the law will be interpreted by stricter standards. However, human rights law also has its drawbacks. It was designed to protect individuals from the power of the state, and even today, the obligations are still on states alone.

There thus remains a problem regarding nonstate actors. It is generally accepted that international humanitarian law binds such actors, and indeed, the criminal courts have drawn no distinction between nonstate actors and their regular counterparts. While some commentators have argued for an extension of human rights obligations to nonstate actors,[50] the extension is not generally accepted, and certainly the main human rights bodies, including the European Court of Human Rights, deal only with cases against states. It is thus possible for a state to find itself challenged for breaches of human rights committed on its territory by rebel forces fighting against it, on the basis that the state failed to take adequate measures to protect its citizens from the actions of the nonstate actor. The nonstate actor is itself immune from such proceedings. This too is an attractive avenue for claimants, as the state will have funds to pay compensation—and is compellable—whereas the nonstate actor is unlikely to be in the same position.

This is an area where there has been considerable development over the last couple of decades, and that development is not over yet. The Court already faces a plethora of cases arising from the international armed conflict between Russia and Georgia in 2008. This will pose many difficulties both in terms of attribution—the extent to which the actions of South Ossetian forces are attributable to either of the two states involved—and the law. To what extent, if at all, will the Court take into account international humanitarian law? Will it continue down its current line in which its judgments are based solely on human rights law with an occasional nod to international humanitarian law where it can be used in support of a conclusion reached?

Where Now?

Where does all this leave the soldier on the ground, the pilot in the plane, or the planner at operational headquarters? Are the law and operational reality beginning to drift apart? The problem starts right at the beginning with the categorization of conflict. The borderline between law enforcement and armed conflict seems now to be highly contentious. In another of the Chechnya cases, which also involved air bombardment, the Court held that as "no martial law and no state of emergency has been declared in Chechnya and no derogation has been made under Article 15 of the Convention . . . the operation in question therefore has to be judged against a normal legal background. . . . The massive use of indiscriminate weapons . . . cannot be considered compatible with the standard of care prerequisite to an operation of this kind involving the use of force by State agents."[51] Surely this was not in fact "a normal legal background." However, it appears that, because Russia had not sought to derogate from the European Convention for the Protection of Human Rights and Fundamental Freedoms, the Court was not prepared to consider this as an armed conflict and, as such, subject to international humanitarian law.

At the same time, international humanitarian lawyers have sought to push the threshold of Common Article 3 to as low a level as possible so as to be able to apply the Geneva law protections contained therein as widely as possible. While this might not be inconsistent with human rights protections in itself, the current wish to bring Hague law down to the same threshold has increased the risk of a collision with human rights law.

The Geneva law protections are generally consistent with human rights law, and indeed, many of the prohibitions contained in Hague law will also be consistent. The problem, however, comes with the permissions under Hague law. Under human rights law, targeting depends on conduct; under international humanitarian law, it depends primarily on status. Under international humanitarian law, a combatant can be attacked at any time simply because of who he or she is. Under the International Covenant on Civil and Political Rights, every human being has an inherent right to life. No one shall be arbitrarily deprived of his or her life.[52] The European Convention goes further. "No one shall be deprived of his life intentionally" except in certain specified circumstances.[53] Even in these circumstances, the force used must be "no more than absolutely necessary." This imposes a far higher standard.

Similarly, an object can be attacked under international humanitarian law if it is a military objective, defined as "those objects which by their nature, location, purpose or use make an effective contribution to military action and whose total or partial destruction, capture or neutralization, in the circumstances ruling at the time, offers a definite military advantage."[54] It is recognized that in attacking such objects, there may be incidental loss of life or damage to civilian objects, but this is acceptable if the expected incidental loss and damage is not excessive in relation to the concrete and direct military advantage anticipated.[55] This again is a very different test from that which would be applied under human rights law.

It follows that for commanders planning air operations, the law sends out differing signals. Under international humanitarian law, the first question is whether the target is a military objective. If so, what is the expected incidental loss of life or damage to civilian objects? Is that excessive in relation to the concrete and direct military advantage anticipated?

But even that is not the end of the line. The commander is still under an obligation to try to avoid, or at least minimize, incidental loss of civilian life, injury to civilians, or damage to civilian objects. He may do this by varying the time or direction of attack or by using a particular weapons system. International humanitarian law thus has a series of hurdles for the commander to overcome before he can launch an attack, whether by air or by other means. Nevertheless, the bottom line is that innocent lives may be lost and property may be damaged quite lawfully. The various tests are looked at primarily through the eyes of the attacker, and he or she will be judged not by an absolute standard but based on the information that was reasonably available to him or her at the time.[56]

Human rights law looks at the situation through the eyes of the victim. If there is loss of life or damage to property, it must be justified under the strict requirements of the law. Whereas international humanitarian law accepts that in war lives will be lost but seeks to minimize the loss of innocent lives, human rights law starts from the premise that any loss of life must be justified to a high standard. The onus is on the attacker and the loss of life itself creates a prima facie case.

However, there is a further issue. The laws of war apply equally to all sides. But why should they? Why should an aggressor be able to take advantage of the law when he is acting illegally from the start? The answer to this is twofold. First, it is usually unclear who the aggressor is. In the Kosovo air campaign, the action was held by some to be "illegal but legitimate."[57] If you draw a distinction between lawfulness and moral righteousness, where does that leave the commander? In most cases, both sides claim to be acting lawfully, as in the case of the Russia/Georgia conflict. In such cases, could both sides set aside the application of the laws until there was a ruling by the International Court of Justice as to who was in the right? Because such a ruling might take years, if it could be achieved at all, it would effectively mean that armed conflicts would be fought without laws.

Secondly, many of the laws of war are for the protection of combatants. These include the provisions dealing with the sick and wounded and prisoners of war. It is not the foot soldiers who start wars. Why should they have their protection reduced because of the actions of senior politicians? Soldiers (and even airmen) do not normally have degrees in international law to enable them to argue on the finer points of legality under the *ius ad bellum*, but if their country was denied the protection of the law, they would be the ones who would suffer.

A Comprehensive Approach for the Path Ahead

So what can be done to resolve this uncertainty? Between 2004 and 2010 a series of expert meetings took place under the auspices of the Harvard Program on Humanitarian Policy and Conflict Research. These resulted in the production of the *Manual on International Law Applicable to Air and Missile Warfare*.[58] This impressive work seeks to draw together the various strands of international law applicable to air and missile warfare into a coherent whole. It is not law in itself, but it is hoped that the *Manual* will assist in clarifying an undeveloped area of law. The *Manual* contains 175 "Rules," with commentary attached to each that explains the background and reasoning.

However, even this *Manual* had problems with two of the most difficult areas: non-international armed conflict and the relationship between international humanitarian law and human rights law.

With regard to non-international armed conflict, a decision was made to maintain the position that the "black-letter Rules" would be applicable only to international armed conflict, but the Commentary for each Rule would explain whether and to what extent the Rule extended to non-international armed conflict.[59] Events in Chechnya and Libya in March 2011 have shown that air warfare is not limited to international armed conflict, so this compromise is helpful.

Insofar as human rights law is concerned, the Introduction to the Commentary states: "There was discussion of human rights law without agreement. Most members of the Group of Experts believe that it has only minimal bearing on air and missile warfare in international armed conflicts because the law of armed conflict is *lex specialis*."[60]

While this may be correct, it raises the question of the meaning of *lex specialis* and, in particular, the relationship between human rights law and the law of armed conflict in non-international armed conflicts. Human rights bodies, in particular the European Court of Human Rights, seem reluctant to give regard to the law of armed conflict in such cases and prefer to deal with the matter on the basis of human rights law, with perhaps a cursory glance toward the law of armed conflict.

While the *Manual* is helpful, it takes the debate only so far. The wider debate on the interrelationship between international humanitarian law and human rights law goes far beyond air and missile warfare. Thus, the Group of Experts was probably wise not to try to tackle it in any depth.

However, that does not mean that the debate does not exist, and it is unfortunate that some on both sides seek to minimize the importance of it. There are substantial differences in relation to the conduct of hostilities—Hague law—and these should be recognized. There are good arguments on both sides. For example, it may be dangerous to allow status-based targeting in low-intensity non-international armed conflicts where law enforcement methods are still possible. Most Common Article 3 conflicts are best dealt with under the law enforcement paradigm insofar as the use of force is concerned. The Geneva law provisions of Common Article 3 are fully compatible with human rights law, so the two legal systems can live together quite comfortably. Indeed, Common Article 3 will assist in laying down a baseline standard where states seek to derogate from human rights law as they may be permitted to do.

On the other hand, it is near suicide to expect armed forces to operate in a law enforcement mode in high-intensity non-international armed conflicts, such as the Sri Lanka conflict or those in Iraq and Afghanistan.

The International Court of Justice did not assist when it looked at the position and stated in Delphic terms that

> As regards the relationship between international humanitarian law and human rights law, there are thus three possible situations: some rights may be exclusively matters of international humanitarian law; others may be exclusively matters of human rights law; yet others may be matters of both these branches of international law. In order to answer the question put to it, the Court will have to take into consideration both these branches of international law, namely human rights law and, as *lex specialis,* international humanitarian law.[61]

What the Court did not say is where the dividing line is. A pragmatic solution may be to consider that in low-intensity armed conflicts, insofar as the use of force is concerned, human rights law should take precedence, with international humanitarian law providing the fallback baseline position. On the other hand, in high-intensity armed conflicts, the positions would be reversed. However, this general rule would require considerable nuancing. As I have outlined, Geneva law and human rights law are generally compatible, so there would be no problem extending Geneva law down to the Common Article 3 threshold. Similarly, Hague law prohibitions could be also applicable, though only where they did not conflict with human rights standards. This would mean, for example, that use of force in such low-intensity armed conflicts, whether by land, sea, or air forces, would be governed by law enforcement and human rights standards, not by the standards of international humanitarian law.

On the other hand, in high-intensity armed conflicts—essentially those meeting most of the Additional Protocol II standards, although not necessarily requiring the involvement of state armed forces[62] the use of force would be governed by international humanitarian law standards. Thus, status-based targeting and attacks where the expected incidental loss of life or damage to civilian objects was not excessive in relation to the concrete and direct military advantage anticipated would be permitted. Human rights law would not be excluded in such circumstances, but it would need to be interpreted in the light of international humanitarian law.

Conclusions

So how would this affect aerial bombardment, and where do ethics come in? First, history teaches us that once a weapons system or method of warfare has been invented, it cannot be uninvented. At the turn of the nineteenth and twentieth centuries, attempts were made to stifle air bombardment at birth. They did not succeed. We have to accept that aircraft and air warfare are here to stay. Indeed, with the advent of drones and other pilotless vehicles, the chances must be that there will be an increasing emphasis on air warfare. The attraction of an air campaign, particularly if the enemy has limited ground-to-air assets, is self-evident. It is unrealistic to expect the genie to go back into the bottle.

The second point to be drawn from this historical review is the need for pragmatism. It can be argued that early attempts to regulate air warfare such as the 1923 Hague Draft Rules of Aerial Warfare were simply ahead of their time. The technology of the day did not permit compliance while waging an effective air campaign. The result was that the Rules were ignored. The law of armed conflict accepts the fact of war and therefore needs to be pragmatic in its application.

This applies as much to ethical principles as to the law. As was shown during the Second World War, an unrealistic legal regime cannot be protected even by ethical principles. However, where the legal regime is realistic, ethical principles can play a major role in fine-tuning that regime. The current legal regime, as expounded in the Harvard *Manual on International Law Applicable to Air and Missile Warfare*, may not be as restrictive as many would like. However, that is where ethics comes in. Hague law is, and has always been, largely reliant on the good faith of the participants. The law only provides the skeleton. It is ethics, chivalry, the warrior ethos—call it what you like—that puts the flesh on that skeleton.

Law and ethics are necessarily intertwined. As stated earlier, the law of war without ethics is an empty shell, while ethics without law is an aspiration. This was recognized in the earliest stages of the codification of the law of armed conflict in the famous Martens Clause, which first appeared in the Second Hague Convention of 1899.[63] In its 1977 and current version, found in Article 1(2) of Additional Protocol I, the text states: "In cases not covered by this Protocol or by other international agreements, civilians and combatants remain under the protection and authority of the principles of international law derived from established custom, from the principles of humanity and from the dictates of public conscience."[64]

When teaching soldiers, I reduced this to simple terms: if in doubt, do what you think is the right thing to do. The law can only take the decision-making process so far. In the end it comes down to doing what is right.

Part II

Interpreting, Criticizing, and Creating Legal Restrictions

CHAPTER 5

Clever or Clueless?
Observations about Bombing Norm Debates

CHARLES J. DUNLAP JR.

                For those favorably disposed toward the overthrow of oppressive regimes occasioned by the momentous events in the Arab world of early 2011, the news of the effectiveness of air attacks against the forces of a dictator threatening brutality toward his own people is a welcome development. In the wake of NATO intervention against the regime of Muammar Gaddafi in Libya in March of that year, the *New York Times* cited an example of what allied air power had accomplished. It quoted "a rebel spokesman using the name Aiman" who described how government tanks and artillery had been firing into the besieged city of Misurata "until three waves of airstrikes forced them back. 'After the airstrikes, things have been quiet,' he said by telephone."[1]

    Indeed, the use of aerial bombing in support of what has been called the Arab Spring" caused one enthusiastic (overenthusiastic?) analyst to proclaim that Operation Odyssey Dawn (the name of the Coalition air operation) is providing a "new lease on life for humanitarianism" by vindicating the "fragile responsibility-to-protect norm."[2]

    The Arab Spring generated virtually a tsunami of interdisciplinary discussion about the meaning, effect, and utility of the use of force at the call of the United Nations in the furtherance of humanitarian goals.[3] Much of that inevitably addresses the technology of war, of which airpower is frequently seen as the most sophisticated expression. It is imperative, therefore, that the enormous intellectual firepower of the academy be applied to this effort, as practitioners too often lack the time or the environment for the kind of considered reflection that best produces introspective analysis and thoughtful guidance for going forward.

The purpose of this essay is to attempt to facilitate the potential contribution of the range of academics, philosophers, theologians, nongovernmental organization representatives, and others outside the armed forces and government whose views are vital to this important dialogue. It will seek to complement and widen the expertise of discussants by highlighting certain technical information about airpower and by presenting something of a military perspective. Why? Because in this writer's experience the debate about the legal and ethical issues of bombing can become obfuscated when, for example, the terms of reference are confused and, especially, the technology of modern air warfare is misunderstood.

To illustrate, sometimes interchanges get bogged down in lengthy polemics whose predicates seem to assume that the means and methods of air warfare—not to mention the doctrine and methodologies for its application—were somehow frozen in place circa 1945. In other instances, the interpretation of *today's* bombing norms is too often sourced in popular understandings of Cold War deterrence strategies. Reference to Vietnam-era bombing practices can likewise become mired in circular discussions of the war's wisdom—a discussion that has intrinsic value but that is nevertheless of limited help in meeting the challenge of devising contemporary bombing norms.

It is true that some commentators have delved into more recent conflicts, especially those in Iraq and Afghanistan. Typically, the narrative produced is disapproving as to the use of the air weapon. Such criticism can be well reasoned, insightful, and productive. However, occasionally it is less useful than it might have been because the armed forces in general, and the U.S. Air Force specifically, too often have done an inadequate job of informing the public of relevant factual information.

Of course, sometimes the hostile reviews are merely an expression of an ideological agenda that uses seemingly generic critiques of advanced military technology as a stalking horse for further attacks on the real objective—that is, the defense policies of the United States. Those who level such attacks are not the hoped-for audience of this chapter.

Rather, this chapter is for those who are open-minded in their views, and—in any event—want to ensure that their arguments take into account as much factual data as possible and consider the widest set of views. Even if one remains convinced of one's own critique after evaluating the information and perspective this analysis tries to provide, one can do so with renewed confidence and conviction from having considered and rejected an alternative presentation. It is the absence of awareness of such alternative views that this essay seeks to address.

Discerning What Really Happened and Happens

One of the most frustrating aspects of any discussion about legal and ethical norms applicable to airpower is the role history plays in the debates. Unfortunately, the conversation is not always as fully informed as it should be. It is true that some historical studies about airpower are so technical that they can be difficult for those whose expertise lies in another discipline to fully assimilate, but there are also readily available sources that are cogent, concise, and easily digestible. Perhaps the finest example of the genre is Phillip Meilinger's short book *Airpower: Myths and Facts,* which is available online.[4] Meilinger works to dispel many misunderstandings about airpower, including, for example, those concerning World War II aerial bombardments.

Dr. Rebecca Grant's short essay about the much-maligned Kosovo air operation is another easily accessible perspective that counters many of the misconceptions regarding that campaign. One need not accept her conclusions per se to appreciate that correcting some widely reported *factual* errors (e.g., that no NATO aircraft flew below 15,000 feet)) is important in considering the legal and ethical norms that were actually observed.[5] Offering alternative readings of what is assumed to be the "history" of past bombing operations may help illuminate contemporary discourse.

There are several more extended examples of relatively recent scholarship that provide needed perspective for some classic issues of air warfare. For example, with reference to the bombing of Dresden—which is frequently used as a bumper sticker of sorts to denigrate airpower—historian Frederick Taylor observes that the number of civilian deaths from the strike "still wrenches at the heart six decades later" but adds:

> This does not mean that the Allied bombing of Dresden cannot be justified. Dresden was not an "open city," but a functioning enemy administrative, industrial, and communications center that by February 1945 lay close to the front line. . . . The bombing of Dresden was not irrational, or pointless—or at least not to those who carried it out, who were immersed deep in a war that had already cost tens of millions of lives.[6]

Taylor's observation does not make Dresden some kind of recommended template for a current approach to bombing. It simply shows that historians continue to argue over these controversies. Some point out, for example, that the overall bombing campaign—for all its many faults—did have the effect of imposing a huge burden on the Nazis' ability to wage war. Among other things, they were obliged to divert "two million people, 55,000 anti-aircraft guns [and] 20 percent of all ammunition" to the air defense effort.[7] Were it not for the allied air offensive, says historian Richard Overy, Nazi "frontline troops might have had as much as 50 percent more weaponry and supplies."[8]

Nor is this to suggest that bombing should be conducted against an otherwise unlawful target simply to force an adversary to defend it. To the contrary, enormous effort was focused—with real success—at disrupting a bona fide target (Nazi war industries), and military effects of great importance were produced by that effort and sacrifice (more airmen were killed in the Eighth Air Force in Europe alone than the Marine Corps lost in all theaters during the entire war).[9] Such information may be useful to offset the oft-heard assumption that the World War II air campaign was immaterial to the outcome of a war and therefore that all air warfare must be similarly ineffective.

Perhaps the real lesson is that with respect to a highly technological means of warfare such as aerial bombardment, the value of historical examples is necessarily temporally limited. This may be why airmen tend to look at history somewhat differently than perhaps others do. In an otherwise sneering and bitter denunciation of air force students attending the service's war college, academician Daniel J. Hughes has something of a point when he accuses the officers of "having little interest in theory and history, which they frequently regard as [made] irrelevant by advances in technology and military capabilities."[10]

Airmen *are* keenly aware of how dramatically the irrefutable laws of physics affect their machines; a better machine typically will defeat an opposing pilot, however talented, and this shapes an airman's mindset.[11] There is good reason for this concern. In 1945, the B-29 Superfortress was regarded as the most fearsome air weapon ever built, as it progressively devastated Imperial Japan. Yet less than six years later, the propeller-driven bombers were slaughtered in Korea by then state-of-the-art Russian-built MiG jet fighters.

Technological evolutions, especially those of the past very few years, have revolutionized the dynamic of air warfare. Information-age technologies have wrought several extraordinary changes, two of which are especially important to bombing norms: the emergence of *precision* strikes[12] and *persistent* intelligence, reconnaissance, and surveillance (ISR) capabilities.

Regarding precision weapons, consider that during World War II bombs *on average* would land within perhaps 1,200 feet of their target. Today, to be rated a precision weapon a "munition must hit within three meters, or less than ten feet" of the target.[13] These are the kinds of "smart" weapons used in contemporary conflicts, not the "dumb" bombs of other conflicts. For example, "close to 100 percent of all weapons carried and employed by aircraft in Afghanistan are of the precision type."[14]

Equally or more important is the revolutionary impact of persistent ISR enabled by the development of a variety of long-loiter aeronautical vehicles and more powerful sensors.[15] Today, ISR platforms can keep some battlespaces under near-constant surveillance, and this has significant implications. For example, *USA Today* reported that in the air attack that killed al-Qaeda operative Abu Musab al-Zarqawi in 2006, it took "600 hours of surveillance by a Predator drone to track Zarqawi and a matter of minutes for an F-16 to drop the bombs that killed him."[16] In 2008, journalist Mark Benjamin reported on the options such technology gives decision makers:

> The Air Force recently watched one man in Iraq for more than five weeks, carefully recording his habits—where he lives, works and worships, and whom he meets. . . . The military may decide to have such a man arrested, or to do nothing at all. Or, at any moment they could decide to blow him to smithereens.[17]

Interestingly—and somewhat counterintuitively—overhead surveillance can sometimes provide superior situational awareness to that obtained by soldiers on the ground. One published report points out, for example, that "despite the distance, the real-time video feeds [provided by aircraft] often give [remote air controllers] a better vantage point than an Army unit has just down the street from a group of insurgents."[18] Thus, it is simply inaccurate to believe that with today's capabilities civilians are necessarily put more at risk by air operations than by ground operations because of a dearth of relevant intelligence.[19]

The revolutionary impact of technology is not limited to weapons and aircraft; it has also radically changed the *process* by which operations are planned and carried out. In order to harness the potential of the information age, the air force has constructed advanced combined air operations centers (CAOCs) filled with technologies that facilitate not only that application of force but also the observance of legal and ethical norms. For example, *US News & World Report* noted that in the CAOC:

> the center painstakingly plans its strikes, says an officer in the targeting team. Analysts calculate the size of bomb fragments and the distance they travel from the strike site, using detailed maps and video footage to

gauge potential for human casualties and property damage. In another area, analysts don 3D glasses to read maps that show precise heights of palm trees and the walls of any given compound to help determine "collateral concerns."

Similarly, the *New York Times* described it this way:

> At the air operations center, targeting specialists spend hours before each mission measuring distances from the potential strike zone to the nearest house, building, mosque, school or hospital. . . . Vast numbers of public, religious and historic sites make up a computer database of no-strike zones. Special goggles are worn while reviewing digital images compiled from surveillance aircraft and satellites to give a detailed, three-dimensional view of the target area.
>
> The bombs themselves are chosen carefully and sometimes modified. Some designed for air burst are instead programmed with a delayed fuse to bury themselves before exploding, thus reducing the blast range. One sort of bomb has even been loaded with less explosive, filled instead with concrete, to cause great damage where it hits but no farther.[20]

The *Times* also noted that air force lawyers vet the targets to ensure that the proposed bombing conforms "to a complex body of military law, including the Geneva Conventions, acts of Congress and court decisions."[21] Those specially trained lawyers, who are on duty in the CAOC around the clock, use a variety of computerized analytical and communication tools to conduct sophisticated evaluations of all aspects of the air operation and to provide real-time advice as required.[22]

In any event, the result of this blending of law and innovative technologies is that even a Human Rights Watch analyst was obliged to admit that "in their deliberate targeting, the Air Force has all but eliminated civilian casualties in Afghanistan."[23] Having said all this, asymmetries can arise over differences as to what law applies in determining legal and ethical norms.

Sorting out the Legal Cacophony

Today those concerned with the legal parameters of aerial bombing and with the law of war in general, are presented with a confusing cacophony of treaties, declarations, agreements, and protocols, along with claims of binding international law that exists only as custom.[24] The situation is further complicated by the fact that the preeminent airpower nation, the United States, is not a party to some of the leading international agreements that many (if not most) nations have acceded to.[25] This can vastly complicate efforts to discern applicable legal norms.

A good example is Protocol I to the Geneva Conventions, which contains much-discussed rules about the protections of civilians.[26] Even though it is not a party to Protocol I, the United States conducts its operations in a way that seems to indicate that it accepts the bulk (but not all) of it as customary international law. Regrettably, the publication of the U.S. Department of Defense's long-awaited *Law of War Manual*, which was anticipated to elucidate this and many other issues, has not, as of this writing, yet occurred.[27]

Until it does, the best and most widely accepted compilation of existing international law applicable to bombing is the *HPCR Manual on International Law Applicable to Air and Missile Warfare* issued by Harvard's Program on Humanitarian Policy and Conflict Research.[28] Although this *Manual* is not without controversy, it is the product of a six-year effort that included many leading international experts from both legal and military disciplines. Notably, it bills itself as a "restatement" of existing law and emphatically does not purport to create any new norms.

What is remarkable about the slim volume is that it reveals that relatively little law is explicitly limited to air and missile warfare. This would suggest that air operations should get no more scrutiny—and perhaps less, given the relatively small numbers of civilian casualties they cause—than other kinds of fires and military operations. Yet they do. James Baker, a former member of the U.S. National Security Council, illustrates a common perception:

> Air power is more susceptible to legal and policy adjustment than ground combat, in light of the variations in means and method of attack available through variation in munitions, delivery azimuth, angle of attack, aim point, fuse, and explosive, all amplified with the assistance of computer simulation.[29]

In many respects, what Baker says is true (and this explains why airpower can be discreetly applied). However, characterizations like his are also apt to create an expectation of perfection that is unattainable with virtually any weapon, given the proverbial fog and friction of war. In fairness, it is often the military's own actions, such as the distribution of videos showing bombs falling precisely down the airshafts of enemy buildings, that exacerbate the assumption of infallibility, which in turn distorts discussions of the application of force via aerial bombing.

Misperceptions about the law also create misunderstandings. The St. Petersburg Declaration of 1868 is an illustration of the mischief that can ensue when the legal applicability of a particular norm is misapprehended.[30] The United States is one of the nations that were not part of the St. Petersburg convocation and it has never agreed to the convocation, which is not, in any event, part of customary international law. To the extent that it still retains vitality, it binds only nations who are a party to it.

Of most relevance to this discussion is the Declaration's Preamble. It states, in part, that "the only legitimate object which States should endeavor to accomplish during war is to weaken the military forces of the enemy; that for this purpose it is sufficient to disable the greatest possible number of men."[31] Scholars often cite this wording to support the premise that the purpose of war is simply to kill the opponent's military personnel. Yet the "object" of war is not, per se, to "disable the greatest number of men" in the adversary's military forces, as some suppose. Rather, as the great military theorist Carl von Clausewitz explains, war is "an act of violence to compel our opponent to fulfill our *will*. . . . Violence . . . is therefore the *means;* the compulsory submission of the enemy to our will is the ultimate object."[32] The destruction of the enemy's military forces is one way of achieving that end, but in the view of many strategists, it is an imperfect means of doing so. Sun Tzu, for example, argues in his classic *The Art of War* that "supreme excellence consists in breaking the enemy's resistance without fighting."[33] Experience shows that the erosion of the "will" of an adversary through the *indirect* effects of aerial bombardment on civilians is a key element of victory in modern war.[34]

The St. Petersburg Declaration also stimulates misunderstandings about the status of civilians in war. Contrary to what many may think, the law of armed conflict (LOAC) makes no judgment as to the moral culpability of individuals in defining that status. Instead, it adjudicates status based on certain objective factors—for example, membership in the armed forces of a belligerent and, in the case of persons otherwise considered civilians, whether they directly participate in hostilities. To the extent they do, they are targetable to the same extent as military personnel are.[35]

None of this turns on the ideological proclivities—or the absence of the same—of any person, combatant or civilian. Under the LOAC, a civilian may not be *directly* targeted—even if he or she embraces the most loathsome, odious ideologies and actively promotes the same (short of direct participation in hostilities). Thus, it is incorrect to indiscriminately apply the label of "innocent" to civilians in the context of LOAC. Civilians may be guilty of any number of moral or even legal breaches yet still enjoy immunity from being directly targeted. They are not necessarily, however, *morally* "innocent" civilians.

The late Daniel Boorstin, the former librarian of Congress and formidable Pulitzer Prize–winning scholar, took this concept a bit further. He insisted that Americans in particular suffer from the "Myth of Popular Innocence" that is expressed in the "touching American unwillingness to believe ill of human majorities."[36] Boorstin points out that in reality Hitler, Stalin, and Saddam Hussein could not have carried out their evil deeds without the cooperation of much of the populace. Boorstin adds that civilian societies are not helpless victims of unscrupulous leaders, as "history proves that ruthless rulers can be removed by popular will."[37] In fact, despite having powerful internal security forces, the Soviet Union collapsed when confronted with a determined people's movement. Much the same can be said of the fall of Arab autocrats in the spring of 2011.

It is not surprising, therefore, that there is no legal or, indeed, moral imperative to spare the sentient adult population of a belligerent from the vicissitudes of war, short of refraining from direct targeting. To be clear, while it is plainly wrong to target civilians or to conduct military operations—bombing or otherwise—for the "sole or primary purpose of spreading terror among the civilian population,"[38] it is nevertheless also fully expected—and tolerated—that military operations could have psychological and other unpleasant consequences for civilians.

For example, consider the profound sadness of those who have lost a family member serving as a soldier in the nation's army. Moreover, even the indisputably legitimate destruction of purely *military* objects can impact the civilian population responsible for replacing them. As Dwight Eisenhower said, "Every gun that is made, every warship launched, every rocket fired signifies, in the final sense, a theft from those who hunger and are not fed, those who are cold and not clothed."[39]

More specifically, no imperative of law or ethics prohibits the sentient adult population from suffering certain *indirect* costs of war. For example, as a matter of law, collateral damage from an air attack "does not include inconvenience, irritation, stress, fear or other intangible conditions caused to the civilian population."[40] Accordingly, while the blockades, "no-fly zones," and the destruction of "dual use" infrastructure no doubt impose hardship on civilians, it is hardship that is legally and morally permissible.

Perhaps the most severe penalty civilians must suffer is being killed or injured as a result of an attack on a genuine military target. Every legitimate military operation seeks to avoid such losses, yet the fact remains that the law, properly applied, recognizes that incidental civilian deaths in connection with an otherwise bona fide attack on a military target are acceptable and often *expected*.

This crucial principle (along with the practical reality that it is typically impossible in the midst of military operations to determine which civilians are authentically "innocent") illustrates why the notion that combatants are legally or morally obliged to take more risk than those holding civilian status is so deeply flawed. Apart from the fact that nothing in international law imposes such a requirement, simply because a human being chooses to serve his or her country in uniform *should not be* a rationale to value that life less. It is wrong to convert an ethic of a disposition toward public service into a norm that licenses the devaluation of the lives of those who choose to serve as combatants. The life of the civilian—"innocent" or not—is not intrinsically more worthy than that of the combatant.

Another problematic tendency among some well-meaning advocates is the unbridled assumption that treaties and other restrictions on specific weaponry are an unqualified good. For example, consider the Ottawa Convention's prohibition on anti-personnel land mines.[41] The United States—which is not a party to the Ottawa Convention—has in its inventory the GATOR mine system, an air-deliverable weapon that contains self-neutralizing anti-tank and anti-personnel mines.[42] It may be used as a "runway denial" weapon in that it can scatter mines on an enemy airfield to make it temporarily unusable to hostile forces without actually destroying it. Nations who are parties to the Ottawa Convention, however, cannot use the weapon. What is the alternative for them? Destroy the runway with conventional high-explosives—*but* this will make it unavailable to both postconflict humanitarian relief flights and economic reconstitution.

Many experts are discovering that the well-intended prohibitions on chemical and biological weaponry are having the perverse effect of limiting nonlethal and low-lethality weapons that might otherwise be developed.[43] For example, although riot control agents have great potential to limit deaths and injury, they are forbidden as a method of war by the chemical weapons convention.[44] Such results are garnering criticism. Harvard Law professor Gabriella Blum cites the prohibition on the use of riot control agents to question the morality of "the law's current absolutist stance [that] prevents parties in conflict from lawfully pursuing actions that might lessen the harms of war."[45]

Regardless, challenging science to come up with solutions to military problems that avoid the law's "current absolutist stance" can take things in a direction some may not have expected. For example, as discussed above, it is generally unlawful to use nonlethal tear gas as a means or method of war, and most countries do not permit the use of anti-personnel landmines. What then, are the options left to air operation commanders tasked with such challenges as neutralizing the use of caves by enemy forces?[46]

One solution would be to scatter anti-personnel mines around the entrances. Since that option is foreclosed to parties of the Ottawa Convention, commanders from states that are parties to the Convention may need to entomb a cave's occupants by blowing up the entrance—a disconcerting action that is not necessarily at odds with international law. Another solution, provided by science, is thermobaric weapons.[47] According to one description, the thermobaric bomb is "among the most horrific weapons in any army's collection . . ., a fearsome explosive that sets fire to the air above its target, then sucks the oxygen out of anyone unfortunate enough to have lived through the initial blast."[48] Again, these consequences merely illustrate that when the focus of a treaty is on a certain weapon, as opposed to *effects*, the result can be unintended and a source of concern.[49]

Science does not always produce solutions more troubling than the problem it seeks to solve. Consider cluster munitions, another much-maligned—and much misunderstood—weapon. They are bombs "that release a number of smaller submunitions intended to kill enemy personnel or destroy vehicles" in a given area.[50] Because the submunitions contain relatively small explosives, they are very useful in attacking such targets as anti-aircraft guns on a dam, snipers on a hospital roof, or even factories producing weapons of mass destruction, whose devastation by more powerful explosives might put civilians at greater risk as toxic materials entered the atmosphere.

Yet today, despite the development of advanced technologies that reduce the failure rate of submunitions,[51] many nations either prohibit or severely limit their use.[52] One way science is helping to address legitimate concerns about these weapons is by developing a "newer generation" of cluster bombs that are more accurate and "sensor-fuzed submunitions [that] are designed to sense and destroy [military] vehicles without creating an extensive hazard area of unexploded submunitions."[53]

This raises another issue that is inexplicable to some in the armed forces: that is, the recent obsession of many academicians with remotely piloted aircraft (RPAs)—often inaccurately referred to as "drones." Plainly, RPAs are not "autonomous," as some seem to believe, although, as suggested above, weapons that "autonomously" sense certain characteristics of targets as they home in on them have been in the inventory for decades.[54] There are certainly legitimate issues about RPA use, but—again—it is not clear why these particular systems should be considered differently from other means of using force.[55] Some writers seem disturbed that RPAs are operated at a distance, but throughout the history of warfare combatants have always sought to apply their weaponry from a range beyond their opponents' capability.[56] There is nothing in law or ethics that requires a combatant to give an opponent a "fair fight" in the sense of exposing himself or herself to being killed. Indeed, as General George Patton succinctly (albeit indelicately) put it "the object of war is not to die for your country but to make the other bastard die for his."

One might rightly argue that regardless of what the law may permit, ethical norms demand a higher standard. Undoubtedly, there is a clear relationship between law and ethics. According to historian Geoffrey Best, "It must never be forgotten that the law of war, wherever it began at all, began mainly as a matter of religion and ethics . . . It began in ethics and it has kept one foot in ethics ever since."[57] Nevertheless, few would dispute the idea that law typically represents the baseline of consensus about behavioral norms.

With norms expected to have global application, such as those governing armed conflict, it should be clearly understood that seemingly universal principles are actually less universal than many believe. This is why, for example, the Harvard *Manual on International Law Applicable to Air and Missile Warfare* was described above as a "slim volume." In an increasingly globalized world, interpretations of ethical norms can vary widely, making consensus rarer. Richard Falk is one of many scholars who argue that the proposition that "all persons and people aspire to the same human rights" is just factually untrue.[58]

In any event, ethicists and others need to be especially cautious about calling upon members of the armed forces to apply ethical norms at variance with the law. Service members are not obliged to obey patently illegal orders, but military law does provide that "the dictates of a person's conscience, religion, or personal philosophy cannot justify or excuse the disobedience of an otherwise lawful order."[59] Given that there are many different moral philosophies, the dangers of allowing personal moral norms to trump the law are readily apparent.

For example, some conscientious people believe that abortion amounts to the murder of an unborn child; the law, however, permits abortions under certain circumstances. Thus, force is not permitted to "defend" the unborn. Put another way, in the military context uniformed personnel are obliged to follow the law in conducting operations, and reliance upon their personal interpretations of international norms is done at their peril. Apart from liability for disobedience of orders, even well-meant efforts to "improve" upon the law in situations that do not involve the disobedience of orders can still have dire consequences for those the law is meant to protect, as is discussed below.

Airpower in Counterinsurgency

Until very recently, conventional wisdom was that airpower was either largely irrelevant to counterinsurgency (COIN) operations or affirmatively counterproductive. This is the philosophy that seemed to infect Field Manual (FM) 3-24, the much-celebrated publication by, among others, General David Petraeus.[60] Published in late 2006, it rapidly became "The Book" on COIN.[61] From the perspective of the air weapon, it is significant that FM 3-24 relied heavily upon a study of a rather narrow band of COIN campaigns, mostly from the Cold War era, that largely preceded the information revolution that was so influential in the development of precision weaponry and persistent ISR.[62]

According to a 2006 RAND study emphasizing those conflicts, "air power has been used in a less-visible supporting role" mainly because historically, "insurgencies do not present opportunities for the overwhelming application of the air instrument."[63] As a matter of fact, given the airpower technologies available to those operations of more than a half-century ago, it is not especially surprising that airpower was limited to a mainly supporting role.[64]

More complicated, particularly in the context of a discussion of aerial bombardment, is that FM 3-24 embraced a view of COIN that eschewed violence in favor of a "population-centric" strategy[65] that sought to win hearts and minds[66] through nonviolent nation-building and other developmental projects. While not shunning force entirely, FM 3-24 rapidly became perceived as advocating a "softer approach that won allies" after it was implemented in Iraq in 2007.[67]

In evaluating this "softer approach," it is worth noting that the contributions to FM 3-24 came from what was called an "odd fraternity" of "representatives of human rights nongovernmental organizations and international organizations, academic experts, civilian agency representatives, [and] journalists."[68] Among other things, the resulting document called upon a counterinsurgent to serve variously as a "social worker, a civil engineer, a school teacher, a nurse, [and] a boy scout."[69] Steven Coll described the new doctrine this way in the *New Yorker:*

> [FM 3-24 is popular] among sections of the country's liberal-minded intelligentsia. This was warfare for northeastern graduate students—complex, blended with politics, designed to build countries rather than destroy them, and fashioned to minimize violence. It was a doctrine with particular appeal to people who would never own a gun.[70]

Unsurprisingly, airpower—especially in its strike role—was marginalized into a five-page annex in a nearly 300-page document. And that brief reference discouraged its use.[71] What FM 3-24 did not take into account was the dramatic technological revolution that had taken place in recent years. In the fall of 2007, retired army general Barry McCaffrey observed:

> We have already made a 100 year war-fighting leap-ahead with MQ-1 Predator,[72] MQ-9 Reaper,[73] and Global Hawk.[74] Now we have loiter times in excess of 24 hours, persistent eyes on target, micro-kill with Hellfire and 500 lb JDAM bombs,[75] synthetic aperture radar, and a host of ISR sensors and communications potential that have fundamentally changed the nature of warfare.[76]

Such capabilities were not overlooked by military commanders—including one of FM 3-24's principal authors. Notwithstanding the "softer" persona of FM 3-24, General David Petraeus's COIN operations in Iraq were decidedly "hard." Although he publicly derided the notion of "killing and capturing" as an avenue to COIN success, that is exactly what happened in Iraq following the issuance of FM 3-24. Tens of thousands of Iraqi males were swept up and incarcerated in huge detention camps,[77] and a dramatic increase in "killing"[78] complemented the "capturing" that combined to finally bring the violence under control.

Much of the killing of insurgents was accomplished by airpower. Despite the admonitions of FM 3-24, airstrikes increased fivefold.[79] The results were significant: retired air force lieutenant general Mike Dunn asserts that "90% of the terrorists [who were] killed [were] killed by airpower."[80] Somehow, it seems, this increasing utility of airpower as a COIN weapon got translated into an assumption that airpower was a major cause of civilian deaths.

That was really never the case with Iraq. For example, a 2003 Human Rights Watch investigation of major combat operations in Iraq "found that, in most cases, aerial bombardment resulted in minimal adverse effects to the civilian population."[81] A study published in 2009 in the prestigious *New England Journal of Medicine* entitled "Weapons That Kill Civilians" did complain about airstrikes in urban areas but nevertheless produced statistics that showed that during the 2003–2008 timeframe in Iraq, *only about 6% of civilians* who died as a result of the *conflict* were killed by air weaponry.[82]

There is also little to support the notion that the military success of 2007 is attributable to FM 3-24's "softer" persona. Even though security increased markedly in Iraq in 2007, a 2008 survey of Iraqis found that 61 percent still believed that the presence of U.S. forces made security worse in their country, and of those who thought the security was improved, only 4 percent believed U.S. forces deserved the most credit.[83] Notwithstanding the vital role of airstrikes and the obvious fact that few hearts and minds were won, Iraq is now considered the exemplar of COIN success.

Afghanistan reflected a somewhat similar yet different challenge. In 2001 a unique blend of airpower, small numbers of U.S. special operations forces on the ground, and alliances with indigenous opposition forces enabled the United States to unseat the Taliban in a matter of weeks.[84] In the ensuing years, however, NATO took responsibility, and efforts to prevent Taliban resurgence foundered, as did myriad nation-building projects. In 2009, General Stanley McChrystal was sent to Afghanistan with a charter to stabilize the situation in order to permit a withdrawal of foreign forces.

During NATO's tenure, its approach to airpower can most charitably be described as misguided. In June 2007, NATO announced that there would be no airstrikes if it "knew there were civilians nearby."[85] In 2008, another NATO spokesman declared that no airstrike would take place "if there is the likelihood of even one civilian casualty . . . not even if we think Osama bin Laden is down there."[86] The law certainly does not require such extraordinary measures to avoid civilian casualties—and for good reason.

By replacing the proportionality standard of Protocol I, which permits attacks that cause incidental civilian casualties so long as they are "not excessive in relation to the concrete and direct military advantage anticipated" with a "zero casualty" rule, NATO evidently did not seem to comprehend the wisdom behind the Protocol's approach. In its approach, NATO telegraphed to the insurgents that all they needed to do to protect themselves from air attack was to surround themselves with civilians—and that is exactly what they did. If NATO had followed the Protocol, the insurgents would not have had as much incentive to shield themselves with civilians.

Unfortunately, General McChrystal's decision in June 2009 to further restrict airstrikes proved disastrous for civilians. By June of the year following the implementation of the restrictive rules, Afghan civilian deaths had skyrocketed by 31 percent,[87] and Coalition military casualties likewise rose sharply.[88] Importantly, the astonishing increase in civilian deaths was *not* the result of the airstrikes that did take place. A study released in July 2010 showed that airstrikes were responsible for only a small percentage of the casualties caused by Coalition forces.[89] For example, traffic accidents involving U.S. and Coalition vehicles killed two and a half times as many Afghan women and children as did airstrikes.[90]

When General Petraeus replaced General McChrystal in June 2010, he vastly increased the use of the air weapon. Noting that airstrikes had risen 172 percent by October of that year, a journalist accurately declared that there was "once again a full-blown air war over Afghanistan."[91] General Petraeus appears to have grasped the fact that since the vast majority of civilian casualties are caused by insurgents (some 76 percent),[92] the best way to protect civilians is to kill those who would kill them. The policy of forgoing airstrikes operates to spare insurgents to live to kill civilians.

Statistics proved the worth of Petraeus's strategies. By the end of 2010, he was able to cut the rise in civilian casualties from the 31 percent that had occurred under McChrystal's approach to half that rate.[93] Most remarkably, a UN report released in March 2011 declared: "Although the number of air strikes increased exponentially, the number of civilian casualties from air strikes decreased in 2010."[94]

Nevertheless, airstrikes are often singled out for the proposition that the deaths they cause ipso facto increase insurgent recruitment. Actually, dispositive evidence about insurgent recruitment motivation is scant, and what does exist is subject to varying interpretations. For example, that the Taliban kill by far the most civilians suggests that they believe that doing so furthers their cause in some way. Being Afghans themselves, one would assume they would not conduct operations so deadly to civilians otherwise.

And it appears they may be right. The *Christian Science Monitor* reports that "there is little indication these Taliban indiscretions [causing civilian casualties] have backfired on the movement so far."[95] Another interesting perspective is offered by Afghan expert Jeremy Shapiro. Shapiro believes that the Afghan government highlights civilian casualties to get leverage with the Coalition, but local officials in his experience "tend actually not to be too concerned" with the civilian casualty issue.[96]

Other experts have challenged—convincingly—the notion that RPA strikes spur insurgent recruitment. For example, after conducting on-the-ground research, analyst Christopher Swift found that such strikes simply do not "drive al Qaeda recruiting" in Yemen.[97] With respect to Pakistan, scholars Christine Fair, Karl Kaltenthaler, and William Miller challenged conventional wisdom in an *Atlantic* magazine article entitled "You Say Pakistanis All Hate the Drone War? Prove It" by arguing that while "drone strikes are not very popular among a large section of Pakistani society . . . Pakistanis are not united in opposition to drone strikes."[98] "In fact," the writers contend, "many Pakistanis support the drone strikes."[99]

None of this is to suggest that any death is less than a tragedy; rather, it is simply to dispute the negative inferences about the *military* effect of unintended civilian casualties that is so often attributed to airpower. In truth, a significant amount of scholarship indicates that the physical presence of foreign forces on the ground is the biggest recruitment stimulant for insurgents. COIN expert William R. Polk insists that the "fundamental motivation" for insurgents is an "aim primarily to protect the integrity of the native group from foreigners."¹⁰⁰ And this has proven true in recent conflicts.¹⁰¹

In fact, some of the key underlying premises of FM 3-24 are now being questioned. Jill Hazelton of Harvard's Belfer Center argues in a recent interview that the "conventional wisdom" of COIN—that "the development of healthy, participatory, well-governed states will defeat insurgency"—has never actually worked.¹⁰² She contends:

> Generally, states that succeed in COIN rely on the use of force, offensive and defensive, to destroy the insurgent military threat by military means, and they also provide limited, targeted political accommodations to gain the cooperation of useful political actors within the populace and insurgency. Success in COIN does not require the protection of the populace, good governance, economic development, or winning the allegiance or the loyalty of the great majority of the population. It does not require building up all of the institutions of the state. These goals may be important to the meeting popular grievances in a particular case, or important to the counterinsurgent for a variety of reasons, as with the United States in Iraq and Afghanistan, but the empirical evidence does not show that they are necessary for success.¹⁰³

Importantly, her research also shows that "successful COIN cases include less sensitivity to civilian casualties than the conventional wisdom prescribes."¹⁰⁴ Along the same lines, Francis J. "Bing" West, former assistant secretary of defense for international security affairs and best-selling author, notes that although "our senior leaders say the war cannot be won by killing," the war "will surely be lost if we don't kill more Islamist terrorists and hard-core Taliban."¹⁰⁵

West, a former Marine who has just written a new book on Afghanistan, sees airpower—as both a high-tech surveillance platform and a precision strike weapon—playing a central role in a new strategy that he proposes:

> Push the Afghans to fight their own war. Stop fighting for them. Create the Adviser Corps we have needed for the past ten years. *Our air surveillance is so extraordinary today that we can deploy about 50 advisers per 400-man Afghan battalion and patrol rigorously without* unduly *risking our advisers*. We do not need 100,000 troops. . . . The Taliban needs to mass in order to threaten to retake government control in the urban areas. *Given our air, they cannot mass.*¹⁰⁶

Concluding Observations

As stated at the outset, the potential contribution of the academy to the proper interpretation and development of legal and ethical norms associated with air warfare is tremendous. The erudition—not to mention sense of commitment—of many moral philosophers, theologians, historians, ethicists, legal scholars, and many other disciplinary experts is a reservoir of talent that needs to be tapped. This is especially so when the means and methods of warfare are increasingly complex, especially as related to the air weapon. Better ways to use force, especially in sensitive situations such as those posed by recent conflicts, can be found if *all* the information and perspectives are considered by the functional experts working in harmony, if not always in agreement.

Of course, no amount of discussion will impact those whose moral tenets reject the concept of the just war.[107] It may surprise some, but those in the armed forces—especially those who have seen the horrific consequences of war firsthand—are often the ones most opposed to the use of force. Senior military officers spend their entire life around young people who comprise the vast majority of the armed forces; to them, they are real people who they often know in a personal way. And they are acutely aware it is these same young people they must send into harm's way and that many of them will not return or will return much different than they went.

Sensitivity to this reality is important. An illustration is the infamous incident when UN ambassador Madeleine Albright asked Chairman of the Joint Chiefs of Staff General Colin Powell: "What's the point of having this superb military you're always talking about if we can't use it?"[108] Such a casual depersonalization of those who will be expected to go into harm's way and, if necessary, pay the ultimate price is deeply offensive to those in uniform. As the BBC reports it, General Powell—who it describes as "this most military of politicians who has watched men die"—answered Ambassador Albright by icily observing that "American GIs are not toy soldiers to be moved around on some global game board."[109] Cavalier references to the troops—and the risks they take—is dangerous ground.

There is also a lesson for those ready to ascribe nefarious motives to military professionals. Consider the case of Philip Alston, the UN special rapporteur who claimed in a study of targeted killings[110] that because RPA operations can be conducted "entirely through computer screens and remote audio feed, there is a risk of developing a 'PlayStation' mentality to killing."[111] Of course, no such evidence of that exists. To the contrary, what we do know is that those operating these systems take their responsibilities extremely seriously, to the point of suffering psychologically because of it.

Dr. Peter Singer, the Brookings Institution researcher who authored the book *Wired for War*, about high-technology weaponry, found that

> in the beginning we feared that drones may make the operators not really care about what they're doing. But the opposite has turned out to be true. They may almost care too much. We're seeing higher levels of combat stress among remote units than among some units in Afghanistan. We found significantly increased fatigue, emotional exhaustion and burnout. Drone operators are more likely to suffer impaired domestic relationships, too.[112]

Dr. Singer explained this phenomenon by noting, among other things, that a "remote operator sees the target up close, he sees what happens to it during the explosion and the aftermath. You're further away physically but you see more."[113] His conclusions dovetail with earlier reports in the *New York Times Magazine* about the stresses conscientious RPA operators suffer.[114] As one air force official put it, RPA operations are "a deeply, deeply emotional event. It's not detached. It's not a video game."[115] In fact, the *New York Times* reported in early 2013 that RPA pilots suffer stress disorders just as those in combat do.[116]

That Professor Alston would make such a serious accusation without offering evidence rightly raises questions about the rest of his analysis and conclusions. All of this is yet one more illustration of the importance of thoroughly understanding the systems *before* engaging in speculation about the motives of the professionals involved. The reality is that academics can get their facts wrong. For example, for years many critics of the bombing campaign in the 1991 war with Iraq alleged that the destruction of infrastructure and the subsequent economic sanctions were responsible for the deaths of 500,000 Iraqi children. Yet new scholarship convincingly argues that this allegation is a myth.[117]

Furthermore, few things frustrate military professionals more than critics who do not bother to learn about systems they disparage. With respect to RPAs in particular, retired air force lieutenant general David Deptula insists that the "truth is, RPA are the most precise means of employing force in a way that reduces collateral damage and minimizes casualties" and argues that the "critics don't understand the reality of 'drone' operations."[118] Deptula points out that

> the persistence, situational awareness, and degree of control possible with an RPA allows for the immediate suspension of lethal engagement if circumstances change or questions emerge—even after a weapon has

been released or launched. RPA are networked aircraft and their data can reach any spot on earth in less than two seconds.

Hence, in addition to the hundreds of operational, maintenance, and intelligence personnel, many lawyers and senior leadership are directly involved with RPA lethal engagements. That kind of oversight is rarely, if ever, the case with the use of manned aircraft or with boots on the ground or sailors at sea. The power of our intelligence networks allows RPA essentially to carry around their own command and analysis center and legal counsel as an integral part of their payload.

Similarly, the methodology of a much-touted study by Stanford and New York University that was critical of RPAs was deconstructed by subsequent analysis.[119] More meticulous studies find that "drone strikes are associated with decreases in the number and lethality of militant attacks in the areas where strikes are conducted."[120] When the "lethality of militant attacks" is eroded, innocent civilians benefit. This is especially important given the reported decline in civilian casualties from RPA missile attacks. While incidents still occur, the New America Foundation reported in June 2013 that only four civilians (as opposed to seventy-eight militants) died in fourteen strikes in Pakistan.[121] In any event, one scholar recently concluded:

> In the end, drone strikes remain a necessary instrument of counterterrorism. The United States simply cannot tolerate terrorist safe havens in remote parts of Pakistan and elsewhere, and drones offer a comparatively low-risk way of targeting these areas while minimizing collateral damage.[122]

It is also useful for anyone who wants to influence policy to exercise caution in attacking the motives of military personnel for this reason: doing so will not resonate well with the American people. Not that anyone should shy away from criticism where criticism is due—which is often the case. But notwithstanding periodic scandals and incidents of terrible misconduct, polls show that the armed forces as an entity remain the most trusted institution in American society.[123]

Such public approval is in stark contrast to that afforded an institution such as the UN, which is not held in nearly as high esteem. More particularly, military leaders are exceeded only by nurses in the public's positive estimate of their honesty and ethics.[124] This may be worth considering when judging the decisions of military commanders. They often must make difficult life-and-death decisions in an extremely compressed time frame and do so based on imperfect information produced in the chaos of battle.

That is why, for example, international law does not judge command determinations as to what constitutes "excessive" civilian casualties based only on knowledge gained in hindsight.[125] Even the U.S. Supreme Court has concluded that "it is difficult to conceive of an area of governmental activity in which the courts have less competence" than the "complex, subtle, and professional decisions" military officers must make.[126] Although penetrating analysis of the conduct of military personnel should be made, reticence in ascribing mens rea in the first instance could be useful.

Finally, no amount of discussion will convince those invested in the belief that force has no place in human affairs. Chinese president Hu Jintao recently said that "history has repeatedly proved that the use of force is not an answer to problem. . . . Dialogue and other peaceful means are the ultimate solution to problem."[127] This may be so. But what force *can* do is create the space—and incentive—for dialogue by those not otherwise disposed to engage in it. It is becoming increasingly clear to many experts that to get terrorist or insurgent groups to "cease their pursuit of an objective via armed violence," governments "should focus on the physical attrition of such groups as being the primary contribution of force to gaining such a policy goal."[128]

Of course, no use of force is desirable. Yet in the twenty-first century, bombing with all its flaws—is in many instances better than no action at all. As one commentator put it, "No one, certainly no one in the US military, has ever claimed that a bombing campaign can be carried out without any loss of innocent human life. Yet the use of airpower is essential if America is to be able to prevail at an acceptable cost of both life and treasure for itself."[129] And in the absence of American leadership—and its distinctive airpower capabilities—one wonders what would have happened to the people of Kosovo or, more recently, the people of Libya.

Indeed, the slaughter in Syria has led to calls for an implementation of a no-fly zone.[130] While this is doable, it would likely involve a significant bombing campaign to suppress Syrian air defenses.[131] Many also fear it could draw the United States and its allies into a wider war.[132] It is imperative that no one forget the sheer horror and ugliness of war, whether the result of bombing or any other use of force.

As John Stuart Mill famously observed, "war is an ugly thing, but not the ugliest of things. The decayed and degraded state of moral and patriotic feeling which thinks that nothing is worth war is much worse."[133] That war may be inevitable should spur our efforts to do whatever we can to ameliorate its ugliness, and an aggressive, interdisciplinary rethinking of bombing norms is a logical endeavor toward that end.

It would be a mistake, however, to be overly optimistic. Clausewitz counsels:

> Kind-hearted people might of course think there was some ingenious way to disarm or defeat the enemy without too much bloodshed, and might imagine this is the true goal of the art of war. Pleasant as it sounds, it is a fallacy that must be exposed: war is such a dangerous business that the mistakes which come from kindness are the very worst.[134]

That said, there is indisputably a place for hard-headed realism (if not "kindness") in the development and application of legal and ethical norms. When tied to a genuine understanding of modern weaponry, the strategies for their use, and the people who use them, real progress can be made. Achieving that "genuine understanding" does, however, require a serious investment of time and intellectual energy by all concerned.

The stakes are very high as we look ahead. With respect to airpower, futurists George and Meredith Friedman muse that the images of high-technology precision weapons striking their targets with extreme accuracy carries a "deep moral message," especially "when contrasted with the strategic bombardments of World War II."[135] According to the Friedmans, "War may well be a ubiquitous part of the human condition, but war's permanence does not necessarily mean that the slaughters of the twentieth century are permanent."[136] That is a proposition with which this writer would readily agree.

CHAPTER 6

The American Way of Bombing and International Law

Two Logics of Warfare in Tension

JANINA DILL

In Afghanistan and Iraq, American bombs struck their targets with spectacular precision. Yet both wars took longer and were more costly than expected. They attracted vociferous criticism for the toll they took on the populations under attack, and unambiguous political victory arguably eluded the United States. In today's wars the problem is no longer how to hit what you aim at, but to decide what you should aim at in the first place![1] The question arises with particular acuteness in air warfare, where attacks tend to be preceded by considerable deliberation and often require a choice between objects. Air strikes respond to urgent needs or immediate threats much more rarely than decisions to launch attacks on the ground, which are often made under considerable time pressure and with much less room for the contemplation of alternatives.[2] When faced with an enemy society, what targets should the United States pick for air attack?

The answer lies in the principle of distinction. The understanding that all persons and objects can and must be assigned either to a civilian sphere that is immune from direct attack or to a sphere of legitimate military targets is what "the whole idea of a law of war absolutely depends on."[3] But what does it mean to distinguish in accordance with the law? This chapter shows that there are two fundamentally different notions of what distinction in war ought to look like. One is exemplified in recent U.S. doctrine, specifically air force doctrine, which is inspired by a long line of strategic thinking about air power. The other understanding of distinction emerges from a systematic interpretation of the positive international law that defines a legitimate target of attack as codified in the Protocol Additional to the Geneva Conventions of 12 August 1949, and relating to the Protection of Victims of International Armed Conflicts (Protocol I), 3 June 1977 to the Geneva Conventions (hereafter First Additional Protocol).[4]

The chapter demonstrates that these diverging notions of what it means to properly distinguish in war are indicative of the struggle between two fundamentally different visions for how combat operations ought to be conducted. The law, when interpreted systematically, aims to regulate warfare by allowing only the targeting of that which needs to be open to engagement if a competition between two militaries is to proceed. It envisions warfare to follow a *logic of sufficiency*. The alternative approach to distinction prescribes attacking what helps end the war most quickly and achieve its political goals most directly. It is based on a *logic of efficiency*. The two logics have radically different implications for which parts of a modern society can become objects of attack from the air.

The Legal Definition of Military Objectives

The First Additional Protocol contains the most recent and most elaborate codification of the principle of distinction. Its "Basic Rule," Article 48, spells out the obligation "to distinguish [at all times] between the civilian population and combatants and between civilian objects and military objectives" and to "direct . . . operations only against military objectives." A legitimate target of attack is thus in the first instance military, an object or a person that counts as what the treaty refers to as a "military objective." Both objects and persons (combatants) can be military objectives. This chapter focuses on objects because air strikes have only recently started to include the systematic targeting of persons. For commanders and operators in contemporary armed conflict from the air, distinction is first and foremost a question of classifying objects as either civilian or military.[5]

Military objectives as far as objects are concerned are defined positively in Article 52(2). The rule establishes that those objects may be attacked "which by their nature, location, purpose or use make an effective contribution to the military action and whose partial or total destruction, capture or neutralisation in the circumstances ruling at the time offers a definite military advantage." Two criteria—an "effective contribution to military action" and a "definite military advantage"—hence determine whether an object can be reckoned a military objective. They are closely linked. In reality, it is inconceivable that the engagement of an object that makes an effective contribution to the enemy's military action would *not* yield a military advantage. In turn, the main, though not the only possible, reason why an attack on an object should be militarily advantageous is that the object contributes to enemy military action.[6]

One rare example of an object that fulfills one but not the other of the two criteria is a bridge over a river that if destroyed forces the enemy to use another one that better lends itself to being ambushed. The attack provides a military advantage and renders the first bridge a military objective. That bridge, however, has not yet and possibly never would have contributed to the enemy's military action. It follows that objects that make an effective contribution are a subclass, though arguably by far the largest, of objects whose engagement will provide a military advantage. The criterion of military advantage has the final say when international law draws the line between immune civilian objects and legitimate military targets.[7]

The Connection of Objects to Military Operations

It is the connection of an object to combat operations, those of the enemy belligerent (effective contribution) and one's own (military advantage), that puts an object into the category of military objectives. But how close must that connection be? And what is the right degree of nexus between an attack and the military advantage? The text says that the advantage has to be "definite," which could mean tangible, visible, palpable, or, alternatively, precise, determinate, distinct, or unequivocal. These words describe the quality of the advantage to be achieved—the first group alludes to the fact of the existence, possibly the likelihood of its emergence; the second group refers to the sharpness of the contours of the advantage, as it were. They do not describe the advantage's connection to the attack. Designations such as direct, immediate, prompt, or instant would do so.

Does the other criterion help shed light on the required connection between an object and the enemy's war effort and hence the minimum degree of nexus between an attack and a military advantage for the object to count as a military objective? Contributions to the enemy military have to be "effective," which means actual, real, significant, but not direct. An example of an object whose contribution to the military effort is real and significant, even vital, yet indirect is the food-supplying industry. Since soldiers need to eat, it quite literally sustains the adversary's war effort. By the same token, its engagement ultimately generates a military advantage because hungry forces are less effective. Moreover, complicating the enemy's food procurement ties up manpower and resources that otherwise could be used in support of the fighting.

However, the military advantage arising from an attack on food suppliers does not directly arise from that attack. It is two causal steps rather than one away from the destruction of the object in question. The result of the attack is that the industry is in ruins and food supplies decrease; first causal step, soldiers get hungry; second causal step, military effectiveness declines. In turn, the food-supplying industry is doubtlessly part of a society's war effort. Yet it is two causal steps removed from the enemy's *military* effort, meaning the engagement of the enemy belligerent in hostilities. Compare the food-supplying industry to the arms-supplying industry. An ammunition plant generates an output that soldiers need to fight rather than to live. As a result, the decrease in military effectiveness directly follows its engagement. It is widely accepted that weapons production sites are military objectives. That food suppliers that service the military should be off limits is contested.[8]

Yet in modern society the engagement of many other objects yields potentially significant military advantages that are only two or more causal steps down the line. For instance, the engagement of any production facility or industrial plant supporting the noncombat-related logistics of the armed forces, such as clothing or shoe production, is two causal steps away from a decrease in military effectiveness. Similarly, attacks on import-export businesses or, in general, any object used for a taxable economic activity decrease the belligerent's financial resources in step one and decrease its capacity to procure weapons in step two. Treating manufacturing sites and facilities used for trade, which only indirectly support the military effort, like military objectives is controversial.

Even further removed from the initial attack is the military advantage arising from a neutralization of certain media installations. An attack causes a decrease in propaganda. This in turn may lower civilian morale (step one); civilians, as a result, may contribute less to the war effort (step two). Alternatively, they might seed doubt in or cause worry to their relatives at the front (step two). It is only in step three or even further down the causal chain that this decreases military effectiveness.

Another example is the engagement of public infrastructure—for instance, power plants, communication links, transportation, bridges, or roads—that is of *no* current or future use to the military but whose destruction "inconveniences" civilians, including regime cronies or political leaders. Attacks on such infrastructure likewise require at least three causal steps before they yield a military advantage.

What about schools that enhance the cognitive and physical abilities of boys and young men who, as a result, will be better suited for military service? This example illustrates that the more causal steps separate the military advantage from the initial attack, the less certain it is that the advantage will materialize at all. Due to intervening agency, any extra causal step increases the chances that the initial effects of the attack will be mitigated. For instance, a belligerent government deprived of the ability to broadcast propaganda could hand out leaflets instead or put up signs to "keep calm and carry on." Every step makes the military advantage anticipated from the attack slightly less certain, and at some point it becomes entirely hypothetical.

While the principle that the law excludes schools from direct attack is incontrovertible, the status of many of the mentioned objects is disputed. They fall into three broad categories: 1) objects that procure what soldiers need to live rather than to fight; 2) objects that are crucial for the belligerent state to sustain its productive and extractive capacity; and 3) objects that are crucial for civilian morale and/or the public's support of the government. While some scholars interpret Article 52(2) to allow the engagement of only those objects that contribute to the enemy's *military* effort directly and hence reject the permissibility of targeting objects from the three categories unless in specific circumstances where they actually do directly contribute to combat operations, others consider a connection to the war effort sufficient.[9]

The Commentary to the First Additional Protocol interprets the word "definite" to mean the military advantage must be "concrete and perceptible . . . rather than . . . hypothetical."[10] It thus clearly rejects the inclusion of objects into the category of military objective whose attack will yield a military advantage only if sweeping and uncertain assumptions about intervening agency materialize. Perceptible belongs in the category of words that describe the quality of the required advantage. "Concrete" could refer either to the latter or to the connection between the attack and the advantage. If we consider it to imply a reference to the connection between the attack and the advantage, it would suggest that a high degree of nexus between them is required.

The U.S. Interpretation and "Sustaining" the War

Although the United States is not a party to the First Additional Protocol, right after the treaty negotiations the language of Article 52(2) was adopted into operational law manuals, signaling that the United States accepted the definition of military objectives as binding.[11] Yet in the 1997 "Field Manual on the Joint Targeting Process" a new criterion to assess the significance of a mission emerged: "war-sustaining." The document asks whether an attack or an operation reduces the enemy's ability to "sustain the war effort."[12] The *Joint Doctrine for Targeting* of 2002 draws on the term war-sustaining to explain the definition of military objectives.[13] The term then forms part of this definition in Military Commission Instruction No. 2 of 2003. Military objectives are therein defined as objects that "effectively contribute to the opposing force's war-fighting or war-sustaining capability," as opposed to the Protocol's criterion of "an effective contribution to military action."[14] This new formulation suggests that a direct link to the competition between two militaries is not the only way an object can become a military objective. Another way is to contribute to an enemy's "war-sustaining capability." The *military* advantage that may ultimately arise has only an indirect connection to such attacks, as shown above.

Since 2004 the *Operational Law Handbooks* issued by the Judge Advocate General School start the definition of military objectives with a verbatim repetition of Article 52(2) and then offer this explanation: "The connection of some objects to an enemy's war fighting or war-sustaining effort may be direct, indirect, or even discrete. A decision as to classification of an object as a military objective . . . is dependent upon its value to an enemy nation's war fighting or *war-sustaining effort* (including its ability to be converted to a more direct connection), and not solely to its overt or present connection or use."[15] It is unclear what a discrete connection between an object and the adversary's war-sustaining effort might mean. However, it is perfectly clear that this explanation embraces an indirect connection to the competition between two militaries as sufficient for an object to fall on the non-immune side of the distinction line.

Just how low the required degree of nexus is according to U.S. legal doctrine becomes evident in the *Operational Law Handbooks* of 2006 to 2008, which list without qualification all "(1) Power, (2) Industry (war supporting manufacturing/export/import), [and] (3) Transportation" as military objectives.[16] The broadest conception of military objectives in legal doctrine yet is implied by the permission to intentionally attack "objects that contribute to an opposing state's ability to wage war" that has been featured in operational law manuals since 2010.[17] Ultimately the whole of an enemy's society, with the possible exception of children and the elderly, can be considered to somehow contribute to a state's ability to wage war. The statement includes no specification at all of the required degree of nexus between an object and the actual *military* effort.

Definitions of Progress during Military Operations

Even more fundamental than the question of the minimally required connection between an object and combat operations and the directness with which a military advantage must arise from an attack is the question of how we define military advantage in the first place. What does it mean to achieve progress during the conduct of hostilities? The point of reference used to determine a military advantage could alternatively be the destruction of one object, the thwarting of one operational objective of the enemy, or a discrete step in the process of overcoming the adversary's military forces, such as the destruction of the opposing air defense system. Most commentators agree that it does not have to be a single air strike that provides the advantage. Progress may be the result of several attacks taken together. But what is the ultimate point of reference for the understanding of progress in combat? Is it overall victory?

If victory provides the point of reference for the definition of a military advantage, as military practitioners tend to assume, the question arises whether we mean military or political victory. Clausewitzian wisdom, of course, teaches that the two are never far apart. However, if the political goals of a war are what serve as the point of reference for defining progress, targeting choices might not continue to follow strictly military imperatives. For instance, in a war whose political goal is regime change, the answer to the question of whether military victory (i.e., overcoming the enemy military until the political regime in question can no longer resist) or political victory (i.e., ousting the political regime) is the adequate point of reference for the definition of a military advantage would decide whether such objects as party headquarters, civilian leadership bureaus, and information links between the government and the public are considered military objectives.

The interpretive statement for Article 52(2) introduced by the United Kingdom on the occasion of the negotiations on the First Additional Protocol argues that definite military advantage "is intended to refer to the advantage anticipated from the attack considered as a whole and not only from isolated or particular parts of the attack." The UK's position has been often reproduced and is widely endorsed.[18] The interpretation that "'military advantage' is not restricted to tactical gains but is linked to the full context of one's war strategy"[19] goes only a little further. However, it is on the verge of challenging the position of the International Committee of the Red Cross (ICRC) that "an attack as a whole is a finite event, not to be confused with an entire war."[20]

The U.S. Interpretation and Achieving Political Goals Directly

Since 2003, virtually all doctrinal texts issued by the U.S. Air Force describe targeting as "fundamentally effects based."[21] The doctrine of effects-based operations (EBO) in the operationalization of distinction recommends that targets are to be selected that "contribute *directly* to the achievement of *strategic* objectives."[22] "Strategic" is defined as "the highest level of an enemy system that, if affected, will contribute most directly to the achievement of our national security objectives."[23] The doctrine hence advocates choosing objects as targets that are linked not necessarily to military victory but to the overall strategic—often political—goals of a war. Accordingly, "offensive action [is allowed and welcomed] against a target—whether [it is] military, political, economic, or other."[24] This means that no one can predict what kinds of objects will count as military objectives during EBO because it is the "national security objective," which evidently varies from one war to another, that determines which attacks yield a military advantage.

It is not only the U.S. military that interprets distinction to mean that military objectives are defined according to whether their engagement contributes to the achievement of political goals. For instance, the military goal of Israel's air campaign Cast Lead in early 2009 was to end the ability of Hamas to launch rocket attacks against Israel. This was at times coupled with the broader political goal of undermining Hamas's hold on the Gaza Strip. Using the latter as the point of reference for the definition of military advantage might have led Israel to consider the destruction of government structures and public infrastructure provided by Hamas—which did not directly contribute to the launching of rockets against Israel—as nevertheless providing a significant military advantage.[25]

Another example of the application of EBO, where the genuinely political goal of an armed conflict served as the point of reference for progress, was so-called boutique bombing as practiced during NATO's air campaign against the Federal Republic of Yugoslavia in 1999. NATO exerted pressure on the government of Serbia by attacking the possessions of regime cronies.[26] A very short line of thinking hence connected targeting choices to the overall political goals of Operation Allied Force, one of which was to get Milošević to return to the negotiation table. The effort to engage Serbia's military apparatus was secondary.

Political victory and military victory fall farther apart the more limited the political goals of a war are. The aim of putting pressure on a country's political leadership, in contrast to, for instance, annexing a territory or discovering and disabling weapons of mass destruction, does not require the destruction of the military fabric of the regime's hold on power. In this situation, it seems particularly indicated to avoid the engagement of fielded forces and military hardware and to focus attacks on the leadership directly. Of course, depending on the political goals of a campaign, the objects that need to be engaged in order to directly achieve political victory might be connected to the competition between two militaries only indirectly. The low degree of nexus that is part of the U.S. interpretation of distinction hence follows the same imperative of directly achieving political goals with force. The effect of these two interpretive positions—the political point of reference for the definition of military advantage and the low degree of nexus between objects and the actual military struggle—on distinction is similar: 1) they broaden the category of objects that can potentially be military objectives; and 2) they open up the possibility that the definition of military objectives can be flexible and can track the political goals of a war.

I suggest that these two positions are both indicative of a quest for efficiency in air warfare. Attacking objects whose effects most immediately further one's ultimate goals regardless of whether or not they are also directly connected to the competition between two militaries allows for more efficient warfare. Making the desired political end state that goal is more efficient than attacking what leads to military victory, which presumably then needs to be translated into political victory at the negotiation table or during occupation. Indeed, the doctrine of EBO prescribes that mission accomplishment should be "sought while minimizing cost in lives, treasure, time, and/or opportunities," seeking "to achieve objectives most effectively, then most efficiently."[27]

The Logic of Efficiency

This focus on efficiency seems to make sense in light of the political context of contemporary armed conflict. Given the increased public scrutiny made possible by embedded journalists, social media, and a 24-hour news cycle, the reputational costs of inflicting death and destruction are higher than they ever were in human history. The global rise of human rights norms and the increase in so-called casualty aversion, specifically among Western publics where their own compatriots are concerned, has rendered this issue even more acute. Avoiding notoriously drawn-out wars of attrition, getting major combat operations over with as quickly as possible, and declaring "mission accomplished" are central imperatives for states waging war in the twenty-first century.

In addition, the prohibition on the use of force, which has ended the notion that armed conflict is a regular tool of statecraft, has rendered it crucial that governments keep the political reasons that they are using to claim that this *particular* war is nevertheless necessary, legitimate, and/or legal at the forefront of combat operations. That every air strike in NATO's Operation Allied Force seemed to directly hurt and put pressure on Milošević, whose alleged evilness provided the basis for the air campaign's claim to legitimacy, may have made an otherwise highly controversial military endeavor somewhat more palatable.

In Operation Iraqi Freedom, the claim that the United States was waging war against Saddam Hussein, not the Iraqi people, was meant to be made plausible by the decision to take air power "downtown" early on. This way the Coalition could directly attack the Baath regime's centers of power rather than destroying tanks and killing conscripts in the Iraqi desert. Of course, this also meant that the United States accepted higher numbers of civilian casualties as a result of "collateral damage" than it would likely have incurred during a war more focused on traditional attrition.[28] In any event, the ever-more-tenuous connection of attacks to the competition between two militaries and their more direct relevance for the achievement of final political goals—indeed, the infusion of warfare with political considerations—is paradoxically an imperative that is rooted in our internalization that war is no longer an acceptable continuation of politics by other means.[29]

This draws attention to the fact that seeking efficiency in war is not merely a strategic imperative. It is also one among several imaginable ways of dealing with the undesirability of war. The underlying logic is to get the war over with as quickly as possible. The implicit normative proposition is that wars ought to be sharp so that they may be brief.[30]

Though the interpretation of the principle of distinction that is manifest in U.S. usage accommodates a compelling imperative that belligerents with the strategic offensive face in modern armed conflict—to wage war as efficiently as possible with a view to achieving the political end-state on which the explanation and legitimization of death and destruction hinge—it would be wrong to think that this logic of efficiency is a new invention. In fact, the desire to jump over battle lines, fielded forces, and military hardware to directly attack an enemy's centers of power is as old as air power itself. The ability to circumvent the competition between two militaries and bomb for political or psychological effect is the distinct capacity of air forces, and this ability has inspired air power enthusiasts since Guilio Douhet and Hugh Trenchard to prescribe just that.[31]

Elliot Cohen, a contemporary expert on air power, provides an exemplary argument based on what I refer to as the logic of efficiency when he observes that "the sprinkling of air strikes over an enemy will harden him without hurting him and deprive the United States of an intangible strategic asset."[32] As a result, he argues that "when presidents use [air power], they should . . . hurl it with devastating lethality against a few targets (say, a full-scale meeting of an enemy war cabinet or senior-level military staff) . . . to cause sharp and lasting pain to a military and a society."[33] From this imperative of maximum pain per air strike he infers that "it appears likely that civilian populations or large portions of them will continue to be the objects of terror." He extrapolates his own stipulation about the efficient use of air power to a statement about the nature of war: "In many cases today, war means bringing power, particularly air power, to bear against civil society."[34]

While Elliot Cohen, much like his "air-minded" forefathers, does not seem to recognize the legal rule of distinction at all—any object can potentially be a military objective if an attack hastens the achievement of victory—many military practitioners make more sophisticated arguments that reveal a vision in which combat operations follow the logic of efficiency without defying the principle of distinction and international law in such a flagrant way. On the contrary, the incarnation of the logic of efficiency that is prevalent in contemporary strategic thought is backed, seemingly even inspired, by a particular reading of international law that interprets Article 52(2) to allow the engagement of objects with only an indirect connection to the competition between two militaries if they provide an advantage for the achievement of a war's ultimate (often political) goals.

General Charles Dunlap, for instance, advocates psychological warfare targeting civilian morale.[35] His argument for why this is legal rests on an interpretation of the legal definition of a "military objective" that considers an indirect connection of an object to the competition between two militaries, and he argues that a political, psychological or other advantage is all that is required for an object to be fair game. "Even a passing familiarity with Clausewitz and other strategists—not to mention America's own experience in Vietnam—make it plain that 'undermining the government's political support' does offer a very 'direct and concrete' military advantage especially in today's world."[36] However, the genuinely military advantage from undermining civilian morale arises only a couple of causal steps down the line because the contribution of civilian morale to the military effort is indirect.

The Logic of Sufficiency

What, then, is the vision for warfare that underlies an interpretation of Article 52(2) on the other end of the spectrum, where the degree of nexus is high and the point of reference for the definition of advantage is military in nature? I argue that the alternative logic for regulating war rests on two commands addressed to belligerents: first, sharply distinguish objects that are closely (meaning one causal step) connected to the competition between the enemy militaries from everything else! The latter (the "everything else") is immune from direct attack. Second, sharply distinguish between ultimate goals (political) and intermediate goals (military)! While hostilities are ongoing, belligerents have to bracket their larger political goals when planning how to act—that is, deciding what to attack. I call that "sequencing." Belligerents are commanded to observe a strict sequence of the use of force that has the goal of military victory now and the translation of military into political victory later.

Two assumptions about sufficiency underlie this logic: first, in a significant number of cases it is sufficient—in order for a belligerent to have a genuine possibility of a generic military victory—to engage objects and persons that are closely connected to the competition between the two militaries and thus count as military objectives. In turn, being allowed to attack objects and persons whose engagement provides a genuinely military advantage is necessary for the competition between two militaries to proceed and a possibility of military victory for one side to exist. This means that a high degree of nexus between an object and military action, and hence between an attack and the resulting military advantage, is required. Why? Because it may be sufficient for military success, and a genuine opportunity for military success is all that is promised to any belligerent.

The second assumption of sufficiency is that in a significant number of cases, military victory is sufficient to allow states to subsequently achieve their legitimate political or other goals. The second sufficiency assumption hence rules out war as an effective instrument for the achievement of a wide range of political goals in the first place. This restriction makes sense only in light of the blanket prohibition on the use of force in contemporary international law. The current international legal order rests on a presumption against the use of force as a continuation of politics by other means. If states had a right to use force as a regular expression of sovereign statecraft, the second sufficiency assumption would be incoherent. The assertion that generic military victory is sufficient is hence as much a normative command as an empirical assumption—it is sufficient for the goals that any state ought to be pursuing with force.

Of course this logic of sufficiency does not per se preclude a belligerent from seeking efficiency—that is, trying to minimize costs in (one's own) blood, treasure, and time. But the two logics differ in what they consider the goal over which other factors can be minimized. For the logic of sufficiency, that goal is military progress narrowly defined, with the goal of overcoming the enemy militarily. What the logic of sufficiency defies is the efficient (direct) pursuit of political goals with force. In addition, this logic draws a fairly definite line around objects that have a high degree of nexus to the competition between two militaries and whose engagement yields a military advantage in one causal step. Even if it would be much more efficient to attack other objects to produce military victory, the latter remain off limits. The logic of sufficiency puts both distinction and sequencing (no minimization of time and resources over the achievement of the political goal) before efficiency.

To sum up, the way the logic of sufficiency deals with the undesirability of war is not by getting it over with quickly but by circumscribing it to the maximum extent possible. The normative proposition underlying the logic of sufficiency is not that wars ought to be as brief as possible but that they ought to be limited as much as possible to the competition between two militaries.

Sufficiency versus Efficiency

Are both logics equally supported by what seems to be an international legal regime that is indeterminate in a crucial place, Article 52(2)? No. I maintain that the First Additional Protocol envisions that combat operations will be conducted according to the logic of sufficiency. As the first treaty to regulate warfare in light of the prohibition on the use of force, the First Additional Protocol does not allow appeal to the notion of a just cause or an acceptable political aim. The law regulating the conduct of war (*jus in bello*) is independent from the reasons for resort to force and their legality (*jus ad bellum*). Law that does not allow any appeal to the causes and reasons for which a war is fought as a guide for the conduct of hostilities must work from the assumption that there is a stable (if very abstract) concept of military victory that is valid across most wars, notwithstanding their different political contexts. It is that notion of "generic" military victory that represents the furthest possible point of reference for the determination of a military advantage. In direct contradiction to the interpretation of distinction exemplified in EBO, a systematic interpretation suggests that belligerents' goals have to be achieved through advantages that appeal to this concept of generic military victory only, rather than directly to the political, moral, or other reasons for which they are ultimately fighting.

In turn, without the interpretation of the word "military" as the ultimate point of reference for the determination of an advantage, it would be impossible to draw a line between acceptable reasons for action in war (conduct) and the legal, political, and moral reasons for why a state is using force (resort). But the fact that the First Additional Protocol is premised on the separation of *jus in bello* and *jus ad bellum* is incontrovertible. Its Preamble unequivocally spells out that the Protocol's provisions must be applied "without any adverse distinction based on the nature or origin of the armed conflict or on the causes espoused by or attributed to the Parties to the conflict."

As described above, the internalization of the prohibition on the use of force is one major reason why it seems crucial for belligerents to wage war efficiently. Yet when we read the positive law that embodies the principle of distinction in light of the prohibition on the use of force and the inevitable separation of the regulation of conduct in war from the causes or goals for which belligerents fight, the logic of sufficiency emerges—a logic that defies efficiency in the two ways described above. The law that defines distinction is quite indeterminate. As a result this efficiency-defying logic of sufficiency emerges only in a systematic or contextual reading that interprets Article 52(2) in light of the separation between the regulation of resort to force and conduct in war. This, in turn, may explain why the issue of whether U.S. bombing practices that follow what I refer to as the logic of efficiency violate international law is a controversial one. The struggle between two diverging logics of how wars ought to be fought is not acknowledged. Because air warfare is uniquely conducive to applying force for the direct achievement of political goals, the logic of sufficiency is in particular danger of losing out to the logic of efficiency during air targeting, the comprehensive subjection of U.S. combat operations to international law notwithstanding.

CHAPTER 7

Force Protection, Military Advantage, and "Constant Care" for Civilians
The 1991 Bombing of Iraq

HENRY SHUE

One of the most important and insistent challenges faced during war is making legally and morally defensible judgments about the protection of civilians and the protection of one's own forces when the two are in tension. Another challenge is weighing each of these two against military advantage. The official final report on the Persian Gulf War of 1991 offers the following murky reading of international law: "An attacker must exercise reasonable precautions to minimize incidental or collateral injury to the civilian population or damage to civilian objects, consistent with mission accomplishment and allowable risk to the attacking forces."[1] Does this mean that "we are minimizing civilian losses, once we have done what it takes to accomplish the mission and to avoid excessive risk to our own forces"? If so, it appears to declare first priority to the mission (military advantage), second priority to protecting our own forces, and only third priority to protecting civilians. I shall note some actual results for Iraqi civilians of this kind of thinking in the case of the 1991 bombing, which was followed immediately by the decade of severe sanctions imposed by the war that were primarily managed and enforced by the United States.[2] Then in light of this preface I shall explore some ways of thinking about how to handle the complex triangular balance among military advantage, force protection, and "constant care" for civilians.

The 1991 Bombing of Iraq and the "Invisible War"

Joy Gordon has written the most authoritative study thus far of the effects on civilians of the war and the consequent sanctions that were intended to prevent Iraq from acquiring weapons of mass destruction and were completely dominated by the United States through its de facto veto role in the committee of the Security Council that controlled the severity of the sanctions. Here there is space to glance only briefly at some conclusions in her book plus one important subsequent reassessment. Gordon notes the assessments respected international organizations made of the effects of the bombing: "In February 1991 a team sent by UNICEF and WHO surveyed the damage from the bombing throughout Iraq. They reported that, throughout the country, in urban as well as rural areas, the destruction to power plants and refineries was so extensive that 'this necessitates a careful rationing of the system's remaining fuel supply, which will not last more than five weeks.' Iraq's infrastructure was so devastated that 'Baghdad has no public electricity, no telephones, no gasoline for civilian vehicles, and less than 5% of its normal water supply.'"[3]

Gordon introduces the most famous and influential judgment as follows:

> In March 1991, Martti Ahtisaari, Under-Secretary-General for Administration and Management, visited Iraq with a UN delegation of personnel from the major humanitarian agencies, including UNICEF, WHO, UNDP, FAO, and UNHCR. In his report, Ahtisaari observed that "the recent conflict has wrought near-apocalyptic results upon the economic mechanized society. . . . Iraq has, for some time to come, been relegated to a pre-industrial age, but with all the disabilities of post-industrial dependency on an intensive use of energy and technology." Ahtisaari described in some detail the extensive breakdown in infrastructure, including water purification and sewage treatment, agricultural production and food supplies and distribution, the destruction of the telephone system and all modern means of communication, and the large number of refugees whose homes were destroyed in the air war. He identified the particular urgency of energy needs. Without the production of electricity, he noted, "food that is imported cannot be preserved and distributed; water cannot be purified; sewage cannot be pumped away and cleansed; crops cannot be irrigated; medicaments cannot be conveyed where they are required; needs cannot even be effectively assessed. It is unmistakable that the Iraqi people may soon face a further imminent catastrophe, which could include epidemic and famine, if massive life-supporting needs are not rapidly met."[4]

Far from allowing sufficient aid to be provided, the United States insisted, through the UN Security Council, on the strictest sanctions in history, which Gordon terms an "invisible war" and which blocked the import into Iraq of the parts needed to repair the damage to the energy infrastructure from the 1991 bombing. The exact results of the combination of the extreme sanctions regime and the effects of the bombing continue to be hotly debated.

Unfortunately, attention has tended to focus on a very high official estimate by UNICEF of the "excess mortality rate" of children under the age of five during the next decade, and at the time she wrote Gordon, like most experts, accepted the official estimate: "the majority of the studies over the course of the sanctions regime strongly suggest that, for the period from 1990 to 2003, that figure is at least 500,000."[5] However, Gordon now rejects as exaggerated the figure of 500,000 additional deaths of children under the age of five, believing that in many cases sacrifices by family members may have prevented children's deaths, although sometimes only at the price of malnutrition of some family members, given the undeniably extensive shortages of food and medicine.[6]

That one extremely high estimate of one particular measure of damage has proven groundless does not, of course, mean that nothing much happened. Abundant evidence remains that the combined collapse of multiple elements of infrastructure—water treatment, electricity, health care, transportation, agriculture—had profoundly serious effects on health, even if the specific effects on child mortality rates were not reliably quantified by UNICEF. Obviously the 1991 bombing alone would not have led to the effects that occurred without the sanctions maintained by the Clinton administration from 1993–2000 and continued by the Bush administration until the American invasion in 2003, but the thorough destruction of the energy infrastructure by the 1991 bombing was a necessary condition for the extensive suffering under the war-imposed sanctions over the next decade.[7] Here we need to consider whether that initial destruction was itself permissible.

Laws of Armed Conflict

Having extremely briefly sketched some facts, we next need to look briefly at the most relevant law. As I indicated at the beginning, the challenge is to juggle at least three separate considerations: military advantage, losses to one's own forces, and civilian losses. The complexity can be made manageable if seen through three interlocked Rules, two of which are embodied in the laws of armed conflict (LOAC) and one of which may be emerging.[8] I state the bare Rules here, although it will require the bulk of the chapter to explain adequately what they mean and why they are reasonable:

1. Civilian losses from an operation must not be expected to be excessive compared to the military advantage from the operation.
2. Losses to one's own forces from an operation must not be expected to be excessive compared to the military advantage from the operation.
3. The risk of losses to one's own forces ought sometimes to be increased in order to reduce losses to civilians, but the risk of civilian losses may not be increased in order to reduce losses to one's own forces, except at the extreme.

In order to begin to see what these three Rules mean, we can start with the most relevant elements of the laws of armed conflict. The "Basic Rule" at the beginning of Part III, Section I, "Methods and Means of Warfare," in the Protocol Additional to the Geneva Conventions of 12 August 1949, and relating to the Protection of Victims of International Armed Conflicts (Protocol I), 8 June 1977 (hereafter, the First Additional Protocol) says: "In any armed conflict, the right of the Parties to the conflict to choose methods or means of warfare is not unlimited."[9] This is sometimes described as "perhaps the most fundamental customary principle" governing the conduct of war,[10] but this basic Rule is no more fundamental than the basic Rule of distinction between the civilian and the military.[11] Indeed, the limits on means and methods generally and the limit on choice of military objectives specifically are formulated to avoid excessive losses to civilians. So the two basic Rules are conceptually interlocked: the limits on means and methods are primarily there to limit harm to civilians.

The LOAC construes excessive losses to civilians as one kind of indiscriminate attack. They are defined as follows: "Among others, the following types of attacks are to be considered as indiscriminate: . . . an attack which may be expected to cause incidental loss of civilian life, injury to civilians, damage to civilian objects, or a combination thereof, which would be excessive in relation to the concrete and direct military advantage anticipated."[12] This is the principle that non-lawyers often refer to as the principle of proportionality. A disproportionate attack is one that is expected to cause excessive civilian losses.[13] This is the principle I have simplified into my Rule 1, and this principle requires judging civilian losses relative to military advantage.

The First Additional Protocol spells out the general Rule limiting means and methods in strong terms, specifying the means for avoiding excessive civilian losses as "constant care."[14] Article 57, "Precautions in Attack," requires that "in the conduct of military operations, constant care shall be taken to spare the civilian population, civilians and civilian objects."[15] Constant care is thus specified partly in one Rule focused on choosing the military objectives relative to which civilian losses will be assessed and partly in another Rule more generally about means and methods.

On the one hand, an unqualified requirement of minimization is imposed for choices among alternative military objectives that are anticipated to yield comparable military advantage: "When a choice is possible between several military objectives for obtaining a similar military advantage, the objective to be selected shall be that the attack on which may be expected to cause *the least* danger to civilian lives and to civilian objects."[16] "The least" is obviously the minimum possible in the circumstances. One may never launch an attack on a military objective that will cause more civilian losses if instead one can attain the same military advantage by attacking a different military objective that will involve fewer civilian losses.[17] And as we just saw above in the principle of proportionality, even the least possible civilian losses are still unacceptable if the expected civilian losses would be excessive relative to the anticipated concrete and direct military advantage. The attack must cause the least possible civilian losses that could be caused in gaining that objective and the least possible civilians losses must also not be excessive compared to the military advantage to be gained from attacking that objective. Otherwise, the forces are required to refrain from making any of the attacks being compared. They must find a different military objective that will yield the same military advantage with fewer civilian losses.

On the other hand, the complementary requirement of minimization for choices among alternative means and methods for attacking objectives is qualified by a feasibility condition: "Those who plan or decide upon an attack shall . . . take all feasible precautions in the choice of means and methods of attack with a view to avoiding, and in any event to *minimizing,* incidental loss of civilian life, injury to civilians and damage to civilian objects."[18] Minimization of civilian losses is presented as already second-best to the complete avoidance of them, but the latter is taken to be not feasible.

Thus, all of my Rule 3 except the final clause is impossible to deny if one accepts the laws that explicitly contain Rule 1, because Rule 3 logically follows from them and because the possible emergence, explained below, of Rule 2 alongside Rule 1 cries out for a direct statement of the priority between 1 and 2. If one were to sacrifice additional civilians in order to protect additional troops, the civilian losses would not be the minimum feasible—one would have chosen to protect fewer civilians than one could have in order to protect more of one's forces.[19] That forces must run risks in order to protect civilians has been recognized by Christopher Greenwood: "Although the security of his own forces remains an important part of this calculation, the need to reduce the risk to the civilian population means that a commander may be required to accept a higher degree of risk to his own forces."[20] Less than maximum force protection for the sake of civilian protection is a requirement of the laws of war.

Both, Not Either/Or

Given that situations arise in which the minimization of civilian losses may require members of the military force to take additional risks and therefore possibly suffer additional losses of life, it is not possible to abide by the clear legal requirement that civilian losses must be minimized while indulging any supposed permission that force losses may be minimized.[21] As Rule 3 indicates, the risk of losses to one's own forces ought sometimes to be increased in order to attempt to cause fewer civilian losses. The evil and horror of war lie primarily in the vast extinguishing and blighting of human lives that constitute war as we know it: the deaths and the wounds.[22] The moral limits expressed in the LOAC are by far most concerned with protecting civilians—excluding them from the killing and the crippling. The LOAC say nothing about force protection.[23] I take this to mean only that they simply have a different purpose, civilian protection, and that, wisely or unwisely, they leave force protection to the various militaries, especially their commanders. Consequently, the LOAC contain no explicit expression of my Rule 2, which must accordingly be defended on nontextual grounds. But the grounds seem abundantly obvious.

Force losses can be excessive relative to a civilian life saved, just as civilian losses can be excessive relative to a life of a member of the forces saved. Why? Because all are humans, all are valuable. In this respect, it makes no difference whether they are fighting or not and if they are, which side they are fighting on or whether their side is relatively justified in fighting.[24] Not only is there nothing disreputable about a commander's attempting to protect the lives of the members of the military force, it is a commander's duty to avoid all unjustified risks. This is why Rule 3 needs the language "except at the extreme." The point has been put graphically, if provocatively, by Charles Dunlap, who has said that "citizen soldiers" are "non-combatants in uniforms."[25]

I say provocatively because soldiers after all have not only uniforms but also arms and possibly armor, supportive comrades, training in warrior skills, and often enormous supporting firepower. In fact, one reason why we may, contrary to Dunlap, reasonably expect soldiers to run greater risks than civilians is precisely that soldiers are better prepared in multiple ways to survive the risks, while civilians are typically vulnerable and largely defenseless. Nevertheless, soldiers are equal human beings with the civilians and are often in the lower ranks of some services, if not conscripted literally, then "conscripted by poverty"—that is, entering military service because no better paid work is available. If, as the LOAC requires and as I have proposed in Rule 3, soldiers must bear higher risks than civilians, the reason, whatever it may be, is certainly not that their lives are worth less. And because their lives are not worth less, there ought to be a limit on how many members of the forces may be risked either to achieve a military advantage or even to save a civilian. One could suggest that the obvious practice of force protection is a very early foundation for an emerging Rule 2 in customary law. If so, the importance of the priority expressed in Rule 3 is again highlighted.[26]

We can see, then, that each of two conceivable minimizations must be qualified, but in ways that are different from each other in important ways. On the one hand, it would be irrational to minimize civilian losses literally or absolutely, for two reasons. One is the very general, simple reason that either the minimization or the maximization of any one factor becomes crazy beyond some point: live as long as possible no matter in how much pain? Minimize pain no matter the effect on length of life? Pain is very important, but it is not everything; length of life is very important, but it too is not everything. Too many considerations matter for anyone to be treated as the only thing that matters, which is what strict minimizing or maximizing in effect does. It would be equally nonsensical to literally minimize civilian losses, whatever the costs.

But the important reason is that the costs may be significant indeed: the lives of soldiers. Even though soldiers are much better prepared, trained, equipped, and supported for running risks than civilians are, if greater risks are routinely run, losses will occur. This is required by international law and seems reasonable, but only, I am suggesting, if there is some limit on the loss of life among the forces that can be demanded. At the extreme, which I do not know how to specify with any precision, the price in the lives of the forces could become too great for the life of a civilian—hence, the final clause of Rule 3: "except at the extreme." This means that civilian losses cannot be strictly and literally minimized, with no limit on the price paid in lives by the opposing forces for an additional life saved. I do not believe that anyone actually believes otherwise, even though the letter of the law seems to say otherwise. But the burden of proof always falls on any advocate of the conclusion that enough has been done to protect civilians.

On the other hand, the loss of military lives also could not sensibly be literally or absolutely minimized for, first of all, the same basic reason as in the case of the civilians: that it is almost never reasonable to minimize one consideration no matter the cost in other important values. And more important, international law requires that "constant care" be taken to protect civilians, with the clear implication (embodied in Rule 3) that this can be at some significant—but not, I suggest, unlimited—cost to the military forces.

This is not an idea that is in the least alien to serving force members. Many people, certainly including almost every military person I have ever talked to about the question, think not only that force protection ought not to be maximized but that a fundamental part of a fighter's duty is to take risks to save civilians. What are the armed forces for? An officer at Annapolis told me: "If I had believed force protection ought to be maximized, it would have kept me safer just to become a banker in the first place. I did not join the U.S. Navy to protect myself."[27] So I believe that few, other than perhaps some sycophantic politicians cultivating their constituents, actually believe in maximum force protection, which would presumably mean never sending forces into battle. It is generally thought to be the duty of military personnel to run greater risks when it is reasonable to expect that running the additional risk will substantially reduce the risks borne by civilians. Soldiers, like many personnel in police, firefighting, and rescue work, view the taking of reasonable and justified risks to be at the heart of their jobs. The additional risk cannot, however, be required to pass through some reasonable ceiling, although heroes take it above and beyond that ceiling voluntarily.

Military forces must pursue their missions both with constant care for civilians and with attention to the safety of their own forces. One does not choose between the protection of the civilians and the safety of the troops—one holds on to both, but with a finger on the scales on the side of civilians. The real challenge always is to strike the proper balance.[28] The "strategy" of suicidal infantry charges in World War I and the "strategy" of murderous bombing of civilians in World War II were both morally despicable in their disregard for the value of human life.[29]

The Entire Iraqi Electric Grid as a Military Objective

We are ready now to look in somewhat greater depth at the profoundly troubling case of the U.S. bombing of infrastructure in the Gulf War of 1991, followed by the most extreme international sanctions regime ever imposed after a modern war.[30] The U.S. military's own official account of the war for Congress makes it abundantly clear that the military objective of U.S.

bombing was the Iraqi electric grid *as a whole*. The final report to Congress says: "Disrupting the electricity supply to key Iraqi facilities degraded a wide variety of crucial capabilities, from the radar sites that warned of Coalition air strikes, to the refrigeration used to preserve biological weapons (BW), to nuclear weapons production facilities. To do this effectively required the disruption of virtually the entire Iraqi electric grid, to prevent the rerouting of power around damaged nodes."[31] The bombing seems to have been planned in this way for the sake of some unspecified combination of military advantage and force protection. Disabling the whole system is almost certainly the most efficient way to proceed from the perspective of an attacker concerned with military advantage, force protection, or both. But it is clear from our look at international law that there are limits both to how one pursues military advantage and to what extent one pursues force protection, especially when the price is paid by civilians.[32]

The obvious difficulty is that if one turns off the refrigeration for biological weapons and radar sites by disabling the entire grid, one also turns off the electricity for perishable medicines, perishable foods, water pumping, water purification, sewage treatment, operating rooms in hospitals, and all the rest of civilian life, as noted in the Ahtisaari Report quoted above. The civilian effects are as global as the military effects, and this is readily predictable. And if one disables the entire electric grid with kinetic weapons, the damage cannot be repaired for a long time, even assuming the availability of spare parts; that is, ignoring the likelihood of sanctions.[33] Complete disruption of the electric grid served efficiency, but it did not honor a reasonable interpretation of the laws of war.[34]

The plan for the bombing attacks in 1991 appears to me to have been a clear violation of the First Additional Protocol, Article 57(3): "When a choice is possible between several military objectives for obtaining a similar military advantage, the objective to be selected shall be that the attack on which may be expected to cause the least danger to civilian lives and to civilian objects." Choosing the entire electricity grid as the military objective instead of choosing individual radar sites, biological weapons laboratories, and so forth as the military objectives seems an undeniable case of failing to choose "the objective . . . the attack on which may be expected to cause the least danger to civilian lives and to civilian objects"—an unconditional requirement.[35] The United States chose "the disruption of virtually the entire Iraqi electric grid" with kinetic weapons, presumably because some of the indirect effects of that choice were anticipated to be militarily advantageous and because this kind of attack was safer for pilots (and planes). But other indirect effects of that same choice were devastating for Iraqi civilians. And those devastating effects could reasonably have been expected—some were predicted.[36]

The United States had earlier chosen not to ratify the First Additional Protocol, but the question then simply becomes whether anyone could plausibly argue that Article 57(3) is not expressive of customary law. The most authoritative treatment of customary law for the conduct of war presents as Rule 21 a requirement that is precisely echoed by Article 57(3).[37] Article 57(3) comes close, indeed, to embodying elemental commonsense morality: if one can do something in a more destructive way or in a less destructive way, do it the less destructive way.

Which Effects Should Count?

If the content of Article 57(3) is accepted as binding as a matter of custom and common sense for those not bound by the treaty, one option for defending the decision to immobilize the entire Iraqi grid in 1991 seems to be to deny that the non-immediate effects on civilians count. One of the most important issues affecting judgments about appropriate targeting is how much of the total effect of bombing on civilians ought to be included in the proportionality calculation that determines whether bombing is discriminate. As we saw earlier, the general requirement of proportionality is understood as follows: "With respect to attacks, the following precautions shall be taken: (a) those who plan or decide upon an attack shall: . . . (iii) refrain from deciding to launch any attack which may be expected to cause incidental loss of civilian life, injury to civilians, damage to civilian objects, or a combination thereof, which would be excessive in relation to the concrete and direct military advantage anticipated."[38] One crucial question in the interpretation of proportionality is which expected civilian losses, injuries, and damage ought to be compared to the anticipated military advantage. Common sense would suggest that all significant foreseeable effects ought to be included in the calculation of both military advantage and civilian losses and that military advantage and civilian losses ought both to be calculated within the same parameters.

In the wake of the bombing of Iraq in 1991, Commander J. W. Crawford of the U.S. Naval War College issued a pioneering clarion call for a sufficiently broad and long view of effects on civilians: "Targeting considerations must extend beyond direct effects. Collateral damage, by legal definition, must include a requirement to examine the reverberative effects of military action. Every target set is different, but some targets (like electricity), due to the potential for long-term effects, demand that collateral damage be considered with a significantly broader view."³⁹ Crawford observed that "the devastation suffered by Iraq [in 1991] as a result of systematic targeting of its national electric power system is the best evidence, in the modern era, of the enormous destructive impact that aerial attacks can have on noncombatants."⁴⁰

Within two years of Crawford's 1997 article, the need to be concerned with the reverberation of effects was being emphasized by influential military lawyer Michael N. Schmitt (USAF General, ret.), who first described the general direction of change: "To the extent that the universe of strikeable targets multiplies, so too does the potential for collateral damage and incidental injuries. This actuality . . . will also require greater attention to the attacker's obligation to choose that method or means of warfare least likely to cause collateral damage or incidental injury, while still achieving military objectives."⁴¹ The general point was then developed specifically for what Schmitt called "subsequent-tier, or reverberating, effects" on and "derivative consequences" for civilians: "Understanding of how to apply the rule of proportionality may shift subtly, but meaningfully. If first-tier collateral damage and incidental injury (i.e., damage and injury directly caused by the kinetic force of the attack) become rarer, it is probable that humanitarian attention will increasingly dwell on subsequent-tier, or reverberating effects. . . . Given the increase of dual-use technologies and facilities . . . the risk of subsequent-tier damage and injury may actually increase because of the greater interconnectivity of valid target sets with civilian activities."⁴² Schmitt has continued to develop this position with regard to what he more recently dubbed "knock-on effects": "Further complicating matters is the issue of knock-on effects, i.e. those effects not directly and immediately caused by the attack, but nevertheless the product thereof—it is the problem of the effects caused by the effects of an attack."⁴³ This is a firm acknowledgement that "the effects caused by the effects of an attack" must be taken into account on the side of constant care for civilians.

A subsequent Joint Forces targeting handbook accepts the reading of the norm of proportionality that includes indirect—in fact, highly indirect—effects.⁴⁴ As one would expect in a handbook on targeting, attention is focused on the calculation of military advantage, not civilian effects. However, it would be utterly arbitrary to contend that reverberative/subsequent-tier/knock-on effects are irrelevant on the side of the proportionality balance concerning civilian losses when precisely such effects are what are striven for on the side of the balance concerning military advantage. And in any case the handbook goes on explicitly to list "collateral effects" as one kind of effects that are said to fall within both the primary categories of direct and indirect.[45] The *Operational Law Handbook* for 2010 quotes the article by Schmitt on the nature of knock-on effects and observes: "These knock-on effects . . . *must be considered in the proportionality analysis.*"[46] I take the recognition of indirect civilian effects by both the 2007 *Joint Fires and Targeting Handbook* and the 2010 *Operational Law Handbook* to be evidence that an understanding of the norm of proportionality, interpreted to require including what Crawford in 1997 called "the reverberative effects" of attacks, is gaining a welcome foothold in different types of official U.S. military documents. I would suggest along with Crawford, Schmitt, and others that this is an understanding of proportionality that ought to change permanently.[47] To ignore reverberative effects is to deny a big chunk of reality. Taking constant care for civilians is one of the basic purposes of any legitimate warfare, and civilians ought to be protected against major indirect and longer-term effects—for example, the health and nutritional effects documented in Iraq in the 1990s—as well as direct and immediate ones.

"Constant Care" and Force Protection, Once Again

Another possible ground on which to try to maintain that what appears to be the obvious straightforward reading of Article 57(3) cannot be the correct interpretation would be the claim that it is militarily necessary to take actions such as the disabling of the radar sites only by disabling the entire electricity system—that the military objective had to be the grid as a whole—because the risk to the forces in such situations would be excessive if individual military installations were made the military objectives. Each law of war and its general interpretation must grant minimum weight to both military necessity and moral limit, which are respectively represented here more concretely in part by force protection and constant care for civilians. Twenty years ago it was certainly much more efficient in the use of the attacker's resources to proceed wholesale. It would have taken more sorties, more weapons, and more pilots to attack radar sites and anti-aircraft weapons one by one than to disable them all at once by depriving them all of central electricity.[48] It would also have been somewhat more risky to establish full air superiority more gradually over a longer period of time during which one's planes and crews would have continued to be at risk because of the not-yet-disabled sites, which could be firing surface-to-air missiles at the planes attacking the individual radar installations. So the risk to forces—primarily to pilots and planes—would have been somewhat higher if the whole electrical grid had not been chosen as the military objective. But would the risk have been prohibitively, or even very significantly, higher?

Clearly not. One could claim that it was necessary to do the task in one fell swoop rather than one military installation at a time only if one gave "necessity" a reading that would permit the absolute maximization of force protection—that is, the utter minimization of risk to the lives of one's military forces. Any concrete situation depends on the relative strength of the technological capacities of the opposing sides. The United States had vast technological superiority over the Iraqi forces in 1991, and consequently the risks to attacking aircraft, while utterly real, were manageable. Since it was reasonable to expect (and had been predicted by the Defense Intelligence Agency)[49] that the kinetic disabling of the entire electric grid would wreak civilian havoc—quite possibly the years of basic paralysis that it did in fact usher in under the sanctions that the United States adamantly insisted on— the likely additional losses for the attacker from not proceeding indirectly by way of individual installations as its military objective would have needed to have been remarkably substantial before one could plausibly argue that it was genuinely necessary to proceed in a manner that was so destructive to civilians over the longer term. And on the other side of the balance, any plausible conception of constant care for civilians would clearly militate against the long-term deprivation of potable water, unspoiled food and medicine, sanitation, emergency medical care, and the other services of modern life that all obviously depend on electricity in an industrialized society. I cannot imagine how the interpretation that prohibits making the entire electrical grid the military objective could fail to be customary law. In addition, it is a requirement of the First Additional Protocol and of commonsense morality.

If we leave aside the present-day possibilities for cyberwar as a possible third option that was not available in 1991, the choice was between targeting the grid as a whole, as was done, or bombing individual military installations one by one. More discrete targeting would have involved some additional risks to pilots while the air defenses were eliminated piecemeal, but as far as I can see, those additional risks to the forces would not have begun to approximate the additional risks to civilians from destruction of the entire grid. If this is correct, the U.S. bombing of Iraq in 1991 is a clear case in which military advantage, force protection, or both received too much priority and constant care for civilians received far too little. It is impossible to provide an algorithm specifying how to get the balances right, but the American bombing of Iraq in 1991 provides a clear object lesson in how to get them wrong.

Chapter 8

Civilian Deaths and American Power
Three Lessons from Iraq and Afghanistan
RICHARD W. MILLER

In Iraq and Afghanistan, the United States has engaged in military violence by means including extremely extensive bombing, in a broad sense that includes firing missiles and launching powerful projectiles from tanks and artillery, as well as dropping bombs. I will argue that the practice of American bombing in these countries and of the monitoring of its toll provides evidence for three related claims of deep interest to those who seek to reduce civilians' suffering in war. The first claim is explanatory: to the extent that changes in bombing practices from earlier wars have affected the vulnerability of civilians, the changes have reflected strategic rationales for pursuing U.S. global power by violent means. The other two claims are moral. This strategic driving force has produced large excesses in unjustified civilian deaths due to U.S. military violence, including bombing broadly construed, in Iraq and Afghanistan. Shaped by strategic interests, the monitoring of the toll of military violence among civilians has been biased toward the goal of counterinsurgency in ways that distort moral assessment and add to the suffering of uncontroversially innocent civilians.

An Explanatory Framework

The first claim, about the explanation of changed bombing practices, would not withstand scrutiny if it is understood as a cynical ascription of pervasive callousness. To avoid such misunderstandings, I will briefly sketch the general thesis that I will apply in explaining the changes, distinguishing it from crude simplifications.

The thesis concerns the pursuit of American power. In the making of U.S. foreign policy, when violent foreign initiatives are assessed, guesses about what will best promote U.S. global power are decisive, and avoidance of consequent deaths of people in developing countries does not have substantial countervailing influence independent of that goal. The rich data concerning deliberations at the top, including the Johnson tapes, the *Pentagon Papers*, the Nixon tapes, the Bush-Scowcroft memoirs, and Bob Woodward's books lend much support to this claim. Of course, those engaged in the planning of U.S. foreign policy believe that maintaining U.S. preeminence benefits humanity. But there is no evidence in that rich record that the moral question of whether what serves this goal is always worth its human costs has significant negative importance in foreign policy deliberations. Indeed, this is a doubt that dare not speak its name. These presuppositions of acceptable discourse in foreign policy making, together with the prerequisites for electoral success, the patriotic emphases of American culture and education, and the content and framing of news reports, all work to sustain a project of domineering influence, sometimes violent, on the part of the U.S. government, in which foreign devastation does not have independent dissuasive force remotely corresponding to its strength as a negative moral reason.[1]

In arguing that changes in U.S. bombing practices have been shaped by this project of those who make U.S. foreign policy, I will not defend a simplistic cynicism that denies that American moral repugnance at foreign suffering due to American violence reduces violence and saves lives. Indeed, that repugnance will play an important role in the attribution of the changes to strategies of power.

First of all, U.S. combatants typically want American attacks to be morally conscientious, a commitment that makes a difference. The My Lai massacre was ended by a career officer, Hugh Thompson, piloting a helicopter gunship, who prepared his helicopter patrol to fire on Lieutenant Calley's detachment if they did not stop murdering civilians (on orders that some had already disobeyed despite Calley's threat of court-martial). When reconnaissance pilots in Vietnam became aware that reports of no young men in the fields could lead to the incineration of villages, they sometimes departed from protocol in their reports to save civilian lives. In the narratives of the 2003 invasion of Iraq in Evan Wright's book *Generation Kill*, many of the Marines take their power of life and death over Iraqi civilians to create a special responsibility to seriously consider less lethal alternatives and assume some extra risks to reduce civilian endangerment. They have contempt for Marines who neglect this responsibility, including the lieutenant colonel who commands them, and sometimes ignore orders that would endanger civilians without sufficient cause.[2] These are not just admirable features of fighters whose lives are in peril, but factors in the calculus of American power. What offends the ethic of soldiers demoralizes them. Frequent communication of their moral disgust to the home front would contribute to opposition to their war. I will treat the attitudes of conscientious combatants as life-saving constants, but not important independent sources of the change in bombing practices.

Moral repugnance in the home country at the civilian toll of foreign wars is also an important fact for strategies of transnational power. Significant and growing moral disenchantment makes it hard for a great power to inspire willingness to make sacrifices to advance its foreign initiatives. It makes young people less likely to devote their talents and energies to careers in the planning and pursuit of national power. The great change of course in America's war in Vietnam was Lyndon Johnson's acceptance of the new consensus of his Senior Advisory Group, favoring rapid de-escalation and a negotiated settlement on terms that were likely to lead to the collapse of the Saigon regime. These key advisors from the foreign policy establishment thought victory would be militarily attainable but only at excessive cost in domestic morale. In the crucial deliberations, Cyrus Vance said that "the war was bitterly dividing the country," in announcing his conversion from insistent support for the war effort. George Ball "emphasized, as I had done many times, that the war was demoralizing the country and that we had to get out."[3] In mournful retrospective, he was to describe this demoralization as "the poisoning of the minds of some Americans toward their own government."[4]

Outrage among non-Americans at civilian casualties is also a vital input to the calculus of American power. Their moral repugnance at U.S. military initiatives makes it hard to secure and maintain allies. In the foreign countries in which the United States makes war, moral outrage at death, maiming, and dispossession of civilians caused by American violence can increase and consolidate support for America's enemy. Granted, this is only one effect. Such violence can also make it impossible for the enemy to impose the peaceful order that is most people's most fervent wish, make many people fearful of the enemy's presence, even make them flee to territory secured by the United States and its local ally. Reluctance to inflict collateral damage can create too much safety for the enemy. Still, sound strategic reasoning must take seriously those costs of moral outrage of civilian survivors.

Civilian Safety and the Goal of Power

When this explanatory framework is applied to American warfare in Iraq and Afghanistan, it yields the best explanation of changes in U.S. bombing practices affecting civilians. I do not mainly have in mind changes in announced norms, since these have not been large. Truman's initial description of the Hiroshima bombing as an attack on "a Japanese Army base" targeted because "we wished . . . to avoid so far as possible the killing of civilians" is not just startling testimony to the power of unmitigated gall, it testifies to Americans' past desire to think of their military as showing due care for enemy civilians, even in the midst of a global war of epic proportions, claiming huge sacrifices.[5] While the U.S. military's adaptations of Genevan rules of necessity and proportionality have included new prescriptions, they have been extremely elastic, too elastic to constrain attacks that have a strategic rationale. Still, the scale of destruction and monitoring has substantially changed. The destruction of cities and towns in World War II and the Korean War and the carpet bombing of the Indochina War have not been characteristic of American warfare since.[6] The procedures for monitoring proposed airstrikes in Iraq and Afghanistan for collateral damage are certainly new. So there is something to explain.

The strategic pursuit of enduring goals of American power, learning from mistakes and adjusting to new circumstances, is one plausible explanation. Moral outrage at the civilian toll of the Indochina War provided a weighty reason for de-escalation and ultimate withdrawal. Rather than breaking the will of America's enemies in Vietnam, the bombing had, on balance, strengthened their opposition and enraged Vietnamese against the United States. The terrain and technology of America's more recent wars have permitted more discriminating attacks, facilitating reduction of such costs. At the same time, the swiftness of air strikes and their devastating force have required control by set routines, enabled by new information technology. These uses of new technology in outrage-reducing ways have been accompanied by new military doctrines, also tending to reduce civilian jeopardy. The doctrine of pursuit of definite, limited, clearly attainable goals that contained the carnage of the First Gulf War was explicitly a response to the failures of Vietnam. The heightened reluctance to cause civilian casualties in Afghanistan is, clearly, a response to the failures of Vietnam and the travails of the U.S. occupation of Iraq. The proverb-crafting of the 2006 *Counterinsurgency Field Manual* continually reinforces the strategic rationale, as in the renowned saying, "An operation that kills five insurgents is counterproductive if collateral damage leads to the recruitment of fifty more insurgents."[7]

Still, this evidence of the role of strategic concerns does not sustain the cutting edge of the strategic explanation, the claim that the moral concern to reduce civilian casualties on the part of those shaping U.S. initiatives had no significant independent force in reducing the deadly impact of bombing on civilians. The crucial test is whether the United States has departed from the practice of making the best bet for advancing U.S. interests when this has conflicted with norms of due care for civilians. The strategic explanation passes this test. Its confirmation also provides evidence for the moral claim that U.S. destruction by military means, including bombing in the broad sense, has inflicted a large toll of unjustified civilian death.

Toward the end of the first Gulf War, grave damage to civilian lives through destruction of the Iraqi infrastructure was regarded as a possible way to deprive Saddam Hussein of support and encourage his overthrow. The decision not to march on Baghdad—due to strategic fears of unmanageable disorder—made this route to regime change especially appealing to the United States. The danger of this tactic to civilians was foreseeable and foreseen. On January 22, 1991, six days after war was launched against Iraq, U.S. military intelligence had reported, "Increased incidence of diseases will be attributable to degradation of normal preventive medicine, waste disposal, water purification/distribution, electricity, and decreased ability to control disease outbreaks. Any urban area in Iraq that has received infrastructural damage will have similar problems."[8] Since the danger would be the relatively slow though ultimately widespread devastation of a public health crisis, the reputational costs could be expected to be smaller than those of obliteration bombing. Yet the intentional use of bombing in order to engage civilians in such a process, gravely endangering their lives, is just as objectionable. No doubt, the United States would have preferred to stimulate regime change by civilian desperation unaccompanied by death. Similarly, Truman would have preferred to induce panic and despair through the mere appearance of mass death in Hiroshima followed by ultimate resurrection. This does nothing to make his choice morally permissible.

Responding to the mere chance of a strategic gain through the violent engagement of civilians in grave harms, the United States made its choice. Precision-guided weapons destroyed the power stations on which refrigeration, water supply, and sewage treatment depend and bombs destroyed the main Baghdad sewage treatment plant.[9] Planners of these attacks explained to a *Washington Post* reporter that they were a deliberate effort to strike "against 'all those things that allow a nation to sustain itself.' . . . to let people know, 'Get rid of this guy and we'll be more than happy to assist in rebuilding.'"[10]

The maintenance of that pressure through economic sanctions was a bipartisan project, flawlessly relayed to the Clinton administration. In 1999, a well-publicized UNICEF survey of childhood mortality in Iraq noted that there would have been 500,000 fewer deaths of children under five if pre-sanctions trends had continued.[11] A UN committee, deliberating in secret, in which each representative had veto power, controlled Iraqi imports under the sanctions regime. In her 2002 analysis of documents leaked by appalled UN staff, Joy Gordon found that the United States, "only occasionally seconded by Britain," had, since 1991, "blocked most purchases of materials necessary for Iraq to generate electricity, as well as equipment for radio, telephone and other communications. . . . For example, Iraq was allowed to purchase a sewage-treatment plant but was blocked from buying the generator necessary to run it. . . . In September 2001 nearly one third of water and sanitation and one quarter of electricity and educational supply contracts were on hold. In early 2001, the United States had placed holds on $280 million in medical supplies, including vaccines to treat infant hepatitis, tetanus, and diphtheria, as well as incubators and cardiac equipment."[12] Asked in a 1996 television interview whether gains from the sanctions were worth the price of the deaths of 500,000 children, Secretary of State Albright answered, "I think that is a very hard choice, but the price, we think, the price is worth it."[13] In choosing bombing targets and managing the consequences, the United States had shown the insignificant independent force of moral compunction about civilian lives in overriding the strategic pursuit of transnational power.

The invasion that overthrew Saddam Hussein poses another obstacle to the attribution of significant force to moral norms in constraining strategic rationales that have shaped bombing practices. Despite the vetting of some attacks for collateral damage, the civilian toll of the invasion was quite impressive and did not correspond to dire dangers to American soldiers. Reports of civilian deaths during the invasion in major news sources added up to about 6,000.[14] (This is about the same as estimates of how many Iraqi soldiers were killed.[15]) One hundred nine U.S. troops died in combat in the invasion.[16]

The narratives of *Generation Kill* suggest one partial explanation. Although the morally responsible Marines show life-saving restraint, there are no significant disincentives keeping exuberant killers such as Captain America, a platoon commander who is "highly rated" with "stellar fitness reports," and Encino Man, a company commander with "an admirable service record," from excessive responses of rapid, devastating violence to the many unclear signs of possible danger flickering out of the fog of war.[17] Even more seriously, there are significant career incentives for higher officers to issue aggressive rules of engagement in response to these ambiguous signs, in pursuit of rapid victory. Appalled by their shooting of two shepherd boys when their lieutenant colonel issued needlessly aggressive rules, one Marine walks away from his lame self-justification muttering, "'Protection of forces' my f—— a——. Why did we make that pell-mell f—— rush? So a colonel could score a few brownie points with a general.'"[18]

An additional causal factor in the disproportionate toll is described by the artillery officers who complain of their need to rely on imprecise cluster munitions rather than more discriminating projectiles in the 2003 Human Rights report, *Off Target*.[19] Although clearly contemplating invasion for a long time, the United States government had not bothered to create a discriminating arsenal before quickly launching its invasion in time to avoid the heat of the Iraqi summer.

A systematic account of the culture and psychology of restraint and excess among Americans who killed in Iraq would add to the inventory of causal factors. Some combatants were admirably conscientious. Some seem to have been psychopaths, as one would expect in such a large armed force. But videos of missile launches instantly destroying vehicles and whole buildings in close air support during the occupation, such as the Wikileaked video posted in the *Guardian* on October 22, 2010, seem to indicate something in between, inadequate compunction in highly lethal responses to ambiguous cues. The terse pilots' reports of outcomes (in the case of the *Guardian* video, pretty clearly false) provided no basis for monitoring and mitigating the killing of civilians.[20] It would be revealing to move beyond anecdotes to understand how particular subcultures of killing produced inadequate restraint.

These inquiries into the situations of those who did the killing would make it clear that pervasive callousness toward Iraqi civilians was not the cause of the disproportionate civilian toll. By the same token, they ought to shift concern to other echelons. The structuring of incentives, the preparation of discriminating means of killing, the regulation of rules of engagement, and the orientation of military cultures are the responsibility of people higher up the chain of command, above all, the commander in chief. That is the plain dictate of the morality of due care and the strict meaning of the Geneva Conventions, which are directed at parties to a conflict, without singling out combatants in the field. At higher levels, strategic zeal might have been and was not independently inhibited by constraints of due care.

The consequences of this zeal extended far beyond the initial invasion. The 2006 *Lancet* epidemiological study entails about 76,000 deaths among Iraqi males younger than fifteen or older than forty-four and Iraqi women attributable to Coalition forces in the forty months after the invasion began.

The study notes an increase in the total of violent deaths due to Coalition forces in each of the three years in this period.[21] After the overthrow of Saddam, the carefully planned retaking of Fallujah from a local insurgent government in the fall of 2004 was the most lethal and best-reported operation, a further test of the balance of strategy and compunction.

Shortly before the assault, a U.S. reporter in Baghdad noted that the U.S. command had decided that Fallujah was "a city where Iraq's nationwide insurgency must be dealt a deathblow to convince other insurgent-held cities to capitulate before the January elections. . . . [According to an Iraq analyst at the Institute for Strategic Studies in London,] 'The logic is: You flatten Fallujah, hold up the head of Fallujah and say, 'Do our bidding or you're next.'"[22] This logic was observed.

The retaking of Fallujah began with weeks of bombing that killed women, children, and noncombatant men as well as insurgents.[23] Well before the full-scale assault, health supplies ran out, and the U.S.-installed government refused to provide more.[24] Before the main assault, U.S. airstrikes destroyed a clinic in central Fallujah and a nearby storehouse for medical supplies.[25] When the assault began on November 8, the first goals were to seize Fallujah General Hospital and nearby bridges connecting the hospital with Fallujah proper.[26] U.S. forces prevented doctors and ambulances from entering the city.[27] Shortly afterward, another clinic was bombed and destroyed.[28] These were strategically apt ways of guaranteeing that severely wounded insurgents would die a terrible death. They sealed the fate of many civilians, too. The U.S. command estimated the remaining population at fewer than one-fifth of normal residents, a number amounting to 40,000.[29] Independently reported estimates were in the range of 30,000 to 50,000.[30]

For over a week, after the main assault began, death fell on those left in the city, mainly from the air, including 2,000-pound bombs and other munitions from skies "so crowded with US military aircraft that they are layered in stacks above the city. . . . 'We call it the wedding cake. It's layered all the way up[,]' [said the leader of the ground targeting effort]."[31] The *New York Times* reported that when a group of 300 refugees were detained by U.S. soldiers as they tried to leave the city, "the men were tested for any residue left by the handling of explosives. All tested negative but they were turned back."[32] For eleven days after the U.S. command declared almost complete control over the city, the Red Crescent and other relief agencies were kept out.[33] After the first day of delivering aid, the Red Crescent said that it "feared that more than 6,000 people could have died in the assault and thousands of families are in critical need of assistance."[34] In the next six weeks, a team from Fallujah General Hospital found more than 700 bodies in some of the vast rubble of the wrecked city, more than 550 of them women and children, the rest mostly elderly men. While residents had also buried their dead, with the approval of the U.S. command, after the retaking of the city, these deaths went uncounted.[35]

More compunction about civilian deaths might have slightly softened the convincing deathblow that was the apt strategic goal in Fallujah, a risk well within the constraints of due care. This risk was not taken.

In Afghanistan, the toll of collateral damage due to U.S. violence has been fairly small, and constrained by close monitoring. In recent years, civilian deaths caused annually by U.S. forces seem to have been in the hundreds, not the thousands, in a country of thirty million people.[36] That strategic considerations dictate constraints is an explicit and adamant refrain of the *Counterinsurgency Field Manual*, in presenting the reigning military doctrine, brought home in such pithy sayings as the one about creating fifty insurgents through collateral damage and also in more measured precepts ("inappropriate or indiscriminate use of airstrikes can erode popular support and fuel insurgent propaganda. For these reasons, commanders should consider airstrikes carefully during COIN [counterinsurgency] operations").[37] When the dictates of strategy and morality coincide, this is cause for celebration. But the superiority of the strategic explanation still casts a shadow on the military operation in Afghanistan.

Unleashing lethal disorder has costs to innocents as severe as bombing. In the first two years after the invasion of Iraq, the increase in civilian deaths from criminal activity was essentially the same as the civilian deaths due to Coalition actions, in Iraq Body Count's tabulation, claiming 36 percent as opposed to the Coalition's 37 percent among the 24,865 reported deaths that were tallied. In Afghanistan, the U.S.-NATO command has increasingly relied on the strengthening of local militias, reviving the deadly congeries of warlords. Whatever its costs to innocents, this shift can be part of an effective counterinsurgent strategy, using disorderly means to prevent the imposition of order by the rebels. Similarly, counterinsurgency in Iraq was advanced by the decapitation of the leadership of Moqtada al-Sadr's militia in Sadr City, converting it to a gang of thugs. The COIN manual notes, "Throughout history, many insurgencies have degenerated into criminality. . . . Such disintegration is desirable" (p. 20).

The civilian toll of such tactics pales in comparison with the impact of continued violent civil conflict on subsistence, nurturance, and reconstruction. In this way, above all, the pursuit of American power endangers civilians in Afghanistan. Whatever prolongs war in Afghanistan continues the incineration of the fabric of life that has made the country a global leader in suffering, continually worst or near-worst in such measures as life expectancy and under-five mortality.[38] Faced with that prospect of prolonged violent disorder as U.S. forces surged, three-quarters of respondents in a large nationwide Afghan opinion survey in 2010 supported negotiations with the Taliban; "an agreement to stop the fighting [that] ceded control over certain provinces to the Taliban" was supported by large majorities in the provinces most likely to be ceded, 63 percent in Helmand, 58 percent in Kandahar.[39]

Despite the gains to civilians of peace through compromise, it would impose a strategic loss on the United States, since substantial success of lightly armed insurgents battling U.S. forces reduces the credibility of U.S. military might. Testifying to the Senate Armed Services Committee the day after the Afghan surge was announced, Defense Secretary Gates noted, "What makes the border area between Afghanistan and Pakistan uniquely different . . . is that this part of the world represents . . . the historic place where native and foreign Muslims defeated one superpower and, in their view, caused its collapse at home. For them to be seen to defeat the sole remaining superpower in the same place would have severe consequences for the United States and the world."[40] The goal of global power that shapes American military practice endangers civilians by discouraging compromised endings that can reduce civilian suffering.[41]

On the question of perseverance, current counterinsurgency doctrine takes a strong stand. The COIN manual begins by positing the maintenance of support for continued war as the central challenge of American counterinsurgency. Unable to defeat the United States in the field, enemies "try to exhaust US national will, aiming to win by undermining and outlasting public support."[42] This makes the shaping of a narrative for U.S. consumption that discourages withdrawal a central task, though one that the manual reserves for officers of high rank. "At the strategic level, gaining and maintaining U.S. public support for a protracted deployment is critical. Only the most senior military officers are involved in this process at all. It is properly a political activity."[43]

Defining the Barbaric

My argument that changes in American bombing practices have been due to rationales of power has also supported a moral judgment that the pursuit of American power has produced large excesses in morally unjustified civilian suffering due to American military violence in Iraq and Afghanistan. So far, I have not examined a crucial aspect of the monitoring of this toll in those conflicts—the criteria used to categorize victims as civilians and the standards for acceptable collateral damage. My final claim is that the classification of victims as civilians and the judgments of collateral damage by the United States and supportive agencies are biased against insurgents in ways that endanger uncontroversially innocent civilians—above all, through their impact on the politics of perseverance in war.

The tallies of Afghan civilian casualties with serious international standing are issued twice a year by the United Nations Assistance Mission in Afghanistan (UNAMA). The reports include a glossary with a description of the category tallied as civilians. According to one such entry, the category includes "Any person who is not taking a direct part in hostilities. It includes all civilians as well as public servants who are not used for a military purpose in terms of fighting the conflict . . . as well as political figures or office holders. . . . It includes persons who may be civilian police personnel or members of the military who are not being used in counter insurgency operations and not taking a direct part in hostilities including when they are off-duty."[44]

Applying their criteria, UNAMA states that at least 2,179 civilian deaths resulted from violence of the Taliban and their allies in 2012, noting as especially objectionable 565 incidents of targeted killing of civilians. UNAMA's criteria include people who participate in the central government's endeavor of imposing the order that the insurgency seeks to overthrow. However, U.S.-NATO forces do not seek to exempt from attack such insurgent analogs as Taliban political commissars, shadow governors, or those mainly engaged in enforcing the writs and judgments of the Taliban in the territories they control. Their death or injury plays no role in tolls of civilian targets or civilian collateral damage by U.S.-NATO forces, as tallied by UNAMA, the United States, or most others concerned with civilian suffering in the Afghan war.

This practice of classification is not inevitable in civil conflicts, even if it is the invariable practice of the counterinsurgent side. For example, during the struggle against apartheid, the African National Congress denied that it practiced terrorism and described the morally protected category in its bomb-making manuals in these terms: "The action can be directed against government personnel, police and soldiers, spies, agents, stooges and informers, but not innocent bystanders of any description."[45] This category provides a different lens for viewing Taliban attacks and their toll. In the extensive detailed descriptions of incidents in UNAMA reports, the vast majority of deaths caused by the Taliban and their allies are due to suicide attacks, assassinations, and the use of improvised explosive devices (sometimes land mines planted in roads) directed at troops, their suppliers, police, officials, or locally prominent political supporters of the regime. A typical report on the state of the conflict by the Afghanistan NGO Safety Office, an information center in Kabul funded by Norway, Switzerland, and the EU, describes Taliban violence in these terms: "Attacks remained focused on the IMF [International Military Force] and Afghan Government, targeting personnel, resources and enablers . . . and expanded to include significant attacks on Private Development Companies for their role in supporting the IMF."[46] This is not a focus on those who do not take part in the imposition of the regime the insurgents oppose. It roughly corresponds to the U.S. focus in targeting the insurgency.

Despite the focus of their targeting, the Taliban cause death and injury to many who do not participate in the imposition of the political order they oppose, mostly harming as a side effect of attacks on those who do participate or a side effect of efforts directed at impeding their contribution, say, by mining roads. This collateral damage is much more common than collateral damage from U.S.-NATO attacks. But here, another standard for moral assessment of killing deserves more careful use. The usual basis for assessing the morality of collateral damage inflicted by the United States is not just its size but its military "necessity," largely determined by the extra risks that the attackers would have to take to pursue their goals by means that are less dangerous to those in the morally protected category. Given the resources available to the Taliban, the same standard would permit more consequent death. They do not have drones and missiles. If they carefully monitor improvised explosive devices to ensure that they are activated only by military convoys, they are very apt to be killed when U.S. troops are massively present.

The ANC definition and the Afghanistan NGO Safety Organization assessment would not, by any means, make the Taliban just. They seek to violently impose an unjust order. Sometimes, they violently attack people who are not vulnerable according to the ANC definition. (ANC militants sometimes did this as well.) But in the special dimension of moral assessment summoned up by concern for killing and maiming of civilians, the Taliban are much less bad when the difference between the ANC definition and the UNAMA definition is clearly marked, and highly asymmetric warfare is taken into account in judgments of collateral damage.

The neglect of these distinctions encourages the portrayal of the Taliban as not simply violent and unjust but utterly barbaric. After all, if they targeted people who played no role in imposing Kabul's rule on the same scale as they target participants, or if they routinely gravely endangered these uncontroversial civilians without excuses of military necessity, they *would* be utterly barbaric. Their barbarism would make them inappropriate candidates for participation in a political settlement conceding them substantial political authority, for example, authority in their base in the Pashtun countryside.

Unfairness to those engaged in unjust violence is no grave matter, as such. However, the counterinsurgent slant in standards endangers those who uncontroversially merit special moral concern, those who do not take part in any violent imposition. The slant makes the inclusion of the Taliban in a political settlement look hopeless or immoral, provides a reason to prolong the three decades of war that have incinerated the fabric of life in Afghanistan, and creates false appearances that counterinsurgent violence does not cause disproportionate harm by exaggerating the moral urgency of its goal.

A generalized version of this judgment of barbarism is the basis for the celebration of the new guidelines for counterinsurgency by Susan Sewall, director of the Carr Center for Human Rights Policy at Harvard, in her preface to the COIN manual. She repeatedly presents as an obvious truth the view that all the insurgencies that the U.S. fights are barbaric, making relentless counterinsurgency a moral need. Qualms about immoral tendencies of counterinsurgency are appealing, but such "logic . . . only rewards insurgents' intolerable behavior." The manual describes a "way to win against barbaric insurgents" without descending into barbarism. "To gain some perspective on the COIN manual's intentions, we need only consider insurgents' eagerness to kill civilians."[47]

Ascriptions of barbarism have a very long history in the justification of war. Roman conquests were said to subdue barbarians. Vitoria, the classic defender of humanitarian military intervention, justified the Spanish conquest as subduing barbarians. John Stuart Mill justified the subjugation of native states that he managed from East India House, together with France's brutal colonial wars in Algeria, on the grounds that "a civilized country cannot help having barbarous neighbors. . . . It either finds itself obliged to conquer them, or to assert so much authority over them, and so to break their spirit, that they gradually sink into a spirit of dependence on itself."[48] So long as those who scrutinize America's wars for undue endangerment of civilians leave the standards applied to insurgents unexamined, they will contribute to a familiar justification for violence that gives rise to many civilian deaths.

Pressing the case that strategic needs, not deepened conscientiousness, explain changes in bombing practices can create hard feelings based on misunderstanding, by seeming to contemptuously dismiss the moral concerns of American combatants. Criticizing the one-sidedness of standard means of assessing civilian tolls can provoke even harsher exasperation at apparent solicitude for violently unjust insurgents. Because of these limitations, the lessons that I have derived have little role to play in electoral politics, much less in deliberations in foreign policy planning. All the more reason to advance them in other parts of the complex division of labor in which people of good will seek to protect humanity from violent harm. Without these warnings about the causes, consequences and monitoring of American military violence, the current trend in reduced civilian casualties due to American bombing can create false appearances that gravely endanger innocent noncombatants.

Part III

Constructing New Norms

CHAPTER 9

Proportionality and Restraint on the Use of Force

The Role of Nongovernmental Organizations

MARGARITA H. PETROVA

The core of international humanitarian law (IHL) is an attempt to strike a balance between the necessity of waging war effectively and doing so in a humane manner that both limits wanton destruction and protects innocent civilians against the ravages of war. We see this consideration in the first international treaty on the conduct of war, the 1868 St. Petersburg Declaration, which tries to find a threshold at which "the necessities of war ought to yield to the requirements of humanity." More than a century later the principle of proportionality was codified in Additional Protocol I to the Geneva Conventions in Article 51(5)(b), which prohibits attacks "which may be expected to cause incidental loss of civilian life, injury to civilians, damage to civilian objects, or a combination thereof, which would be excessive in relation to the concrete and direct military advantage anticipated." This is considered a customary rule of international law, and violation of it is considered a war crime.[1]

Yet exactly what the proportionality principle means and how it is to be applied in practice evades clear answers.[2] Apart from the difficulty of balancing two incomparable values—civilian lives and damage versus military advantages—there is no agreement on what the two sides incorporate and how far the temporal and spatial limits of the military and humanitarian effects extend. The Final Report to the Prosecutor by the Committee Established to Review the NATO Bombing Campaign against the Federal

Republic of Yugoslavia summed up the sentiment prevalent among legal scholars: "It is much easier to formulate the principle of proportionality in general terms than it is to apply it to a particular set of circumstances."[3] On the one hand, given the latitude in interpretation, authors have argued that traditionally, military necessity has been privileged over humanitarian concerns[4] and that proportionality has often been used to justify excessive force.[5] On the other hand, it is seen as a tool for restricting legitimate military action whose liberal interpretation could result in prosecution of conscientious military commanders—an evaluation that motivates calls that it be abandoned altogether.[6]

Although the principle is indisputably vague and has allowed both critics and defenders of the same military action to rest their arguments on it, in this chapter I show how nongovernmental organizations (NGOs) have successfully mobilized it as part of their efforts to ban a specific category of weapons, cluster munitions, and how in turn the achieved prohibition might provide grounds for further limitations on the use of explosive force within populated areas. Whereas civilian immunity declares an absolute prohibition of attacks against civilians and an aspiration that civilians be protected at all times, proportionality guides how much civilian protection is achieved in practice when military action does cause unavoidable civilian harm. I argue that NGOs have used principles of proportionality (and discrimination) from IHL as a basis for their arguments but that these principles alone were not sufficient to achieve a prohibition. During the time when NGOs relied exclusively on legal grounds in seeking better civilian protection, their demands were limited to calls for more reliable cluster munitions and the prohibition of their use in populated areas. IHL considerations helped place the issue on the agenda of international talks at the Convention on Certain Conventional Weapons (CCW), but they didn't produce new regulations. A breakthrough came when NGOs and supportive states reframed the debate in humanitarian terms and raised political considerations about the cost of civilian casualties in a stand-alone negotiation process for a new treaty that Norway launched in 2007.

In the course of their campaign, NGOs have challenged the military for its failure to gather the information about the effects of weapons on military targets and civilians that is indispensable for assessing proportionality. Then they used a two-prong strategy that first and foremost highlighted the suffering of civilians and secondarily questioned the military utility of the weapons. NGOs have sought to expand the temporal span during which incidental harm to civilians is calculated while raising to the level of indispensability the bar of military utility the weapons have to surpass in order to be acceptable. The statement that the humanitarian costs of the weapons "far outweighed" their military utility reframed the proportionality principle and became a key argument of this campaign.

The chapter is organized as follows: first, I provide a theoretical framework that focuses on the norm development and issue framing that guides the empirical research, then I present existing legal views about the proportionality balance at the early stage of the NGO campaign. I follow that with an examination of the discursive processes that led to a redefinition of the balance between military and humanitarian interests, and I conclude with some reflections about the implications of the efforts of NGOs to restrain the use of force by looking at a recently launched initiative to limit the use of explosive weapons in populated areas.

Norm Development and Issue Framing

The point of departure for this chapter is the constructivist argument that norms[7] provide the broad contours of what is permissible, expected, and right. However, they are not constant and don't always provide unambiguous guidance for action. Although there are some "foundational" norms that are relatively stable over time (e.g., those enshrining the rights and freedoms of states and individuals), conflict among them and controversies about their application to specific cases spur the development of norms. The meaning of existing norms is defined, expanded, or narrowed and new norms emerge in a process of analogical reasoning and incremental building on old norms.[8] This is what I argue happened with the legal norm of proportionality. On the one hand, it, together with the distinction principle, was used as a building stone for a prohibition on cluster munitions. On the other, the process of employing the proportionality principle has clarified and arguably expanded its meaning to include in the calculation the long-term effects of weapons on civilians, thus placing more weight on humanitarian considerations compared to military ones.

Understandings of what the principle of proportionality entails and how it applies to cluster munitions changed to a large degree as a result of how NGOs framed debates about the humanitarian impact and military effectiveness of cluster munitions. NGOs have been the "signifying agents actively engaged in the production and maintenance of meaning"—a process referred to as "framing" that results in the production of "collective action frames" that help "organize experience and guide action."[9] Framing performs an interpretive function, but it does so by simplifying complex phenomena or focusing on some of their aspects with a view to mobilizing constituents, gaining the support of bystanders, or demobilizing opponents.[10]

Snow and Benford identify three core framing tasks: "diagnostic framing" (identifying a problem and attributing blame), "prognostic framing" (prescribing a solution), and "motivational framing" (urging others to effect change). A number of tactics are used to perform these tasks. At the stage of "diagnostic framing," I pay attention to "frame amplification,"[11] which highlights the humanitarian costs of weapons and focuses on the victims and their suffering. Regarding "prognostic framing," I examine two tactics NGOs have used to simplify the issue: advocating a ban as a solution to the humanitarian problems and advocating placing all types of cluster munitions under the same category irrespective of their design and functional differences. Finally, I concentrate on "motivational framing," which includes frames targeted at the general public that highlight the severity and urgency of the humanitarian problems, and NGO frames aimed at opponents that address military effectiveness in an attempt to make NGO goals resonate with military establishments. In the latter effort, NGOs used what I call "counterframe deflation"—countering and downplaying the arguments of opponents,[12] for example military arguments about the necessity and utility of cluster munitions. This tactic included questioning the credibility of the military regarding the evidence it used to back its claims about the weapons' effectiveness and enhancing the credibility of NGOs by drawing support from military officers. It was used especially in countries where military opposition to the ban was strongest as government policymakers seriously considered support for a treaty.

Legal Views and Military Practice

In a 2002 legal analysis of IHL and explosive remnants of war, Greenwood states that "[the proportionality] principle has usually been discussed by reference to the harm caused to civilians by the impact of munitions and the immediate damage and destruction caused by those munitions. The dangers posed by unexploded munitions appear to have been considered rarely if at all."[13] While acknowledging that a known risk from weapons use should be incorporated in the analysis, he argues that "the risks posed by [explosive remnants of war] once the immediate aftermath of an attack has passed are too remote to be capable of assessment at [the time of attack]," thus privileging a view that only short-term civilian costs be included in the proportionality calculus. Similarly, Boothby suggested that in applying the principle to cluster munitions one has to take into account their "expected" effects on civilians but not those that are reasonably foreseeable; the dud rates of cluster munitions, for example, should not be considered in the proportionality calculation given all the unknown factors that might intervene between the moment of use and their eventual impact on civilians.[14]

On the other hand, legal scholars who are part of the NGO community have argued that the postconflict effects of weapons needed to be part of an analysis of proportionality.[15] Still, early in the cluster munitions campaign even the International Committee of the Red Cross (ICRC) was cautious about what the proportionality rule implied: "Traditionally, the formulation has focussed on incidental damage to civilians and civilian objects likely to occur *during* an attack. It remains to be determined, however, the extent to which the long-term effects of unexploded munitions must be taken into account."[16] In the course of CCW discussions, the inclusion of dud rates into proportionality assessments was examined. Charles Garraway argued that "if it is known that there is likelihood, or even a certainty, that an attack will leave behind some explosive remnants of war, then that must be factored into the equation," although the question "How remote does the collateral damage have to be to require it to be taken into account?" remained open.[17] Previously it has been shown that U.S. military doctrine failed "to incorporate 'indirect collateral damage' into a proportionality analysis,"[18] and NATO and the United States were criticized for not including the aftereffects of cluster munitions in the assessment of collateral damage.[19]

Thus, at least until 2005, legal analyses predominantly asserted that long-term weapons effects are not included in calculations of proportionality because of their remoteness and uncertainty, while U.S. military practices adhered to a limited view of the temporal base for assessing collateral damage.

The Presumption of Military Necessity

Until recently there has been an agreement among military officials, state representatives, technical experts, and even NGO members about the significant military utility of cluster munitions. The military utility was perceived as an obstacle to achieving restrictions on their use, let alone a ban, even a few months prior to the launch of the negotiation process for their prohibition.[20]

The military has often rated the military efficiency of cluster munitions highly. In Vietnam, the expectation was that "these weapons could give us a quantum leap on the enemy."[21] In the 1991 Gulf War ground-launched cluster munitions were seen as "the decisive battle winner."[22] After the 1999 Kosovo campaign, it was claimed that "experience . . . demonstrated the importance of Combined Effects Munitions [cluster bombs CBU-87]."[23]

Similarly, state officials have argued about the high effectiveness of cluster munitions. For example, UK representatives have stated that "[cluster munitions] are still the most appropriate air-delivered weapons in many situations because of their ability to destroy enemy assets dispersed over an area.... The area effect capability of air-dropped cluster munitions is not matched by current precision weapons, or by large unguided unitary bombs.... Artillery-launched cluster munitions will maintain a crucial capacity in the suppression of area targets for a long time to come."[24]

U.S. officials have continuously insisted that cluster munitions "serve indispensable military purposes" and "provide distinct advantages against a range of targets and can result in less collateral damage than unitary bombs."[25] Similarly, French officials have argued that cluster munitions are "peerless in their efficiency" for neutralizing ground targets[26] and that to decide "to do without them would mean to accept an important reduction of land defense capabilities."[27]

An independent expert advising NGOs and governments on cluster munitions concluded in 2000 that "the trend towards increasing use of submunitions seems unlikely to change in the foreseeable future. From a military point of view, they offer unmatched cost-effectiveness."[28] In 2003, another expert even claimed that cluster munitions were a "battle winning munition."[29]

Early on, challenging the military utility of the weapons was not part of the NGO strategy. Rather, NGOs perceived military utility as a roadblock to future restrictions. While a few NGO reports questioned the military effectiveness of cluster munitions, even their conclusions were circumscribed. For example, a Landmine Action report acknowledged that "when the cluster munition works, it works. If it were possible to manufacture a variant with no propensity for failure and causing long-term danger, the military effectiveness of cluster munitions would not be in question."[30] Similarly, Pax Christi admitted that "there is no doubt about the military utility of cluster weapons against large mechanized forces" and that "[they] are certainly not 'out of fashion' even in advanced Western forces."[31] Even in 2006 when the campaign to ban cluster munitions was gathering force internationally, a former NGO member summarized prevailing views that "unlike antipersonnel mines, the military utility of these weapons is more readily and widely recognized."[32]

From Military Necessity to Humanitarian Impact

At the inception of the cluster munitions campaign, NGOs faced formidable opposition based on entrenched perceptions about the military utility and necessity of the weapons. As in the land mine campaign before it, the main

NGO strategy for overcoming this opposition has been to create a counter-discourse that highlights the enormous civilian suffering those weapons cause—the so-called humanitarian "frame amplification."

The first step in this regard was raising awareness by gathering evidence on the magnitude of the problems the weapons caused civilians. NGOs gathered data on the humanitarian cost of cluster munitions, documenting the numbers of civilian casualties and the socioeconomic problems resulting from their use.[33] They also concentrated on gathering information about the rate of failure of submunitions in actual battle use. This information supported their arguments that manufacturer data on failure rates were misleading and that the large percentage of submunitions that remained postconflict turned the weapons into de facto land mines. Although the scale of the cluster munitions problem was more limited than that of land mines, NGOs managed to convey its severity and urgency by calling it a "humanitarian crisis in waiting" and a "looming disaster" unless timely measures were taken.[34]

In addition to piling up statistics about the numbers of unexploded submunitions lying on the ground and numbers of civilian casualties, NGOs sought to present the human face of the victims and make visible the immediate and lasting impact of the weapons on their lives and bodies.[35] Their publications portrayed stories of horrific suffering of innocent victims and impediments to reconstruction efforts in war-torn societies affected by cluster munitions. And they made sure that the people affected by the weapons would be present at the negotiations of the new treaty to remind state delegates of the human lives behind the statistics. This brought an effective emotional power to the NGOs' arguments and helped single out the weapons by associating them directly with the suffering they had caused.

The tactics of amplifying the humanitarian problems, the suffering of victims, and the sense of crisis and urgency were targeted at both the media and the public and at state negotiators involved in the CCW talks, and they were vital for issue framing.

From an IHL perspective, zooming in on the weapons' humanitarian impact changed the postulated balance between military interests and humanitarian costs, even if perceptions of military utility might not have been affected. Since 1999, NGO reporting on the deleterious effects of cluster munitions, especially after the end of conflict, raised awareness of the problem and directed attention to the applicable law. NGOs argued that the use of cluster munitions might be breaching the proportionality rule. Early on, Human Rights Watch challenged their proportionality: "In some circumstances, the long-term harm to the civilian population of cluster bomb use may outweigh the short-term military benefit."[36] Over time, Human

Rights Watch has consistently argued that because of the area coverage and postconflict problems, use of cluster munitions "in or near populated areas is almost always disproportionate, and thus illegal, because of the harm their duds inflict on civilians over time relative to the military advantage sought."[37] Other NGOs have also continuously advocated that the foreseeable long-term effects should be factored in the proportionality calculation.[38]

The ICRC directly challenged the view that long-term effects of cluster munitions were not foreseeable and could not be evaluated for proportionality. It argued that "in light of the experience gained from the use of cluster munitions in past conflicts and the work of governments and organizations to address them . . . the application of the proportionality rule must now include the extended impact of submunitions," emphasizing that when cluster munitions are used in populated areas the long-term consequences of unexploded submunitions are "readily foreseeable."[39]

Thus, NGO reports documenting the humanitarian impact of cluster munitions and their failure rates, which far exceeded those declared by manufacturers and the military, made it difficult to argue that the long-term effects of the weapons could not be foreseen. A report for the CCW argued for including the expected long-term effects by pointing to the wealth of information already gathered by humanitarian groups. Indeed, according to the report, NGOs showed "the inevitability of civilian damage from large numbers of unexploded submunitions deployed in residential or agricultural areas."[40] Thus, even when the use of cluster munitions in populated areas could be seen as indiscriminate at the time of attack because of their wide dispersal, the argument often was that they would be disproportionate because of postconflict casualties from duds. In 2006, CCW states parties acknowledged "the foreseeable effects of explosive remnants of war on civilian populations as a factor to be considered in applying the international humanitarian law rules on proportionality."[41]

Yet even though the principle of proportionality opened an opportunity for NGOs to argue that the humanitarian effects of cluster munitions outweighed their military utility, it didn't provide sufficient grounds for banning them altogether. Indeed, Human Rights Watch and the ICRC, the two organizations most closely focused on interpretation of IHL, have for a long time been cautious not to advocate a total ban on cluster munitions, calling mostly for a ban on use in populated areas and restrictions on inaccurate and unreliable munitions.[42] The dynamics within the NGO campaign changed with the decision of Handicap International to advocate a ban in an effort to reinforce its domestic public campaigns. It managed to pass a ban in the Belgian Senate in 2005 and succeed in making Belgium the first country to ban cluster munitions in early 2006, which made it more difficult for NGOs to call for anything less than what had already been banned by a state.[43] In 2005, Landmine Action strengthened its UK campaign and highlighted the impossibility of pursuing meaningful actions on cluster munitions based on IHL if states made no concrete assessments of the military utility and humanitarian effects of the weapons. It argued that UK practice "systematically gives preference to the military at the expense of increasing risk to the civilian population."[44] Although Landmine Action refrained from calling for a ban on cluster munitions at that time, it declared that the window of opportunity to show that "existing mechanisms of [IHL] can serve the purpose for which they were developed" was closing.[45]

The window of opportunity for legal discussions at the CCW was closed after Israel's massive use of cluster munitions in South Lebanon in the summer of 2006 and Norway's decision to lead a negotiation process outside of the CCW forum. In August 2006, Landmine Action called for a total ban. While Human Rights Watch was still not calling for a complete ban at that point, it declared that cluster munitions pose "unnecessary and unacceptable harm to civilians."[46]

The focus on "unacceptable" humanitarian harm came to inform the stand-alone process of international negotiations Norway launched. The Oslo Declaration that states had to endorse in order to participate in the negotiations established its humanitarian frame of reference—its goal was to conclude a treaty prohibiting "the use, production, transfer and stockpiling of cluster munitions that cause unacceptable harm to civilians" by 2008, but it left ambiguity regarding which weapons posed such an "unacceptable harm." The organizers characterized the process as an advancement of IHL by banning a weapon whose "humanitarian and political consequences— long after the conflicts have ended—by far outweigh their usefulness."[47] Although the proportionality balance between humanitarian and military concerns was the foundation on which the process built, the emphasis shifted from narrow legal and military interests, which often prevailed (especially within the CCW) to political considerations focused on the humanitarian issue. Importantly, this was reflected in the choice of the wording of the Declaration: it sought a ban on cluster munitions that cause "unacceptable" harm to civilians instead of using the legal terms "disproportionate" or "excessive" civilian damage. Defining what "unacceptable" meant became the core of negotiations, and as they unfolded the criteria used became more and more detached from the balancing exercise and evaluation of military needs. Instead, showing that cluster munitions caused humanitarian harm and did not function as advertised became sufficient to consider their effects "unacceptable." To the extent that a balance was sought, it was heavily tilted toward the humanitarian side. This marked a break with legal interpretations of proportionality in favor of stronger civilian protection. Although to arrive at that point it was necessary to go through a proportionality argumentation, the discursive loop ultimately led to a more absolutist understanding of civilian immunity.

In this respect, NGOs were key players showing which weapons were causing humanitarian damage and hence had to be banned. A report prepared by Norwegian People's Aid, the Norwegian Defense Research Establishment, and Jane's Defense expert Colin King was crucial in providing the evidence that even advanced submunitions that touted a reliability rate of 98–99 percent, which many states wanted to exempt from banning, had failed to function properly and had led to civilian casualties in Lebanon in 2006, thus effectively arguing that they were causing "unacceptable harm."[48] By framing the debate in humanitarian terms, NGOs and their state allies shifted the burden of proof that certain weapons work properly and provide high military utility to those who were claiming so.[49] While this approach has been said to exemplify a supplanting of traditional IHL reasoning with a precautionary principle,[50] the outcome still rested on showing the magnitude of civilian costs first, and there are bases in IHL upon which it builds. In clarifying the meaning of proportionality, the ICRC's *Commentary on the Additional Protocols of 8 June 1977* states that in cases of doubt about the proper balance, "the interests of the civilian population should prevail." Furthermore, even a military advantage "of great importance" cannot justify "very high civilian losses": "incidental losses and damages should never be extensive."[51]

Hence it could be argued that when severe civilian harm results from the use of cluster munitions, a prima facie case exists that the proportionality rule is breached, no matter what military advantage is being sought. However, despite the merits of the principle, in practice the lack of clear-cut application guidelines has resulted in weapons malfunctions and civilian casualties in conflict after conflict, including the war in Lebanon in 2006. Israeli justifications of the use of cluster munitions by claiming that they comply with the law once again underscored the fact that reliance on legal constraints is insufficient to ensure civilian protection and a ban of the weapon itself was needed.[52] While all along NGO campaigners have justified the need for restrictions on cluster munitions by pointing to their disproportionate and indiscriminate effects, especially in urban areas, eventually the call for a ban came from more activist-oriented NGOs, who extended the proportionality logic to a political argument about "unacceptable humanitarian costs."

From Military Effectiveness to Disutility and Obsolescence

Not only have NGOs managed to reframe the debate from security into humanitarian terms, they have also sought to undermine arguments about the military utility of the weapons in a process I call counterframe deflation—challenging and downplaying military arguments. As early as the land mine campaign, NGO members had noticed that the "armed forces will not respond to the emotional issue" of humanitarian suffering and deemed it necessary to address issues of military use per se. This approach has also been used in campaigns to ban cluster munitions.[53]

As in the case of humanitarian effects, where NGOs have consistently demanded that civilian casualty data be gathered and evaluated together with battle damage assessments,[54] they have pointed out the lack of studies of the military utility of cluster munitions, an absence that made balancing military advantages and humanitarian costs impossible.[55] They have also argued that in practice the anticipated military advantage of attacks has often been limited. Indeed, NGOs have asserted that in no conflict have "[cluster munitions] of any kind . . . been a critical positive combat factor for the user force"[56] and questioned "if there were instances, either for a specific engagement or for a conflict as a whole, in which the military could not have accomplished its objective without cluster munitions."[57] Thus, from challenging the effectiveness of the weapons in specific strikes, NGOs moved to denying their indispensability in conflicts overall. And they have repeatedly highlighted the heavy toll unexploded submunitions take on friendly forces, a factor that diminishes their supposed military value.[58]

In addition to this direct challenge to the military effectiveness of the weapons, the latter was linked in many cases to the argument that the use of cluster munitions undermined the larger political goals of today's military campaigns to win "the hearts and minds" of local populations and that the high political cost of cluster munitions thus limited their military utility. NGOs repeatedly argued that cluster munitions were relics of the past designed for the battlefields of the Cold War rather than current operations in urban environments. Once again, political arguments factored in the legal analysis and arguably influenced how proportionality was viewed.

Thus, challenging the military utility of the weapons became part of the campaign to delegitimize them. But it came only after the NGOs had convincingly demonstrated the humanitarian cost. Under the scrutiny of NGOs, perceptions about the political costs of cluster munitions changed. At this stage, when momentum behind a prohibition had gathered force, studies of the weapons' military utility were undertaken with a view to making a ban that was justified on humanitarian grounds palatable to the military. As in the U.S. campaign to ban land mines,[59] this step was made in order to portray proponents of the ban as responsible statesmen taking into account military requirements and not as pacifists on a crusade to disarm their militaries.

Importantly, this tactic was most visible where military opposition was an obstacle to achieving a ban. Once the Oslo Process started, it was deemed necessary to show that the prohibition on cluster munitions would not critically undermine military capabilities. Although campaign efforts in the United States were for the most part weak, challenging the utility of these weapons came to the fore in the UK, the major military force that was (reluctantly) participating in the negotiation process. Landmine Action worked actively with parliamentarians to raise the visibility of the problem and put pressure on the government to support the international ban process. In debates in the House of Lords, it was persistently argued that cluster munitions had no military utility in post–Cold War conflicts, in which protecting the civilian population and winning support were of the utmost importance.[60] However, the arguments in this case focused not so much on challenging the military utility of the weapons but rather on refuting their utility after weighing it against their humanitarian and political costs.

In a reprise of a generals' letter published in support of the land mine ban in the United States, nine senior British military leaders published an open letter to the secretary of state for defense under the title "Cluster Bombs Don't Work and Must Be Banned" on the eve of the final negotiations of the Convention on Cluster Munitions in May 2008. They argued that cluster munitions caused casualties among the civilians and soldiers, that they were "battlefield losers," and that prohibiting them would "establish a new benchmark for the responsible projection of force in the modern world." Instead of letting soldiers use these indiscriminate weapons, the UK government should equip them with the "right" weapons, such as precision-guided munitions.[61] NGOs and their parliamentarian allies continued to portray cluster munitions as "outdated relics of the Cold War" of "no real military gain."[62]

The NGO Cluster Munition Coalition (CMC) also questioned the military effectiveness of the weapons and started referring to them as "outdated" and "obsolete." It focused on challenging the "unique" military utility of cluster munitions and placing doubt on claims about their effectiveness. CMC argued that in recent conflicts they had been used as "a weapon of convenience against a wide range of targets" rather than for delivering specific advantages against particular target sets, as the military argued. In addition, it argued that the tasks performed by cluster munitions could well be fulfilled by other weapons.⁶³ Similarly, campaigners argued that "the soldiers . . . know these weapons are outdated" and that "good soldiers should have good weapons"⁶⁴ and accused states that were reluctant to full-heartedly support the Oslo Process of being "more interested in protecting their obsolete cluster munitions than [in] protecting civilians."⁶⁵

In a further attempt to clarify the issue of military utility, the ICRC organized an expert meeting in April 2007.⁶⁶ At it, NGOs continued their charge that if militaries had no reliable data on the failure rate of submunitions, it was not possible to assess either their humanitarian cost or military utility. The meeting did not issue a concrete position on military utility, but it pointed to the lack of reliable information and the need for further study of this aspect, once again casting doubt on the assertions and credibility of the military in this respect.

Finally, in an attempt to provide the missing military analysis of cluster munitions and show that banning them would not be an irresponsible act, the Norwegian Ministry of Foreign Affairs tasked the Norwegian Defence Research Establishment to undertake a study of their military utility. The subsequent report significantly raised the bar for the military utility of cluster munitions: "Cluster weapons do have a satisfactory or adequate effect against most targets. Under certain conditions the effect is quite good. However, no evidence has been found to claim that such weapons are far better than their alternatives to the extent that they are *indispensable*. . . . Cluster weapons do not constitute an *irreplaceable* capability on the battlefield. Alternatives exist, although in some cases they may be less effective than cluster weapons. . . . Thus a prohibition of cluster weapons will not mean that a set of *unique* capabilities is lost."⁶⁷

The conclusion was that although cluster munitions had better military effectiveness than unguided unitary bombs, a combination of precision-guided bombs and advanced types of cluster munitions such as sensor-fuzed weapons delivered higher military utility despite their much higher economic cost.⁶⁸ The study undermined exaggerated and unsubstantiated claims about the "unrivaled" military effectiveness of cluster munitions against certain types of targets. However, it should be remembered that very few weapons could pass the test of being indispensable and unique. If that were a criterion for obsolescence, and hence prohibition, the military would find itself with very few weapons left. But the report strengthened the position of states and NGOs advocating a ban by showing that it was backed by military analysis and enjoyed the support of the military establishment. Importantly, it demonstrated that cluster munitions were neither indispensable nor unique, thus opening the door for arguments that they were dispensable and outdated. Thus, the strategy of challenging unsubstantiated military assertions and raising the standards of military utility that cluster munitions had to meet—from being useful under certain circumstances to being "unique" in their capabilities—led to a diminishment of their military value in the proportionality balance. Clearly, the discursive reframing of the military side of proportionality parted with strictly legal criteria and elevated the debate to the political plane.

Conclusion

I have shown how the NGO campaign on cluster munitions has built upon the principle of proportionality in making its case for why these weapons should be banned. Although NGOs have used the proportionality principle to stigmatize the weapons, IHL alone was not sufficient to put a halt on the use of cluster munitions. A more political, public approach that went beyond legalistic interpretations and at the same time tried to reshape them was required. Employing the IHL language of proportionality, NGOs have both influenced the application of the principle and then departed from it to achieve a higher standard of civilian protection. While the latter has been the goal all along, to get there NGOs initially trod the IHL path.

By raising the political costs associated with the use of specific weapons, NGOs have tried to shift the balance of military decisions. As more actors with diverse perspectives acquire a voice, the interpretation of legal provisions inevitably changes. While scholars have portrayed this trend as NGOs misinterpreting the law and mixing political and legal criteria,[69] it may be both impossible and undesirable to keep new opinions from informing law development. The achievement of the Convention on Cluster Munitions needed an additional political stimulus because of the indeterminacy and divergent applications of the law, but the process leading to the new treaty has also clarified some legal aspects of the humanitarian side of the proportionality balance.

After the Convention on Cluster Munitions was adopted, NGOs from the CMC started thinking about new humanitarian challenges. Two organizations launched initiatives on weapons-related issues that drew upon the cluster munitions campaign—Human Rights Watch focused on strengthening the law on incendiary weapons, while Landmine Action has set its sights on restraining the use of explosive weapons in populated areas.[70] Although both organizations tried to use the cluster munitions approach by emphasizing "unacceptable civilian harm," Human Rights Watch started along the usual path of bringing challenges to existing law within the CCW, while

Landmine Action opted for a political approach that was detached from the IHL framework.

The main arguments about incendiary weapons have been about avoiding "needless and unacceptable civilian suffering" and proportionality. Given the "substantial humanitarian costs," there should be "a clear presumption that the risk to civilians of using such weapons substantially outweighs the concrete and direct military advantage expected from an attack."[71]

Similarly, the impetus for the explosive weapons initiative was civilian protection. Data on armed conflict showed that 90 percent of civilian casualties occurred in populated areas,[72] and explosive weapons killed more people per incident than other weapons.[73] The campaigners' idea was to stigmatize the use of explosive weapons in populated areas as "unacceptable" while voicing skepticism about "any expectation that interpretations of [IHL] will result in progress."[74] Various ideas for an appropriate framing of the issue have been proposed—from contrasting the unacceptability of these weapons in domestic settings where governments are accountable to their citizens with their acceptability when victims are noncitizens to IHL principles of distinction to indiscriminate area effects within populated areas to the long-term psychological impact on individuals and the effects on civilian infrastructure.[75] While NGO proponents have remained guarded about the prospects of IHL provisions to successfully deal with the problem,[76] there seems to be continuous resort to IHL thinking that shows up in supportive state and UN statements on explosive weapons that have focused on their indiscriminate, long-lasting, or disproportionate effects.[77] Although involved NGO researchers insisted on stigmatization and employing various frames, the language used was close to that of IHL.[78] In 2011, Action on Armed Violence (formerly Landmine Action) issued a report that underscored the indiscriminate nature of explosive weapons in populated areas.[79] The way NGOs voiced their concerns became reminiscent of an implicit proportionality argument that highlighted the humanitarian costs and questioned the military side of the equation: "The harm done to civilians by these explosive devices is unacceptably high. At the same time . . . governments have failed to justify their military value"; "governments have a responsibility to measure what they're doing and [to] make a serious, genuine evaluation before they undertake a strike about whether this is going to be unacceptable in terms of civilian damage."[80]

Thus, whereas campaigners against explosive weapons have tried to dispense with the IHL route and adopt a position of the ethical voice that stigmatizes problematic military practices that IHL has continuously failed to restrain, they have found themselves being "sucked back into IHL" language while searching for appropriate frames.[81] Indeed, even if officials who refer to the issue don't mean to invoke specific IHL legal standards,[82] they still imply a broad proportionality measurement of civilian and military interests. This could indicate that an IHL frame may be necessary if the issue is to gain traction. As David Kennedy has argued, the IHL vocabulary of proportionality and military necessity has become dominant, and "it is difficult to image how *else* one would talk about the use of force."[83] While humanitarians face a constant dilemma between applying the cost-benefit analysis of humanitarian law and opting for an absolutist position defending civilian immunity tout court, the case of cluster munitions may be a template for resolving this tension by first using the proportionality principle, then reshaping understandings of what is at stake in the balancing act, and eventually transcending the principle in favor of civilian protection.

Importantly, the Convention on Cluster Munitions reinforces the rule against indiscriminate attacks and includes criteria that are to be used as guidelines to "avoid indiscriminate area effects and the risks posed by unexploded submunitions." These guidelines could be applied to explosive weapons and targeting practices that have the worst effects on civilians. If this and provisions of the Convention on Cluster Munitions that require information to be gathered about victims of cluster munitions are extended to explosive weapons, they could contribute to restraint on the use of explosive force in populated areas and provide a clearer picture of the civilian suffering that results from it.[84] Although building upon those principles will not lead to changes in state practice overnight, it may be a stepping-stone on the way to curbing and possibly stigmatizing the excessive use of explosive weapons (especially high-tonnage and unguided bombs, inaccurate artillery rockets, and the targeting of dual-use facilities) where civilians bear the brunt of violence. Efforts to enhance civilian protection may still be pursued best by relying on a balancing exercise that keeps humanitarian considerations in focus without losing sight of the other side of the scale.

CHAPTER 10

Toward an Anthropology of Drones
Remaking Space, Time, and Valor in Combat

HUGH GUSTERSON

Compare the following two accounts of battle. The first comes from Book III of Homer's *Iliad*:

> When they had thus armed, each amid his own people, they strode fierce of aspect into the open space, and both Trojans and Achaeans were struck with awe as they beheld them. They stood near one another on the measured ground, brandishing their spears, and each furious against the other. Alexandrus aimed first, and struck the round shield of the son of Atreus, but the spear did not pierce it, for the shield turned its point.
>
> Menelaus next took aim, praying to Father Jove as he did so. "King Jove," he said, "grant me revenge on Alexandrus who has wronged me; subdue him under my hand that in ages yet to come a man may shrink from doing ill deeds in the house of his host." He poised his spear as he spoke, and hurled it at the shield of Alexandrus. Through shield and cuirass it went, and tore the shirt by his flank, but Alexandrus swerved aside, and thus saved his life. Then the son of Atreus drew his sword, and drove at the projecting part of his helmet, but the sword fell shivered in three or four pieces from his hand, and he cried, looking towards Heaven, "Father Jove, of all gods thou art the most despiteful; I made sure of my revenge, but the sword has broken in my hand, my spear has been hurled in vain, and I have not killed him."

> With this he flew at Alexandrus, caught him by the horsehair plume of his helmet, and began dragging him towards the Achaeans. The strap of the helmet that went under his chin was choking him, and Menelaus would have dragged him off to his own great glory had not Jove's daughter Venus been quick to mark and to break the strap of oxhide, so that the empty helmet came away in his hand.
>
> This he flung to his comrades among the Achaeans, and was again springing upon Alexandrus to run him through with a spear, but Venus snatched him up in a moment (as a god can do), hid him under a cloud of darkness, and conveyed him to his own bedchamber.

The second account is from Jane Mayer's 2009 *New Yorker* article on the U.S. deployment of drones over Pakistan:

> On August 5th, officials at the Central Intelligence Agency, in Langley, Virginia, watched a live video feed relaying closeup footage of one of the most wanted terrorists in Pakistan. Baitullah Mehsud, the leader of the Taliban in Pakistan, could be seen reclining on the rooftop of his father-in-law's house, in Zanghara, a hamlet in South Waziristan. It was a hot summer night, and he was joined outside by his wife and his uncle, a medic; at one point, the remarkably crisp images showed that Mehsud, who suffered from diabetes and a kidney ailment, was receiving an intravenous drip.
>
> The video was being captured by the infrared camera of a Predator drone, a remotely controlled, unmanned plane that had been hovering, undetected, two miles or so above the house. Pakistan's Interior Minister, A. Rehman Malik, told me recently that Mehsud was resting on his back. Malik, using his hands to make a picture frame, explained that the Predator's targeters could see Mehsud's entire body, not just the top of his head. "It was a perfect picture," Malik, who watched the videotape later, said. "We used to see James Bond movies where he talked into his shoe or his watch. We thought it was a fairy tale. But this was fact!" The image remained just as stable when the CIA remotely launched two Hellfire missiles from the Predator. Authorities watched the fiery blast in real time. After the dust cloud dissipated, all that remained of Mehsud was a detached torso. Eleven others died: his wife, his father-in-law, his mother-in-law, a lieutenant, and seven bodyguards.[1]

Homer's account is of what we might call archetypal combat. Heroic men meet each other as named individuals who are equals in the field of battle, their chances of living, dying, or being wounded shaped by their strength, skill, and luck rather than by any difference in weaponry. Each can eliminate the other only by getting within arm's reach of his opponent. Their combat takes place in front of an audience, which can judge the combatants' prowess, and the encounter of the warriors is physically intimate: Menelaus grabs Alexandrus's helmet, for example. The warriors are also described as having an intimate relationship with their weapons, which are figured almost as an extension of their bodies. And if there is intervention from above, it comes from capricious gods who intervene to spare a man from his apparently fated death.

Contrast this archetypal scenario of combat with Jane Mayer's description of a drone attack in Pakistan. Instead of a heroic encounter between equal combatants, we have a radically asymmetric situation where the drone operators, far from being combatants watched by an audience, have become the audience that observes the act of death; looking on from high in the sky, they have assumed the position of the gods who decide who will live and die. This encounter still has an intimate quality, but the intimacy has become one-sided and asymmetrical: while Mehsud does not even know he is being observed, the drone operators can see him close up, reclining on the roof of his house on a hot evening, his wife attending to his medical needs. Without even knowing he is in combat, he is killed as if by a god's thunderbolt from the sky.

The international drone market is now worth $6 billion a year.[2] Fifty countries are estimated to have built or purchased drones or, as they are more formally known, unmanned aerial vehicles (UAVs).[3] Of these, only three have used them to shoot at targets on the ground: Israel (the first country to do so), the United States, and the UK. In addition, in 2005, Hezbollah used a reconnaissance drone against Israel.[4] The number of drones owned by the United States is often estimated at about 7,000.[5] The best known are the Predator and Reaper drones, both manufactured by General Atomics. The United States is now also in the process of developing smaller drones such as the Switchback, nicknamed the "assassin bug," which weighs less than six pounds, carries tiny warheads, and can be carried in a soldier's backpack.[6] The United States is also seeking to develop "tiny 'nano' drones, which can fly after their prey like a killer bee through an open window."[7]

U.S. drones are operated abroad by both the CIA and the U.S. military, while the Department of Homeland Security has been testing drones for use within the United States and the Federal Aviation Administration has been directed to plan for the integration of as many as 30,000 drones into U.S. domestic airspace within a decade.[8] The United States has deployed drones over Iran, Iraq, Afghanistan, Yemen, Pakistan, Libya, Somalia, and Syria, and it routinely flies them along the U.S.-Mexican border. The Air Force initially resisted the adoption of drones, according to some analysts, because unmanned, slow-flying aircraft with glorified lawnmower engines were at odds with its "Top Gun" alpha-male culture; this left the CIA as the original bureaucratic sponsor of this technology.[9] The first U.S. drones were developed before World War II for use in target practice. They were then used to collect radioactive samples from U.S. nuclear tests and for aerial surveillance and intelligence gathering in the Vietnam War. Israel was the first country to use drones for "targeted killings," and the United States publicly criticized it for doing so until, after 9/11, it also moved to arm drones and use them for aerial strikes against human targets on the ground.[10] The CIA drone program intensified greatly under the Obama administration.[11] Its targets have largely been alleged jihadists (a handful of whom have been U.S. citizens) in countries such as Pakistan, Yemen, and Somalia with which the United States is not officially at war. The CIA calls its aerial assassinations of senior insurgent leaders who have been individually hunted "personality strikes," but many attacks have also been "signature attacks"—aerial strikes on groups of apparent militants whose individual identities are unknown but who are targeted because their behavior follows patterns that look suspicious to analysts thousands of miles away.[12] "When it comes to signature strikes, say insiders, the decision to launch a drone assault is essentially an odds game: If the agency thinks it's likely that the group of individuals are insurgents, it will take the shot."[13]

The New America Foundation, which has been tracking drone strikes in Pakistan to the extent that this is possible from within the United States, has estimated that as of July 2013, U.S. drones had killed between 2,039 and 3,370 people, of whom between 1,585 and 2,733 were "militants."[14] The United States has consistently maintained that the overwhelming majority of those killed in Pakistan were jihadists. Indeed, one unnamed Obama administration official told the *New York Times* in 2012 that innocent Pakistani victims of drone strikes numbered "in the single digits."[15] However, according to the *Guardian*, which relied on information collected by the British human rights group Reprieve with the aid of Pakistanis in Waziristan, by 2011, CIA drones had killed 2,500 civilians in Pakistan.[16] Counterinsurgency experts David Kilcullen and Andrew Exum estimated in 2009 that the United States was killing fifty civilians for every jihadist it succeeded in killing,[17] while Pakistani journalist Noor Behram, who risks his life by rushing to the scene of drone attacks to photograph the aftermath, estimated that 10 to 15 civilians were killed for every militant the drones kill.[18]

The *New Yorker* describes drone operations thus: "Using joysticks that resemble video-game controls, the reachback operators—who don't need conventional flight training—sit next to intelligence officers and watch, on large flat-screen monitors, a live video feed from the drone's camera.... People who have seen an air strike live on a monitor described it as both awe-inspiring and horrifying. 'You could see these little figures scurrying, and the explosion going off, and when the smoke cleared there was just rubble and charred stuff,' a former CIA officer who was based in Afghanistan after September 11th says of one attack."[19]

Those on the ground, on the other hand, may be unnerved by the constant buzzing of unseen drones above, wondering if they will be vaporized by an unseen missile at any moment. Neil Williams, a volunteer for Reprieve, recalls interviewing Tariq Aziz, a sixteen-year-old boy who died a few days later in a U.S. drone strike. "Tariq and other teenagers at the meeting told Williams how they lived in fear of drones. They could hear them at night over their homes in Waziristan, buzzing for hours like aerial lawn mowers. An explosion could strike at any moment, anywhere, without warning."[20]

The U.S. reorientation toward drones over the last decade has been driven by three factors in particular. The first is a desire to save American lives. If an unmanned plane is shot down or crashes for other reasons, the pilot's life will not be lost, since the pilot will be thousands of miles away from the plane. Military personnel, especially pilots, are expensive to train and are not fungible assets in the way they were in earlier wars. Moreover, as the American public becomes increasingly resistant to the deaths of its servicemen, lowering casualty rates becomes important in maintaining domestic support for (or at least acquiescence in) war.[21] The second factor is cost. At about $5 million each for a Predator drone,[22] for example, these aircraft are much cheaper than F-16s, B-1s, Warthogs, or the Joint Strike Fighter,[23] which are now estimated to cost over $150 million per plane. The third factor is tactical. While manned aircraft are more restricted by fuel constraints and concerns of being shot down while lingering, "drones can hover above a target for up to forty hours before refuelling, and the precise video footage makes it much easier to identify targets."[24] This ability to linger or "dwell" in the air makes it possible to analyze patterns of life on the ground for suspicious features, track people and objects of interest as they move, take action in a more deliberate fashion, and collect damage assessment information after a strike.

In this chapter I want to explore the implications of the rise of the drones for warfare. If we imagine the potentially transformative effects of drones on warfare as an arc, we are in the early stages of traveling this arc, but it is already possible to discern where the journey seems to be taking us. (Here I do not want to give the impression that technology, in this case drone technology, is a sort of independent force that acts upon society, forcing changes in norms and institutions. It is rather a process of the kind that Sheila Jasanoff has described as one of "co-production," in which norms, institutions, and technology interact with one another to co-produce a new military sociotechnical ensemble.)[25] I will discuss ways that drones respatialize war, change its pace, and rework conventional military notions of honor and courage. I will also discuss the implications of drones for the laws of war and for the relationship between war and democratic governance in the United States.

The Respatialization of War

The history of military technology is one in which warriors have sought to kill one another with increasing measures of what Pink Floyd's Roger Waters calls "the bravery of being out of range."[26] Hand-to-hand combat gave way to the bow and arrow, which in turn gave way to the gun, and then came artillery and aerial bombardment, each development increasing the potential distance between warriors. Throughout this centuries-long process in which military tactics and technology co-evolved, combatants have sought to find ways of striking a blow from a distance long enough to ensure that they themselves were immune to a reciprocal blow. However, a vestigial vulnerability has always remained: snipers can be shot or captured, planes can be shot down, and destroyers can be sunk.

The development of drones not only increases the distance from which a blow is struck to thousands of miles, thus finally conferring immunity on the attacker, it also triangulates a formerly dyadic relationship between combatants. In the arc of technological development described above, the weapon that might kill the enemy was moved further away from its target, but the weapon's operator still had to be with the weapon to operate it. Weapon and warrior were more or less coincident in space. Drones have disarticulated the spatial relationship between weapon and warrior. In the words of retired admiral Joe Dyer, drones "allow us to project our power without projecting vulnerability."[27] Not only can people on the ground in Pakistan be killed by weapons so far away that they may have had no idea they were being targeted,[28] but the person operating this weapon (the drone) is thousands of miles away from the missile he launches as well. While the weapon that executes the strike is relatively proximate to the scene of destruction (hovering in the sky above Pakistan in this case), the person who controls it may be on the other side of the planet. A formerly tightly packed and spatially concentrated ensemble—weapon and weaponeer—has been disarticulated.[29] In the words of the narrator of a documentary film on drones, "for the first time in history it's possible to kill someone on the other side of the planet in real time by remote control."[30]

The respatializing dynamics here are profound and asymmetrical. The targets of drone strikes are made to feel deeply trapped in the local, from which there may be no escape, while the targeters inhabit a space of free movement that has become stretched to global proportions. This is a little like the contemporary relationship between labor, which is trapped at the local level by lack of resources and by national boundaries, and globalized capital, which is free to move anywhere in the world at a few keystrokes. In the cases of both military conflict and global capitalism, the freedom to move on a global scale affords an important but not necessarily decisive advantage.

It is tempting to say that because remote control technology allows drone operators to act on the battlefield while being so far away from it, they are effectively outside the battlefield, looking in like gods on the outside. But another way of thinking about this situation is to insist that if a combatant acts from a place, then by definition that place is part of the battlefield. If this is so, the drone operator has not so much removed himself from the battlefield as he has globalized the battlefield, bringing the battlefield, or experiential fragments of it, inside the national boundaries of the homeland. While trapping the targeted adversary in the local level by acting from an unseen distance, the drone operator has, at the same time, enabled shards of that faraway local battlefield to embed themselves in his own experience of the local. Thus the flip side of globalization in this process of respatialization is that the clear boundary between the battlefield and a U.S. combatant's personal and family life is in danger of being erased—not (as for Pakistani militants) because the U.S. combatant's family may be struck at any time by the enemy but because the experience of battle in the other's faraway local cannot be quarantined from the U.S. combatant's domestic life. An article in *Stars and Stripes*, describing drone surveillance of U.S. troops in ground combat, vividly conveys what is at issue here. It begins by pointing out that the same communications technologies that enable a drone operator to act from a distance may also give the jarring sensory experience of being in that distant place:

> "A lot of people downplay it, say 'You're 8,000 miles away. What's the big deal?' But it's not really 8,000 miles away, it's 18 inches away," Gersten said. "We're closer in a majority of ways than we've ever been as a service."

The amount of time spent surveilling an area—sometimes hundreds of hours are devoted to a single mission—creates a greater sense of intimacy than with other aircraft, Mathewson said.

One of the great advantages of drones is the ability to loiter. But that also means crews spend a lot of time watching the destruction of war. . . .

"There's no detachment," he said. "Those employing the system are very involved at a personal level in combat. You hear the AK-47 going off, the intensity of the voice on the radio calling for help. You're looking at him, 18 inches away from him, trying everything in your capability to get that person out of trouble."[31]

On top of the experiential whiplash caused by being physically removed from the battlefield while having the (at least partial) sensory experience of being vividly within it, the proximity of the drone operator's pod to his family creates further confusion, now between the battlefront and the home front, which are separated only by the temporal boundaries of shift work. *Stars and Stripes* quotes a drone operator as follows: "You've just been on a combat mission and half an hour later your spouse is mad at you because you're late to soccer practice. Friction at home can stem from just that simple question upon walking through the door: 'How was your day?'"[32]

Klem Ryan makes a strong argument for a military ethics premised on the self-contained battlefield of the eighteenth century that clearly separated combatants and noncombatants, and he is troubled by the way drones undermine core assumptions of international humanitarian law that "belligerents mutually occupy a distinct physical space in which war is conducted."[33] In line with others who dub drone warfare "PlayStation warfare,"[34] he worries that drones offer a numbingly impersonal and "dissociative" style of killing that makes the taking of life too easy.[35] (Here he takes issue with Charles Dunlap and others who suggest that death on screen can be hauntingly vivid, leaving drone operators with high rates of post-traumatic stress disorder.) While I find the eighteenth-century battlefield to be morally superior to what we have today, it has, together with its anchoring distinctions, long since dissolved under the weight of aerial bombing, land mines, death squads, total war, and nuclear weaponry and its associated genocidal targeting doctrines. Bearing in mind Mary Kaldor's observation that the ratio of civilian to military casualties shifted from 10–90% to 90–10% in the course of the twentieth century,[36] the putative precision of drone warfare may even offer a modest improvement in the odds for noncombatants in war zones, even as it plays havoc with our spatial imagination of the war zone. It is likewise too reductive to say that the processes of respatialization in drone warfare simply distance drone operators from the battlefield, thus making killing easier. It would be more accurate to say that they scramble relations of distance, making them simultaneously more elongated and more compressed in ways that are subjectively confusing and paradoxical. They at once making killing easier and harder, creating a new psychological topography that we are struggling to understand.

This process of respatialization, which removes the U.S. combatant from the battlefield even as it embeds the battlefield in his local world, raises the following interesting question: Is there any reason why Taliban operatives, if they could somehow escape their confinement to their local sphere of action, should not come to Nevada, where the military operates drones at Creech Air Force Base, or Virginia, where the CIA operates its drones, and attack drone operators there? (Incidentally, military "pilots" fly their drones in uniform, while CIA drone operators do not, which, according to the criteria the George W. Bush administration applied in the Middle East, makes them "unlawful combatants."[37]) I have asked my students this question a number of times, and after pausing uncomfortably, they mostly answer that the Taliban would have a right to attack a U.S. drone operator while he is operating a drone in his control pod. But what if the Taliban chose instead to blow up his car on the freeway as he drove home from work—as U.S. drones have done to cars in Pakistan and Yemen on many occasions? Would this be a military operation or an act of terrorism? And what if they sprayed him with machine gunfire at a picnic, killing his wife as well, just as the CIA killed Mehsud's wife and other family members in the excerpt from the *New Yorker* with which this chapter opened?

Retemporalization

The passage from the *Iliad* quoted earlier makes it clear that speed is of the essence in archetypal combat, where a failure of reflexes can lead to instant death, and victory can be won or lost through sequences of events that unfold at high tempo. Military strategist Carl von Clausewitz famously bemoaned the way brilliantly formulated tactics could unravel when confronted with the frictional speed of the unexpected, and French theorist Paul Virilio has commented on how technologies of war have operated at higher and higher speeds as they have evolved through history—up to nuclear warheads that now fall to earth from space at six times the speed of sound.[38]

Drone warfare scrambles time and speed in the same way that it scrambles relations in space. Just as drone warfare traps the target in the vulnerability of the highly visible local while enabling the targeter to withdraw into the invisibility of the global, it speeds war up for the target while slowing it down for the targeter. For most targets the moment of combat is confined to a fraction of a second when they realize, if they do, that an explosive is hurtling at them from the sky at hundreds of miles an hour. They reportedly hear a whooshing noise just before they are dismembered. For the drone operators, this climactic moment of kinetic action may be preceded by hours of lingering surveillance, discussion of targeting options, and careful remote pursuit of the enemy across a variety of contexts and terrains. Because a drone can linger in the sky, unseen, for almost two days, targeting decisions that would have been made in a more jagged, impulsive way in the past can now be slowed down and made more deliberatively. And in war the party that controls the tempo of hostilities has an advantage.

Military Honor, Courage, and Democracy

There is an understandable ambiguity about those who operate drones. Are people pilots if they are not in an aircraft when it is airborne? Are people who are thousands of miles away from the battlefield combatants? Do they deserve combat medals? The U.S. Air Force officially refers to drones as "remotely piloted aircraft," suggesting that those who control them are pilots, albeit unconventional ones, but many others just call these people "operators," and some Air Force pilots deride those who operate drones as the "Chair Force."[39] In February 2013 the Pentagon announced a new Distinguished Warfare Medal, ranked higher than the Bronze Star but lower than the Silver Star, to recognize drone operators for, in Secretary of Defense Leon Panetta's words, "extraordinary achievements that directly impact on combat operations, but that do not involve acts of valor or physical risk that combat entails."[40] Two months later, acceding to fierce opposition from veterans' groups, Panetta's successor, Chuck Hagel, cancelled the medal.[41] In her *New Yorker* article on drones, Jane Mayer, underscoring the ambiguity of the enterprise, puts "flown" in scare quotes and remarks on the incongruity of the fact that drone operators are said to wear flight suits. As for whether they are combatants, she calls them, with unkind irony, "cubicle warriors [who] can drive home to have dinner with their families."[42]

Mayer goes on to discuss critical claims that "unmanned systems, by sparing these combatants from danger and sacrifice, are creating what Sir Brian Burridge, a former British Air Chief Marshal in Iraq, has called 'a virtueless war,' requiring neither courage nor heroism."[43] Journalist Glenn Greenwald says, "Whatever one thinks of the justifiability of drone attacks, it's one of the least 'brave' or courageous modes of warfare ever invented. It's one thing to call it just, but to pretend it's 'brave' is Orwellian in the extreme. Indeed, the whole point of it is to allow large numbers of human beings to be killed without the slightest physical risk to those doing the killing. Killing while sheltering yourself from all risk is the definitional opposite of bravery."[44] Jonathan Schell asks if we can even call something so asymmetrical "war." Discussing the use of air power in general and drones in particular in Libya in 2011, Schell says:

> The balance of forces is so lopsided in favor of the United States that no Americans are dying or are threatened with dying. War is only war, it seems, when Americans are dying—when we die. When only they—the Libyans—die, it is something else for which there is as yet apparently no name. . . . In our day, it is indeed possible for some countries, for the first time in history, to wage war without receiving a scratch in return. . . . The epitome of this new warfare is the Predator drone, which has become an emblem of the Obama administration.[45]

Indeed, the way the Predator and Reaper drones operate is more like hunting than war, since there is no chance that the drone's operator can be killed by his target/adversary. "Armed drones are weapons of assassination, not of war as we know it," says *Washington Post* columnist Eugene Robinson.[46] This is why drone strikes are widely perceived in the Middle East as cowardly, and it is why Nobel Peace Prize winner Jody Williams said, "When you sit in Nevada and kill someone 7,000 miles away, I think there's something unethical or immoral about that."[47] Andrew Exum, a former U.S. Army officer who is now an expert on counterinsurgency at the Center for a New American Security, says, "As a military person, I put myself in the shoes of someone in FATA [Pakistan's Federally Administered Tribal Areas] and there's something about pilotless drones that doesn't strike me as an honorable way of warfare." Invoking the warrior ideals of *The Iliad*, he adds, "As a classics major, I have a classical sense of what it means to be a warrior."[48]

In this regard, the drone is continuous with a long tradition of colonial war-fighting technologies—going back at least to the machine guns with which British and French colonial soldiers mowed down spear-carrying Africans—that ensure that the "natives" die in an unfair fight in considerably larger numbers than the colonial soldiers.[49] In this regard it is important that the drone is not envisaged as a weapon that would be of use in fighting evenly matched adversaries such as the Russians or the Chinese, who would quickly shoot a lumbering Predator out of the sky. It is specifically designed as a way of saving American lives in highly asymmetrical postcolonial counterinsurgency operations—what some in the military refer to as "small wars." The asymmetry of drone operations thus derives not just from the technology itself and the way it absents one party to the combat from the scene of combat, but also from the neocolonial context of combat. In neocolonial counterinsurgency contexts, there is always a massive asymmetry in casualties as outnumbered occupying forces use superior technology to subdue indigenous populations.

In his book *On Suicide Bombing*, anthropologist Talal Asad observes that in the era of drones, American "soldiers need no longer go to war expecting to die, but only to kill. In itself, this destabilizes the conventional understanding of war as an activity in which human dying and killing are exchanged."[50] In a way that recalls the *Iliad*, Asad sees the honorable drama at the core of combat as one in which contending soldiers meet to wager their bodies for a cause. It is the willingness to forfeit one's own life that partly lends meaning to war, and it is the mutual availability of the bodies of combatants on both sides for injury and death that establishes an honorable reciprocity between enemies and affords soldiers opportunities to display the courage that is war's defining virtue. Seen through Asad's lens, the absence of this reciprocity of exposure in drone warfare makes it dishonorable. It also suggests that there is a perverse parallelism, or mirror imaging, between drone warfare and the tactic of suicide bombing adopted by insurgents. Drone operators and suicide bombers, either by absenting their bodies or by preemptively destroying their bodies, deprive their adversaries of the opportunity to capture or kill them, thus undermining the structural reciprocity that conventionally, or at least ideally, defines war. It should thus come as no surprise that the victims of both drone strikes and suicide bombings brand these modes of killing "terrorist" and "cowardly." In contemporary war zones where people on the ground are blown to pieces either by unseen American drones in the sky or unsuspected suicide bombers in their midst, it seems that what we conventionally understand by the word "war" is being torn apart from above and below by American technology and insurgent tactics.

Seen from one perspective, the emerging asymmetry between American casualties and the casualties of the other in the killing fields is a comfortable development. General Charles Dunlap says, "I don't feel better if a lot of Americans die in the effort. I'm ok with all of *them* [i.e., the other side] dying."[51] On the other hand, Jane Mayer quotes a U.S. veteran of the Iraq War who says, "There's something important about putting your own sons and daughters at risk when you choose to wage war as a nation. We risk losing that flesh-and-blood investment if we go too far down this road."[52] While we can understand why political and military leaders would be drawn toward a model of war in which only the other side dies, this veteran's claim is important and bears further exploration.

In a brilliant rereading of the second amendment to the U.S. Constitution and its references to an empowered militia, Elaine Scarry has argued that democratic control over war-making depends on the bodies of citizen-soldiers being placed in jeopardy on the battlefront.[53] She argues that in a democracy, collective consent to war is performed not only at the moment when war is declared by the people's elected representatives but continuously throughout the war as citizen-soldiers agree day in and day out to obey orders that could result in their death. Consent to war is continuous, and it requires the endangering of soldiers' bodies in order for this consent to be meaningful. Pointing to soldiers' mutinies during World War I and in Vietnam, she argues that the ability of soldiers on the frontline to withdraw their consent to place their bodies in danger constitutes a sort of check in a democracy against wars that have become unjust or illegitimate. Or, as Tom Englehardt puts it,

> Think of us as moving from a citizen's army to a roboticized, and finally robot, military—to a military that is a foreign legion in the most basic sense. In other words, we are moving toward an ever greater outsourcing of war to things that cannot protest, cannot vote with their feet (or wings), and for whom there is no "home front" or even home at all. In a sense, we are moving toward a form of war without anyone, citizen or otherwise, in the picture—except those on the ground, enemy and civilian alike, who will die as usual.[54]

Thus, putting Scarry and Englehardt together with other commentators, we can say that there are two ways that drones undermine democratic processes with regard to war. The first is that by absenting the bodies of U.S. citizens from the battlefield and externalizing the casualty cost to the enemy, drones begin to remove the fulcrum of ongoing consent to war. The second issue is that by making war more casualty free, drones diminish the citizenry's interest in the gravest decision an elected government can make and create opportunities for the executive to act alone in ways that expand its freedom of military action while blurring the definitional boundaries of war. In Michael Hastings' words, "The remote-control nature of unmanned missions enables politicians to wage war while claiming we're not at war."[55] Thus the United States is at the moment of writing clearly engaged in a low-level war in Pakistan and Yemen, where its drones are killing people on a regular basis, but this war is undeclared. The Obama administration's military intervention in these countries attracts so little attention and critique in the media and from elected representatives because drone technology enables the intervention to take place without large numbers of boots on the ground and without the steady trickle of flag-draped coffins back to Dover Air Force Base. "We just do it because it's costless to us," says Peter Singer.[56] Some U.S. drone attacks in Yemen have involved the deliberate targeting of American citizens, with no recognizable process of judicial review in which those citizens have been able to contest the evidence against them and challenge their government's decision to summarily execute them from the sky.[57] And in 2011 the Obama administration engaged in what it called "limited military operations" in Libya, largely confining itself to aerial attacks on Ghaddafi's forces, but insisted, against the advice of some of its own senior lawyers, that it did not need congressional approval under the War Powers Act because American troops were unlikely to die and therefore this did not constitute a war.[58] This raises the following question: If U.S. drones were not engaged in warfare in Libya, what were they engaged in? In their public statements about the care and diligence with which targeted individuals' behavior is weighed as drone "kill lists" of "unlawful combatants" are assembled, Obama administration officials implicitly suggest that drones might be engaged in a quasi-judicial process rather than warfare. If so, the subjects of justice are found guilty and punished without any of the protections the U.S. Constitution usually affords the accused.

Advocates of drones, on the other hand, often represent them as tools of war that can comfortably be accommodated to the traditional ethics of war. In a much-cited article, Bradley Jay Strawser has defended drone warfare as not just morally defensible but "ethically obligatory."[59] He argues that in war commanders have a moral obligation to safeguard their own soldiers' lives as much as possible and should thus use drones whenever they can. However, when a conflict becomes so asymmetrical that one side kills while the other dies, we must ask, together with Schell and Asad, whether drones are even engaged in something that can be called war and, therefore, whether the ethics of war offer the best framework for understanding. In her book *The Body in Pain*, Elaine Scarry describes war as an "injuring contest" with the power to settle disputes over meaning through the inscription of pain on the body.[60] She contrasts war with torture, which also involves the inscription of pain on the body, but in a radically asymmetrical context in which there is no contest since one side has the power to inflict pain and the other must endure it. In Scarry's schema, drone violence is structurally more akin to torture than to war, and indeed U.S. officials often say that the purpose of drone strikes is to demoralize and break the will to resist of a target population.

Drones trouble us because their use does not neatly fit schemas of war or justice and they disrupt the balance of power between the executive and other branches of government. Whether seen as instruments of war or the execution of justice, it is clear that they act in ways that slip through the net of checks and balances in the Constitution. They pose a challenge to democratic modes of governance that we ignore at our peril.

Peering over the Horizon

Mary Ellen O'Connell discusses the troubling possibility, which some commentators see as more or less inevitable, that future smart drones will be given full autonomy to select and kill human targets on their own.[61] There are other ways that we seem to be moving toward what Nick Turse calls a "drone-eat-drone world."[62] It would be good to have a full discussion of these possible future scenarios now, while there is still time to avert them, although, as O'Connell notes, the international community is better at limiting technologies once they mature than in banning or restricting them beforehand. As we look to the future, we can expect drone technology to proliferate as countries such as Russia, China, India, and Saudi Arabia start using drones against people who are as much of a nuisance to them as jihadists are to the United States. U.S. police departments are also beginning to acquire drones, though only for surveillance missions at present.[63] At the same time we can expect the range of drones available for use to mushroom, so that the classic lumbering drones in the sky will be joined by backpack drones and even smaller drones.

As drones become cheaper and smaller, we can expect nonstate actors to acquire them. After all, the technology is not very advanced (some of it is available off the shelf), nor is it prohibitively expensive. For $300 anyone can now buy their own drone, controlled from an iPad, from Amazon.com,[64] and in Washington, DC, where I live, the local amateur drone user group already has 448 members.[65] At a 2013 conference on drones, Rosa Brooks pointed out that in a world where drones and the Second Amendment coexist, it would be easy to equip such drones with crude weaponry.[66] And according to Peter Bergen, "It is only a matter of time before well-financed drug cartels acquire [drones]."[67] NGOs (nongovernmental organizations) are also beginning to buy their own drones. For example, the World Wildlife Federation has announced that it will use drones to track rhinoceros poachers in Africa.[68] (This raises the possibility that in the future drones could be armed with guns or lasers to kill or disable poachers who are out of the reach of thinly stretched law enforcement on the ground.) Although at present the American people seem to think of drones as an American monopoly that gives the U.S. military a sort of godlike power to strike the enemy from the heavens, it will surely not be too long before a terrorist organization or a future Adam Lanza (the shooter at the Newtown Elementary School) uses a crude drone to attack American citizens. Then the sense of immunity and invincibility Americans currently derive from drone technology will be transmuted into one of vulnerability as Americans move, at least on occasion, into the subject position Pakistanis and Yemenis occupy with regard to drones. (This will recapitulate the narrative arc of the nuclear arms race, where the U.S. monopoly over nuclear weapons was shattered by the Soviets after only four years and the American sense of nuclear cockiness was replaced by a feeling of gnawing terror and vulnerability.)

The emergence of drones has opened a Pandora's box, the contents of which we are only beginning to explore.

CHAPTER 11

What's Wrong with Drones?
The Battlefield in International Humanitarian Law

KLEM RYAN

Barack Obama was sworn in for his second term as U.S. president on the 20th of January 2013 which, as it happened, was also Martin Luther King Day. The coincidence led a journalist to quip that Obama's speech heralding his second term should have been titled "I have a drone."[1] The remark nicely encapsulates the dramatic escalation in the use of drones by the United States that has occurred within the last few years, an escalation that shows no sign of abating. The practice of waging wars using drones and thereby removing one belligerent's combatants from the "battlefield" is a process Hugh Gusterson aptly describes as the "respatialization" of war.[2] Wars fought with drones are no longer confined to a specific area of conflict; anyone anywhere at any time is potentially a target in a "global battlefield" if they are perceived as a threat to U.S. national security. Senior representatives of the U.S. government have been very candid on this point, asserting that the "battlefield versus non-battlefield distinction is," in their assessment, "growing stale."[3]

Proponents of the United States' drone program argue that the extended reach and accuracy of drones is simply a refinement of technology that represents a commendable advance in aerial warfare practice by making possible more ethical and legally compliant behavior.[4] Critics counter that drones have led to a wave of extra-judicial targeted killings and enabled a relentless expansion of U.S. military operations into new regions that include

Afghanistan, Pakistan, Libya, Yemen, and Somalia. Much of the debate has focused on whether drones are a weapon that can be reconciled with existing international humanitarian law (IHL). While there is currently no consensus on this matter, it seems that in the past two to three years the view of drone proponents—that they are compatible—has been increasingly accepted, particularly in the United States, accompanied by some concerns (murmured sotto voce) about the secrecy of the drone program and the specific details of some high-profile cases.[5] Adding to this impression, UN Special Rapporteur Philip Alston's report on drones (which he drafted in response to concerns about their increasing use), argues that from the perspective of IHL, "a missile fired from a drone is no different from any other commonly used weapon, including a gun fired by a soldier or a helicopter or gunship that fires missiles."[6]

I believe, however, that the problem with drones for IHL lies far deeper than Alston's analysis suggests. Even if we concede that drones resemble other weapons systems—such as helicopters and long-range missiles—these other weapons did not lead to claims, as drones have done, that the battlefield (and, relatedly, conventional war) is now largely obsolete.[7] The respatialization (or perhaps we might even say *de*spatialization) of the battlefield that drone use has facilitated and encouraged, I argue in this chapter, undermines important assumptions of IHL (as it is conceived in the Hague and Geneva Conventions), namely that belligerents mutually occupy a distinct physical space in which war is conducted.

In IHL, the battlefield is a space separate from the civilian sphere where the military representatives of the belligerent parties assume the identities of combatants and the privileges and costs that accompany this identity. Historically, the creation and maintenance of this distinct physical space was essential for making possible rules of restraint under conditions of conflict. These rules both moderate the behavior of combatants toward one another (by, for example, allowing for humane treatment of the wounded and the practice of taking prisoners) and limit the inclusion of civilians in hostilities. Additionally, the creation of a mutually occupied physical space in which war was conducted was viewed as central to expediting the conclusion of the conflict by encouraging a negotiated settlement between the belligerent parties.

Aerial warfare has always strained the mechanisms of IHL for limiting war. The large-scale bombing campaigns of the Second World War revealed the terrible propensity of air war to expand the zone of conflict beyond the comparatively narrow battlefields of earlier wars. This problem did not go unnoticed. Because bombers lacked the accuracy required to reliably distinguish military from civilian targets and their widespread use tended to generate indiscriminate results, bombing campaigns were stigmatized as a weapon of terror, even as their role grew in importance during the war. They were tolerable—in the eyes of the Allied command and its supporters, at least—in the exceptional circumstances of the fight against fascism and Japanese imperialism. In other words, the battlefield as a space distinct from other spaces was still conceptually coherent and morally significant, even if all belligerent parties frequently broke the boundary.

Drones, in comparison to the indiscriminate bombing campaigns of earlier wars, understandably appear to represent an advance in restraint given their comparative accuracy. However, this perspective is misleading. In undermining the premise that a battlefield is a specific place mutually occupied by belligerents, drones represent not a technological advance consistent with existing IHL but rather a decisive break with conventional limited war, even though the precursors of this break were clearly evident in earlier weapons technology. The rapid and seemingly ceaseless expansion of the "global war on terror," targeted killings, and "signature strikes" and the general contempt the United States displayed for the Geneva Conventions with regard to the rights of prisoners of war and the wounded are symptoms of this deeper break with the notion of limited war as envisaged in IHL.

To expand on these concerns I start by providing a brief historical background to the development of IHL. I then turn to the importance of the idea of the battlefield as a specific place that belligerents occupy in common and the relationship of this idea to limited war. Finally, in the last sections of the chapter, I relate these concepts to drones.

The Problem of Formulating Effective Rules to Regulate War

It is often asserted that the modern laws of war derive from the classical just war theory of Christian theologians such as Augustine and Thomas Aquinas, which were subsequently expanded and codified by modern legal theorists and jurists.[8] In fact, it is more accurate to say that IHL evolved as the product of a deep disagreement between two competing visions of war.[9] In the first vision—that of the just war theorists—war results from "the iniquity on the part of the adversary that forces a just war upon the wise man."[10] Accordingly, the rights of war—specifically the permission to kill—accrue only to the just belligerents and not to their unjust opponent. This perspective guided the development of Christian doctrine on war and is echoed in the work of influential early modern legal theorists such as Francisco de Vitoria and Hugo Grotius.

The second vision sees war as a duel. Instead of a "just" party seeking to right a wrong in the face of resistance by an unjust opponent, war is viewed as a struggle between contending political communities whose moral status is determined by their adherence to mutually applicable rules of conduct, not by the perceived justness of their cause. Consequently, all belligerents have combatant rights and the associated duty to respect the rules of war. This view also differs from just war theory in that it considers political conflict as a feature of human social interaction that cannot be eliminated, a situation that requires that communities seek rules to contain the effects of such conflicts.[11] The antecedents of this perspective are found in the writings of the ancient Roman jurists, but its reintroduction and formalization in modern European legal discourse traces back to the writings of the fourteenth-century Italian jurist Raphaël Fulgosius. It was subsequently developed by theorists such as Alberico Gentili, Christian Wolff, and Emerich de Vattel. And it is from Vattel that we get the term "regular war" (*guerre réglée*) to describe this approach to limiting war.[12]

Several significant differences for the laws of war arise from these two competing visions, but the most important in the context of this chapter concerns rules for restraining the conduct of combatants during hostilities. Proponents of both views recognize the importance of limiting the harms of war. Aquinas, writing from a just war perspective, states in *Summa Theologiae* that to use "more than necessary violence" even in a just war is "unlawful," whereas "if he repel force with moderation, his defence will be lawful."[13] However, the attempts by Catholic institutions in the Middle Ages to formulate rules in accordance with Aquinas's injunction were largely unsuccessful because of three crucial weaknesses in the just war approach. First, by denying the legitimacy of one belligerent party, a negotiated settlement of the conflict became inherently more difficult to obtain. "Just" belligerents invariably sought the complete submission of their "unjust" opponents, an aim that precluded compromise and negotiation. Secondly, by resisting their "just" opponents, "unjust" belligerents were deemed to have committed an additional wrong that justified greater levels of force being employed against them. Thirdly, a "just" belligerent could not be prevented from using any effective weapon against "unjust" belligerents, which meant that injunctions against the use of specific weapons applied only to "unjust" belligerents and were therefore largely ignored.[14]

The consequence of these weaknesses was that instead of restraining warfare, just war reasoning had the effect of escalating and prolonging wars. "Just" wars tended to become internecine conflicts "waged not only against States and their armies, but also against their people."[15] The problem, as Vattel observes, was that "since each Nation claims to have justice on its side it will arrogate to itself all the rights of war and claim that its enemy has none.... A rightful determination (a décision du droit) of the controversy will not be advanced thereby, and the contest will become more cruel, more disastrous in its effects, and more difficult of termination."[16]

Recognition of these negative consequences of applying just war reasoning (alongside the emergence of the modern nation-state in Europe) provided the conditions for the ascendance of legal theories founded on the regular war perspective in the seventeenth and eighteenth centuries and the subsequent formalization of these theories into international law during the nineteenth and early twentieth centuries.[17] Whereas just war theories advised the just party to exercise restraint, regular war stressed the mutual interest (and indeed the obligation) of warring parties to restrain conflicts. For, as Francis Lieber states in his 1863 code for the U.S. Army, "Modern wars are not internecine wars, in which the killing of the enemy is the object."[18] Instead, the object of regular war was peace, but specifically a peace achieved while minimizing the evils of war for *all* belligerents to the greatest degree possible. Central to achieving this goal is the maintenance of a relationship between the belligerents that, in the words of the jurist Hersch Lauterpacht, makes "possible a measure of intercourse between enemies during the war and some voluntary relationship after it."[19]

There are two important elements of the regular war approach to limited war that I want to briefly note before I turn to drones: the role of the battlefield in relation to combatant identity and the idea of rule compliance through a reciprocal relationship between belligerents.

The Battlefield and IHL

In regular war, and subsequently IHL, the battlefield is conceived of as a physical space linked to a time and event where particular rules and obligations govern the people present in that location. IHL therefore draws on the basic concept of law as being grounded in and dependent on what Carl Schmitt terms "enclosures in the spatial sense," by which the physical control of territory provides the basis of legal authority that is applicable in a particular time and place.[20] Just as law defines boundaries (the scope and rights of private property ownership or the limits of national borders, for example), so too the laws of war seek to define the physical space in which war is to be conducted and to whom the rules apply. Those who are wearing uniforms or bearing arms on a battlefield are reasonably considered to be active participants in the war (unless they are *hors de combat*). Individuals who do not meet these criteria are not. The intent is to create a barrier between the world of the battlefield and other spaces and to enable actors on the battlefield to recognize the appropriate rules (and risks) to which they are subject.

Early theories of regular war operated under an implicit assumption that there were natural and largely immutable physical limitations on the reach of weapons and therefore on the space of the battlefield.[21] It was so obvious to most eighteenth- and nineteenth-century theorists and jurists that war occurred on battlefields that were clearly distinguishable from other spaces that this did not warrant explicit mention. It was only in the late nineteenth and early twentieth centuries, as new weapon technologies developed, especially the development of aerial warfare, that the need to define the battlefield became evident. For without the clear physical distinction of the battlefield, concepts essential for the effective functioning of the laws (such as the identity of combatants and what constitutes a military target) lose their clarity.

The Hague Conventions of 1907 represent the first major overt attempt in IHL to expressly separate combatant and civilian spaces. Some of the Conventions prohibit combatants from deliberately situating military facilities in civilian areas or targeting civilian residences, and some issue the injunction to respect enemy buildings dedicated to science or charitable purposes.[22] These early efforts to delineate combat and civilian spaces were, however, placed under significant pressure by rapid advances in military technology during the two world wars. As the author of the *Commentary on the IV Geneva Convention* observes, these early regulations were "drawn up at a time when hostilities were confined to the area close to the front."[23]

In the postwar effort to bolster IHL in response to these changes there was a rush to draft new guidelines that would more explicitly delineate the battlefield (referred to in later IHL as the "combat area" or "zone") from other spaces. As a part of this process, the battlefield was for the first time formally defined as "in an armed conflict, that area where the armed forces of the adverse Parties actually engaged in combat, and those directly supporting them, are located."[24]

A range of conventions draw on this definition, such as Article 19 of the third Geneva Convention, which requires that captured enemy combatants be immediately removed from the "combat zone" so that they are placed "out of danger";[25] Article 14 of the fourth Geneva Convention, which allows for the creation of "safety zones" and "neutralized zones," both of which are distinct from "combat zones";[26] and Article 15 of the 1977 Additional Protocols, which requires an occupying power in enemy territory to provide humanitarian support in "territories which are no longer combat zones, but are considered to have regained a measure of stability."[27] These

Articles show that the aim of IHL of separating civilian spaces from the battlefield remained an important element of the project of limiting war, even as the evolving technology and practice of war was making such a clear distinction increasingly difficult to maintain.

Dissociation and the Laws of War

The idea of a shared battlefield fills two important functions for the regular war perspective. While it separates civilians from conflict, it also implies a direct relationship between belligerents that furthers the goal of placing limitations on war. Maintaining a dialogue between belligerents and seeking a relationship based on reciprocity, Vattel recognized, form essential conditions for adherence to the laws of war. For while reciprocity and dialogue between belligerents does not guarantee restraint, the absence of them tends to reduce inhibitions to violence. The collective term I use for the implications of this lack of engagement is "dissociation," which comes in at least three distinct forms, each exacerbated as the distance between belligerents increases: dissociation of agents from their violent acts, dissociation of the targets from the source of the violence directed against them, and dissociation of the public from the violence committed on its behalf.

Dissociation of agents from their violent acts has been clearly demonstrated to have negative implications for compliance with the law, as a 2004 ICRC (International Committee of the Red Cross) report on violations of IHL outlines: "Many studies have shown that people find it difficult to kill their fellow human beings at close range and that special conditioning is needed to overcome this inhibition. Conflicts in which recourse is had to advanced technologies which permit killing at a distance or on the computer screen prevent the activation of neuro-psychological mechanisms which render the act of killing difficult."[28]

Furthermore, as it becomes easier to kill the enemy, there is an attendant shift in the reasoning for killing. "The humanity of the other side is denied by attributing to the enemy contemptible character traits, intentions or behaviour: 'We are superior, they are inferior.' 'We are fighting for an honourable and disinterested cause, they are fighting for inadmissible interests and objectives deserving only condemnation.'"[29]

Studies show that as group norms change and violent practices are repeated, the force used against opponents escalates.[30] What would once have been inconceivable first becomes permissible and then acceptable as "social norms, institutions, and culture change in ways that make further and greater violence easier and more likely."[31]

In his discussion of IHL in *The Nomos of the Earth*, Schmitt makes essentially the same point when he contrasts the conduct of conventional land warfare with naval blockades. Whereas land warfare "introduces the necessity of direct contact between the occupying army and the occupied territory," "a blockading fleet has only a negative relation to enemy territory and its inhabitants, because it considers both the land and the people to be nothing more than the goal of a forceful action and the object of the means of compulsion."[32]

Moreover, argues Schmitt, the increasing dissociation made possible by new weapons inevitably leads to the "remoralization" of warfare. For as soldiers pursue the means to annihilate the enemy with greater precision and decisiveness, the enemy must been seen as deserving of the treatment, lest those that use the weapons are forced to confront the dreadful nature of their actions.[33]

Schmitt also points to the second form of dissociation—that of the targets from the source of the violence directed against them—when he writes about the special problems that arise from air warfare: "Independent air war dissolved the connection between the force applying power and the [enemy] population. . . . With air bombardment, the lack of relation between military personnel in the air and the earth below, as well as with inhabitants thereon, is absolute. . . . Independent air war allows neither the one nor the other side a possibility to establish a relation."[34]

From the perspective of regular war, the absence of a relationship between belligerents is significant because it crucially undermines the development of and adherence to conventions. For soldiers to adhere to rules that limit their behavior, those rules must be perceived by the parties to be reciprocal, not merely in the abstract but in mutually demonstrable practices that contribute to the shared goal of less brutal and intractable conflicts.[35] As Schmitt observes, one-sided air warfare makes such a relationship impossible and therefore undermines the basis for limited war.

This spatial separation also contributes significantly to a third form of dissociation: the dissociation of the public from the violence committed on its behalf. Carl von Clausewitz's famous aphorism—"war is simply the continuation of policy with the addition of other means"—is often quoted without recognizing that its importance lies in the insight that wars begin, are waged, and are finally resolved within a political context.[36] While Clausewitz is commonly thought to have advocated "total war," in fact he understood that such an escalation would be unbearable for political communities. As the force directed against an enemy increases, Clausewitz observed, so too does resistance and the cost of pursuing the war. This led him to believe that the terrible cost of war would naturally inhibit escalation and encourage belligerents to carefully assess the aims of the war and seek a political rather than a military solution.[37] If, however, a political community were to be relatively immune to such costs and thereby in an important respect dissociated from the consequences of war (a situation that Clausewitz did not foresee), then an important inhibitor to war would be removed.

Drones and Dissolution of the Battlefield

Drones are at the forefront of the developments in military strategy that are dissolving the barrier between the battlefield and civilian spaces. These developments have largely been driven by the United States, the global leader in the use of weaponized drones in combat since 2001.

Although to date there has been little detailed public disclosure relating to drone operations, it is evident that drones have been central to the United States' military operations in Pakistan, Libya, Yemen, and Somalia and have been deployed extensively in Iraq and Afghanistan. What has emerged from these operations—based on reports on the strikes themselves, information gleaned from government documents, and interviews with individuals involved in their planning and execution—is that drones have significantly shaped the way the United States defines and targets its enemies. The early stated use of drones (under the second Bush administration) was to target and kill "high value" operatives directly associated with the leadership of al-Qaida. However, it has recently become clear that under the Obama administration, drone attacks have been progressively expanded to include a category of persons that the United States labels "militants."[38] According to numerous reports, many attacks against "militants" in recent years have been so-called signature strikes, in which the identity of those targeted is not known but their actions or characteristics (their signature) suggests that they are "hostile" to the United States.[39] The signature characteristics for targets can be as broad as being a military-aged male (late teens to sixties), carrying a weapon, and being present in a region in which people associated with terrorist or militant activity are believed to be.[40]

These targeting practices are intricately linked with the tactical capabilities of drones, specifically their range, endurance, surveillance capability, and, perhaps most importantly, their lack of risk to the user and the concomitant vulnerability of the target. It is clear that without these combined advantages the concept of a "global battlefield" where enemies can be hunted and killed irrespective of their location would not be tenable. What is also clear is that as the battlefield has expanded, the idea of combatant identity at the core of IHL has degraded. When attacks are not time bound or geographically limited—making enemies perpetual targets of force—individuals can no longer clearly position themselves as active participants or nonparticipants in war. By contrast, IHL envisages a battlefield environment where combatants can clearly signal their association with the conflict by wielding a weapon or by formally surrendering or simply leaving the battlefield and returning to their homes. These options simply are not available to combatants targeted by drones, as it is usually not evident that they are being observed or targeted, and even, if they are somehow aware of being observed or targeted, how their actions are being interpreted.[41]

Furthermore, as the battlefield envisaged by IHL (and regular war more broadly) dissolves, the mindset of the just war perspective reemerges. Where the limited wars of IHL are waged between belligerents of equal standing, drone wars are waged against individual "insurgents," "militants," and "terrorists" who are denied the status of combatants. The expansive definition of militant used by the United States follows naturally from the just war rationale that enemies are morally wrong and individually accountable rather than representatives of a community involved in a political struggle and with whom one can have reciprocal relations. This point was clearly illustrated when unnamed U.S. government officials discussing signature strikes with reporters from the *New York Times* argued that their approach to targeting "is one of simple logic: people in an area of known terrorist activity . . . are probably up to no good."[42]

Illustrating the worry that drones may dissolve the necessary conditions for operative rules in war, a 2011 UK Ministry of Defence (MoD) discussion paper on drones raised concerns about the blurring of battlefield boundaries and its implications for reciprocal behavior between combatants: "Is the *Reaper* operator walking the streets of his home town after a shift a legitimate target as a combatant? Would an attack by a Taliban sympathizer be an act of war under international law or murder under the statutes of the home state?"[43]

In traditional conventional wars, IHL has a clear answer to this question. Article 50 of the Additional Protocols specifies that individual combatants may enter civilian spaces for the purpose of visiting their families so long as they do not alter the civilian character of the population.[44] But the idea that combatants can have this privilege (and civilian populations the associated protection) is fundamentally undermined when no delineation between the battlefield and other spaces is recognized.

WHAT'S WRONG WITH DRONES? 217

Drones and Dissociation in War

As I outlined above, at least three distinct forms of dissociation can arise in war, all of which are evident in the use of drones. I consider each in turn.

Dissociation of Agents from Their Violent Acts

Despite arguing that drones are functionally indistinguishable from other weapons systems (as quoted above), Philip Alston, in the conclusion to the 2010 UNHRC report on targeted killings, outlines the main concern he sees with these weapons: "Because operators are based thousands of miles away from the battlefield, and undertake operations entirely through computer screens and remote audiofeed, there is a risk of developing a 'PlayStation' mentality to killing. States must ensure that training programs for drone operators who have never been subjected to the risks and rigors of battle instill respect for IHL and adequate safeguards for compliance with it."[45]

Charles Dunlap, in his chapter in this volume, takes exception to Alston's suggestion that drone operators are at risk of becoming dissociated from killing, calling this "a serious accusation" that is contradicted by the available evidence.[46] To rebut Alston, Dunlap cites Peter Singer's findings that drone operators have displayed comparatively high (in relation to operational units in Afghanistan) levels of "combat stress . . . fatigue, emotional exhaustion, and burnout."[47] Rather than dissociation, he argues, such findings suggest that drone operators experience higher levels of engagement with combat as a result of the time they spend tracking specific targets and the graphic visual detail of aerial strikes provided by the drone's advanced video feed technology.

When we consider the impact of war on human psychology it is clear that the range of responses observed vary; some may find video warfare disassociating, others intensifying. It would not be surprising, therefore, to learn that drone operators have different psychological responses. With this said, as with any new weapons systems, there is a time of adjustment as militaries work out how best to deploy these systems and what is required of the weapon's operators. As experience with drone warfare increases, the recruitment and training processes for drone operators will undoubtedly become ever more effective at identifying and developing personnel suitable for the role.

There are clear historical precedents for such a process. When submarines were introduced into naval service in the early twentieth century, many mariners were found to be psychologically incapable of working in the demanding, confined conditions. Subsequent advances in submarine technology, psychological profiling, and naval training techniques resulted in an intensive selection process for submariners that became more effective at identifying candidates who were suitable for submarine service and eliminating those who are not.[48] It is likely that as with submariners, drone operators will be found who are capable of operating under the particular conditions required.

Indeed, evidence suggests that such a process is under way. In 2012, the *Air Force Times* published an interview with a trainee operator who responded to the question of how she felt she would be impacted by the images of killing by remote by saying that "she has no worries that she will be traumatized by watching as targets she is tracking get blown up. To her, that just means another U.S. or coalition service member's life has been saved."[49]

Her views mirror those of experienced drone operators, as a 2011 U.S. Air Force psychological assessment survey noted:

> In one surprising finding that challenged some of the survey's initial suppositions, the authors found limited stress related to a unique aspect of the operators' jobs: watching hours of close-up video of people killed in drone strikes. . . . "The going-in assumption was that we were placing these guys under a great amount of stress because of all this video feed," said Col. Kent McDonald, the chief of neuropsychiatry at the school of aerospace medicine and one of the study's two authors. In one-on-one interviews with 85 operators, the authors found that many felt a sense of accomplishment in protecting troops on the ground. . . . [The study found that] *4 percent or less of operators were at high risk of developing post-traumatic stress disorder[,]* . . . [whereas] *the percentage of troops returning from Iraq and Afghanistan at risk of developing post-traumatic stress disorder was 12 to 17 percent.*[50]

What did cause stress for the operators, according to the report, was not combat as much as the "long work hours and frequent shift changes due to staff shortages," issues that the U.S. military is seeking to address through the training of more operators.[51]

It appears, then, that, contrary to Singer's conclusion and Dunlap's argument, killing with drones *is* considerably less stressful for operators than combatants in traditional forms of fighting. Furthermore, such killings are readily rationalized when associated with the idea of protecting one's own combatants from harm. As with all new technologies, we will come to understand their impact on users only over time, but if any currently employed weapon has the potential to induce dissociation in its user, undermining IHL's objective of inducing restraint in combatants, it is the drone.

Dissociation of Targets from the Source of Violence Directed against Them

As Gusterson's chapter in this book vividly illustrates, people are usually unaware they are the targets of drone operations; strikes often occur "out of the blue." Indeed, this is a key element of the effectiveness of drones as a weapons system. It is also the source of much of the hostility directed at drones. Death in war often occurs unpredictably for the target—artillery shells and snipers' bullets usually catch their victims unaware. Nevertheless, these weapons entail some local human presence, and those targeted usually know that they are at some degree of risk in a particular location and are relatively safe in others. The sense of risk under drones is far less specific, and consequently all the more pervasive, as a distant enemy watches, evaluates, and strikes anyone anywhere and at any time.

A 2012 report from the law schools at Stanford and NYU entitled *Living Under Drones* provided one of the first scholarly efforts to establish what life is like for those subject to these conditions.[52] The report recounts the paralysis of uncertainty and fear caused by drones and the marked changes in social practices, particularly around communal events such as funerals, for which people are increasingly reluctant to gather lest a drone operator decide to target the event.[53]

The distance between the operator and the target community precludes dialogue that might mediate the situation and provides no common grounding for mutual understanding of the conflict. To each party (but particularly to the target) the other is an alien, not a person with values, concerns, and perspectives with which one must, or even can, engage. As a teenage witness of a drone strike commented to a researcher, "America is 15,000 kilometers away from us; God knows what they want from us."[54] This is a problem that some in the U.S. government have recognized and expressed concern about. A small group of congressmen observed in a letter to President Obama that drones are "faceless ambassadors that cause civilian deaths, and are frequently the only direct contact with Americans that targeted communities have."[55]

In this sense, drones represent Schmitt's concern about the spatial separation of military personnel from their targets in its purest form and the attendant problems for developing mutually applicable laws of war (noted above). When belligerents are thousands of miles apart, the relationship required for reciprocal rules to limit war simply does not exist. We should not be surprised, then, if the use of drones tends to produce hostility and continued violence, as opposed to restraint, engagement, concession, and peace.

Dissociation of the Public from the Violence Committed on Its Behalf

An article in the *New York Times* in 2011 outlined a third concern about drones and the United States' expanding list of wars: "Drones raise questions about the growing disconnect between the American public and its wars. Military ethicists concede that drones can turn war into a video game, inflict civilian casualties and, with no Americans directly at risk, more easily draw the United States into conflicts."[56]

Similarly, the UK MoD report on drones asks, "If we remove the risk of loss from the decision-makers' calculations when considering crisis management options, do we make the use of armed force more attractive?"[57] Interestingly, this question was driven by the recognition of the report's authors of the dissonance between the Clausewitzian conception of war as a naturally limited activity and the contemporary reality of war fought by remote control. Thus they observe that, in Clausewitz's account, "one of the contributory factors in controlling and limiting aggressive policy is the risk to one's own forces. It is essential that, before unmanned systems become ubiquitous (if it is not already too late) we consider this issue and ensure that, by removing some of the horror, or at least keeping it at a distance, that we do not risk losing our controlling humanity and make war more likely."[58]

In other words, one of the most significant concerns about drones, in the MoD's view, is that they effectively sever the (already tenuous) link between force and resistance that historically offered one of the most important mechanisms for limiting war.

The MoD's discussion oscillates between different forms of dissociation, both individual and social, and it appears that both concern the report's authors. But it is with regard to the dissociation of the public from the violence of war facilitated by drones that the authors make their most interesting observation: "The recent extensive use of unmanned aircraft over Pakistan and Yemen may already herald a new era. That these activities are exclusively carried out by unmanned aircraft, even though very capable manned aircraft are available, and that the use of ground troops in harm's way has been avoided, *suggests that the use of force is totally a function of the existence of an unmanned capability—it is unlikely a similar scale of force would be used if this capability were not available.*"[59]

Subsequent events have added weight to this conclusion. In June 2011, President Obama faced criticism for authorizing the United States' participation in the NATO air campaign on Libya in contravention of the U.S. War Powers Act (1973). The act requires the president to seek congressional approval for any use of U.S. military force that lasts longer than sixty days.[60]

Obama's response to this criticism is instructive: He argued that the use of drones in Libya (supplemented by missiles fired from U.S. naval forces in the Mediterranean) did not contravene the act, because as there were "no troops on the ground" and "Libyan forces [were] unable to fire at them meaningfully," the intervention did not meet the definition of "military force."[61] It is clear from this argument why Dennis Blair, former director of national intelligence in the Obama administration, labeled drones as "dangerously seductive."[62]

Even if it is strategically employed to circumvent domestic legal constraints, the argument that the use of drones does not constitute military force as commonly understood has important implications. It clearly illustrates that, contrary to those who argue that drones are merely "just another weapon," the unique tactical capabilities of drones transcend conventional political limitations on war. Moreover, this situation is compounded by the fact that public resistance to the employment of force is also increasingly negated by the operational capabilities of drones, as a 2012 Columbia Law School report notes: "The precision capabilities of the technology . . . provide seeming assurance that as the U.S. expands drone strikes to occur more frequently and in more regions, the strikes are nevertheless carefully limited."[63]

This expanded use of drones has correlated with the widespread reemergence of classical just war concepts of war in public and academic discourse. Only a few paragraphs after the *New York Times* notes a "growing disconnect" between the U.S. public and the wars fought on its behalf that risks "more easily draw[ing] the U.S. into conflicts," the reader is told how these concerns can be set aside:

> "There's a kind of nostalgia for the way wars used to be," said Deane-Peter Baker, an ethics professor at the United States Naval Academy, referring to noble notions of knight-on-knight conflict. Drones are part of a post-heroic age, he said, and in his view it is not always a problem if they lower the threshold for war. "It is a bad thing if we didn't have a just cause in the first place," Mr. Baker said. "But if we did have a just cause, we should celebrate anything that allows us to pursue that just cause."[64]

This is not an isolated opinion; just war arguments are becoming increasingly common in normative theories of war.[65] Bradley Strawser, in one of the most cited articles in the *Journal of Military Ethics*, titled "Moral Predators: The Duty to Employ Uninhabited Aerial Vehicles," presents a clear neoclassical just war argument in support of drones: "If an agent is pursuing a morally justified yet inherently risky action, then there is a moral imperative to protect this agent if it [is] possible to do so. . . . As a technology that better protects (presumably) justified warriors, UAV [drones] use is ethically obligatory, not suspicious."[66]

In other words, even if drones are inconsistent with the underlying concepts of IHL and cause deep resentment and hostility from the populations against which they are deployed, so long as their targets are declared "unjust," qualms about deploying drones are readily assuaged. The risk that this growing disassociation of the public from the costs and consequences of wars fought on its behalf will result in more "just" wars is clear.

Conclusion

Max Huber, then president of the ICRC, observed in 1955 that "if there is continued exploitation of technological developments for military purposes, as has unfortunately been the case since 1907, the existence and the value of what still remains of the laws of war and even of the law of nations in general, will become problematical."[67]

When we talk of violations of IHL it is usually to criticize those who ignore conventions, use disproportionate force, or use inhumane weapons. The problem for IHL posed by weapons such as drones is different. The argument I have advanced in this chapter is that the unique capabilities of drones undermine IHL in a less obvious but more fundamental way. In effectively eliminating the practical distinction between the battlefield and other spaces, drones collapse the key barrier upon which the concepts of combatant identity and distinction rely for their efficacy. Furthermore, the propensity of drones to encourage various forms of dissociation that inhibit the formation of a relationship between belligerents removes important mechanisms for restraining aggression.

Of course, drone proponents will respond that the breakdown of the barrier between the battlefield and other spaces is not logically entailed by drones; nothing precludes drones being used in conformity with IHL in conventional war. This truism is relevant only if we ignore reality in preference for philosophical abstraction. The transformation toward robotic warfare is still in its comparative infancy, but it is gathering speed. It is already evident that the practice of warfare has changed (at least as far as countries such as the United States is concerned), making the existing conventions for limiting war increasingly untenable.

The argument I have outlined expands the notion of what constitutes noncompliance with IHL. New weapons have largely been deemed independent from normative critique; so long as they do not violate the requirement of discrimination or otherwise offend humanitarian sensibilities, no wrong is thought to have occurred. Indeed, not only have conventional weapons largely been excluded from moral critique (at least since the Second World War), the normative debate has never shed the long-held hope of some—as Tammy Biddle documents in her chapter in this volume—that new advances would one day eliminate war itself.[68] But new weapons have not ended war; they have only created new ways of prosecuting them. Drones herald a transformation in how wars are fought, one that risks rendering IHL impotent to impose effective restraints on the conduct of future conflicts.

CHAPTER 12

Banning Autonomous Killing

The Legal and Ethical Requirement That Humans Make Near-Time Lethal Decisions

MARY ELLEN O'CONNELL

Long before the computerization of weapons technology, humanity debated the normative acceptability of new weapons.[1] The invention of the long bow, gunpowder, airplanes, weapons of mass destruction, and so on have all raised moral and legal concerns.[2] Unmanned aerial combat vehicles, or drones,[3] became the focus of debate when the United States used a drone to launch a missile attack that killed several people in November 2001 in Afghanistan.[4] It was the first known use of a drone, operated from a great distance, to kill. As the debate over drones grew, another debate, on the legality and morality of autonomous weapons, intensified.[5] In certain respects, autonomous weapons are as old as any weapon if they are defined as weapons that may be triggered by a target rather than by the user of the weapon. A camouflaged pit, a spring gun, a land or maritime mine, or an improvised explosive device can be triggered by the target without the user being in the vicinity. Such weapons have long been the focus of philosophers and legal scholars. The advent of robots with computer programs that can learn has renewed the debate and deepened the concerns. Advances in artificial intelligence mean that once a robot is constructed and programmed, it will be able to make the decision to attack without additional human intervention.[6] Such an attack could occur at great distance from the time and place of the robot's origin.

In response to these developments, discussion is building toward a norm against the use of fully autonomous robotic decisions to deploy lethal force.

This developing norm is reflected in documents of such diverse origin as the United States Department of Defense[7] and Human Rights Watch.[8] While consensus is building toward the norm, consensus does not yet exist about how to move beyond establishing the norm to winning global acceptance of it. While it is possible to find support in existing law for such a norm, the legal case is based on inference and analogy. Such support may be sufficient to gain wide acceptance, but there is really no denying that the norm would be strengthened by preliminary discussion and eventual negotiation of an affirmative treaty ban on fully autonomous killing. The discussion beginning in the context of a new Protocol to the Convention on Certain Conventional Weapons is a promising start.[9]

In April 2013, United Nations Special Rapporteur Christof Heyns called for a moratorium on moving beyond the design stage in the development of fully autonomous weapons pending the formation of a panel of experts to "articulate a policy for the international community on the issue."[10] Wendell Wallach, chair of Yale's technology and ethics study group, also called for a moratorium in 2013 and for the U.S. president to "sign an executive order declaring that a deliberate attack with lethal and nonlethal force by fully autonomous weaponry violates the Law of War."[11] This chapter proposes a treaty that will say much the same as Wallach's executive order. The negotiation of such a treaty will require the expert input and global debate needed to develop a strong express rule requiring that human beings make any decision for the near-time application of offensive lethal force to other human beings and property. Even without the successful conclusion of a treaty, the negotiation alone could go a long way toward creating a principle of customary international law that bans autonomous killing.

The research for this chapter began in 2010. It is likely that as soon as the chapter is published some of the information contained in it will be inaccurate, given the pace of technological development. Nevertheless, the best time to consider the law and morality applicable to new technology is, arguably, before it is fully operational. Once technology is in use, restrictions tend to be more difficult to obtain.[12] On the other hand, regulating technology still under development poses obvious problems. Every attempt will be made here to focus on the most likely future scenarios. The proposal made in the final section, titled "Norm Building," is designed to apply to certain existing weapons systems, as well as future ones. Thus, the proposal should be relevant regardless of what scientists invent next.

The remainder of this chapter is divided into three parts: It will begin with a brief overview of what we know at the time of writing about autonomous weapons. The discussion will move on to the law governing killing, including the international law regulating lethal weapons. The final part will introduce a proposal for a treaty ban on removing humans too far from any decision to kill. The precise details of such a ban will require international negotiation; this chapter will focus on why such negotiation is imperative.

Autonomous Weapons Technology

In 2010, the scientific community did not yet have a consensus definition of what constitutes a fully autonomous weapons system.[13] By mid-2013, a common definition had emerged. In November 2012 the United States Department of Defense (DoD) issued a directive titled "Autonomy in Weapons Systems." The purpose of the directive is to establish "DoD policy and assigns [sic] responsibilities for the development and use of autonomous and semi-autonomous functions in weapon systems." The directive defines an "autonomous weapons system" as a weapons system that "once activated, can select and engage targets without further interventions by a human operator. This includes human-supervised autonomous weapon systems that are designed to allow human operators to override operation of the weapon system, but can select and engage targets without further human input after activation."[14]

UN Special Rapporteur Christof Heyns refers to "lethal autonomous robotics" as weapons systems "that, once activated, can select and engage targets without further human intervention."[15] Military ethicist Deonna Neal uses a similar definition of a fully autonomous weapon. She calls it a robot "which uses some form of artificial intelligence to guide its decision-making and that is capable of target discrimination and regulating its use of force independently of human 'eyes on target' verification or authorization before it kills someone."[16] The Human Rights Watch definition is also similar: "If a weapon were fully autonomous, it would 'identify targets and . . . trigger itself.'"[17]

These definitions exclude many autonomous weapon systems by specifying that the weapon have the ability to "select and engage targets." The general view is that states do not yet possess fully autonomous weapons systems but are definitely seeking them. Although a few commentators continue to raise doubts about whether scientists can or will develop fully autonomous systems, the weight of opinion indicates it is only a matter of time. How much time is disputed. Current estimates predict that fully autonomous weapons will emerged as early as 2015 or as late as 2050.[18] At a meeting of the International Society of Military Ethics in January 2011, Neal cautioned that "there is disagreement among the engineering community as to whether an autonomous robot . . . can actually be created."[19] Yet a few weeks later at another meeting, a former U.S. Air Force research scientist expressed his confidence that there is "nothing holding us back. The technology is a 'slam-dunk.'"[20] He argued that the United States is not yet fully automating drones because of "cultural resistance," not because of technological hurdles.[21]

Despite any "cultural resistance," there is no doubt that scientists are hard at work on the relevant technology for fully autonomous systems.[22] The 2012 DoD directive shows the level of U.S. involvement in planning for autonomous weapons systems. In the United States, the Committee on Autonomous Vehicles in Support of Naval Operations wrote in 2005 that "the Navy and Marine Corps should aggressively exploit the considerable warfighting benefits offered by autonomous vehicles (AVs) by acquiring operational experience with current systems and using lessons learned from that experience to develop future AV technologies, operational requirements, and systems concepts."[23]

At a press briefing in 2007, a spokesperson for the U.S. Department of Defense's Unmanned Aerial Systems Task Force, Dyke Weatherington, spoke of the need to go beyond the remotely controlled technology in existence today that requires human intervention. For example, in "air-to-air combat—there's really no way that a system that's remotely controlled can effectively operate in an offensive or defensive air combat environment. That [requires] . . . a fully autonomous system."[24] Finally, a Human Rights Watch report released just a few days before the DoD directive was released in November 2012 states, "Some military and robotics experts have predicted that 'killer robots'—fully autonomous weapons that could select and engage targets without human intervention—could be developed within 20 to 30 years."[25]

The breakthrough to fully autonomous weapons has either already happened or will occur in the foreseeable future. As will be argued in the section on norm building, the time to clarify the applicable legal and moral principles is now.[26] We already have a variety of semi-autonomous and passive autonomous weapons such as land mines:

> Indeed, several military robotic-automation systems already operate at the level where the human is still in charge and responsible for the deployment of lethal force, but not in a directly supervisory manner. Examples include: (i) the Phalanx system for Aegis-class cruisers in the Navy "capable of autonomously performing its own search, detect, evaluation track, engage and kill assessment functions" . . .; (ii) the MK-60 encapsulated torpedo (CAPTOR) sea mine system—one of the Navy's primary antisubmarine weapons capable of autonomously

firing a torpedo and cruise missiles . . .; (iii) the Patriot anti-aircraft missile batteries; (iv) "fire and forget" missile systems generally; and (v) anti-personnel mines or alternatively other more discriminating classes of mines (e.g. anti-tank). These devices can each be considered to be robotic by some definitions, as they all are capable of sensing their environment and actuating, in these cases through the application of lethal force.[27]

Efforts to build norms against fully automated killing should begin before the technology is in wide use. A recent UK Ministry of Defense report stated of unmanned aerial vehicles: "Most of the legal issues surrounding the use of existing and planned systems are well understood and are simply a variation of those associated with manned systems."[28] The same could be said of fully automated systems. Yet it is the thesis of this chapter that humanity has not yet taken into account the impact of increasing physical and temporal distance on our legal and moral principles on killing. Development of the capacity to deploy robotic target selection should be accompanied by both an audit of applicable legal and ethical norms and affirmative action toward a central, treaty-based principle to restrict the removal of human beings from the offensive near-time kill decision.

Several arguments already exist against drafting a new rule. One posits that computers will be able to make target selections better than human beings.[29] Another is offered by Schmitt and Thurnher who argue that by definition human beings will not be taken out of actual kill decisions because human beings will build and program robotic weapon systems. Thus, no new rule is needed to ensure that humans are in the loop. They admit that human input could occur long before a robot resorts to lethal force but then they fail to grapple with the implications of this fact.[30] Temporal distance from the kill decision is the critical issue. Most would agree that a computer programmed to kill months or years before an actual operation no longer has meaningful human involvement in the deployment of lethal force. The current law and system of accountability for targeting decisions is built around human involvement, as Schmitt and Thurnher acknowledge.[31] Yet, with no additional evidence, they assert that the system will continue to work adequately even when humans are far removed from the kill decision. It would be irresponsible at best to base legal and moral standards for killing on such an unproven assertion that defies common sense.

Jakob Kellenberger, former president of the International Committee of the Red Cross, has said: "The deployment of such [fully autonomous] systems would reflect a paradigm shift and a major qualitative change in the conduct of hostilities. . . . It would also raise a range of fundamental legal, ethical and societal issues, which need to be considered before such systems are developed or deployed."[32] Kellenberger may or may not be correct that the deployment of fully autonomous systems will be a paradigm shift over semi-autonomous ones. That issue is not as important, however, as his second point about the fundamental legal and ethical principles we have established over centuries. These principles are premised on a closer association between the decision to use a weapon and the death or destruction resulting from that use than is often the case with today's weapon systems. Without legal intervention, scientists may continue to develop robots that take the kill decision ever farther from the human beings who should bear responsibility for making it.

Lawful and Ethical Killing

International law prohibits the resort to lethal force except in limited circumstances.[33] In peacetime, international law permits police forces and other government authorities acting under police rules to use lethal force to save lives immediately. No innocent bystanders may be killed in such operations. Governments may resort to military force when challenged by organized armed insurgents on their territory or when attacked with significant force from abroad. The United Nations Security Council may also authorize the use of force by states to restore "international peace and security."

A state's response to a significant attack from beyond its territory is regulated under the international law of self-defense. The law of self-defense is comprised of Article 51 of the United Nations Charter and additional rules found in the law of state responsibility and the general principles of law.[34] The International Court of Justice has identified these additional rules in a number of important decisions, starting with the 1948 *Corfu Channel* case between the United Kingdom and Albania.[35] Military force may be exercised by a state that is the victim of a significant armed attack on the territory of a state responsible for the attack. The exercise of such force must also conform to the general principles of necessity and proportionality.[36] States may, of course, take defensive action in other circumstances, but such action may not include major military force on another state's territory. States may also resort to force when challenged by an organized armed group within the state that is attempting to overthrow the government or to secede. States appear to tolerate outside intervention by states assisting a government in ending insurgent or secessionist military challenges.

Fighting, whether for internal control, in self-defense, or as authorized by the Security Council, amounts to armed conflict if organized armed groups fight each other with a certain amount of intensity. Missiles and bombs, regardless of how they are deployed, are lawful for use only within the actual fighting of an armed conflict. In other words, missiles and bombs are permissible for use in armed-conflict hostilities only. If a police force were to use bombs or missiles, it would generally be resorting to excessive force. Beyond armed-conflict hostilities, authorities may use only that force necessary to save a human life immediately. Bombs and missiles risk killing bystanders, which means that the current generation of drones deploys too much firepower for lawful use outside of actual armed conflict or in response to a significant armed attack by a state that triggers the right of self-defense under Article 51 of the UN Charter.

This law governing resort to military force is a subfield of international law still referred to as the *jus ad bellum*. The law was reconfirmed by a consensus of all UN members at the 2005 World Summit in New York.[37] The rules on how military force may be used during an armed conflict are found in the Hague Conventions, the Geneva Conventions of 1949 and their Additional Protocols of 1977, customary international law, and, again, general principles (collectively the *jus in bello*.) The rules on conduct of force are the subject of regular review and comment by the International Committee of the Red Cross (ICRC). Also in 2005, the ICRC published a comprehensive review of customary international humanitarian law (IHL) for the two types of armed conflicts for which there are well-developed sets of rules: international armed conflict and non-international armed conflict.[38] It is important to also emphasize that certain human rights principles apply even during an armed conflict.[39] The European Court of Human Rights, the Inter-American Court of Human Rights, and the Inter-American Commission have investigated whether governments have used excessive force and have thus violated the right to life.[40]

These contemporary rules on the resort to and conduct of lethal force, whether in peace or armed conflict, developed most directly from the Just War Theory of Augustine and Aquinas. Augustine drew on Aristotle and Cicero for the concept that peace is the normal state and that violence is justified only to restore peace. Moral philosophers continue to teach that the taking of human life may be justified to protect human life.[41] In other words, the exceptional right to resort to lethal force rests squarely on a justification of necessity. Current law reflects the understanding of what necessity permits as a moral and ethical matter.

Within the *in bello* context, it is becoming increasingly apparent that necessity is the most important guide to regulating the use of force, given the rise of irregular forces. Necessity determines what level of force may be used—military force where enemy combatants may be killed without warning or police-level force necessary to save a human life. Thus, other battlefield targeting rules—proportionality, distinction, precaution, and humanity—are relevant after a necessity decision is made. In the mid-2000s, the ICRC sought to broaden the category of persons subject to intentional targeting because of their status as persons in a "continuous combat function." The ICRC insists, however, that such persons may be targeted only when it is necessary to do so. The standard of necessity depends on the circumstances, whether the situation constitutes armed conflict or not. In the words of the ICRC's Interpretative Guidance on the Notion of Direct Participation in Hostilities under International Humanitarian Law:

> In classic large-scale confrontations between well-equipped and organized armed forces or groups, the principles of military necessity and of humanity are unlikely to restrict the use of force against legitimate military targets beyond what is already required by specific provisions of IHL. The practical importance of their restraining function will increase with the ability of a party to the conflict to control the circumstances and area in which its military operations are conducted, may become decisive where armed forces operate against selected individuals in situations comparable to peacetime policing. In practice, such considerations are likely to become particularly relevant where a party to the conflict exercises effective territorial control, most notably in occupied territories and non-international armed conflicts.[42]

This greater emphasis on necessity is, in effect, a new restriction on the use of lethal force under the *in bello* rules. It is consistent with the overriding obligation to respect human rights, which compels that any close case be decided in favor of peacetime standards for the resort to lethal force.

The law in mid-2013 reflects an ever-greater restriction on resort to lethal force, whether by state against state in self-defense or during an armed conflict by combatants against each other. Still, the lawful use of lethal force requires an exercise of conscience. Even where a president or soldier has the legal right to kill, the decision to do so will ultimately be an act of moral judgment. Moreover, legal scholars know that in rare circumstances, individual conscience may compel action in defiance of law. This chapter argues that the ultimate decision to kill must be made, therefore, by a human being at or very near the time of the lethal impact. Even if scientists develop a computer that can replicate the human conscience, the decision must not be given up to a machine that cannot be held accountable.

However, John Aquilla, executive director of the Information Operations Center at the Naval Post Graduate School, has said, "I will stand my artificial intelligence against your human any day of the week and tell you that my A.I.

will pay more attention to the rules of engagement and create fewer ethical lapses than a human force."[43] Similarly, Ronald Arkin, author of "Governing Lethal Behavior in Autonomous Robots," a study funded by the Army Research Office, believes computer software can be developed to incorporate proportionality, recognition of surrender, uncertainty, and other fundamental concepts to lawful conduct on the battlefield.[44] Anderson and Waxman have also argued that robots will make better, more accurate decisions in life-and-death matters, whether the context is health care or war-fighting.[45]

Others insist that scientists will not be capable of designing a computer sophisticated enough to make lethal force judgments reflecting the principle of necessity. They argue that these judgments will always be subjective decisions that only a human being, not a computer, can make.[46] A similar argument was made when computer engineers first began to predict that a computer could defeat a human being at chess. The counterargument was once that a computer would not be able to make the subjective decisions necessary to outwit a human adversary.[47] Today it appears well within the realm of the possible that computers will be programmed to be capable of doing what experienced battlefield lawyers currently do.

What seems unprogrammable is conscience, common sense, intuition, and other essential human qualities. Accountability is another challenge that seems impossible to overcome in the case of autonomous killing.[48] Current systems for holding individuals accountable for killing require a certain mens rea (mental intention), something a computer does not have. Without accountability, the importance of norms about the use of force would likely diminish.

It is already proving too easy to kill with robots.[49] Giving up the decision entirely to a computer program will truly lower the barrier and remove, literally, the humanity that should come to bear in all cases of justifiable killing. From the perspective of law, morality, and strategy, it seems essential that a human being who has training and a conscience and who may be held accountable should always make the awesome, ultimate decision to kill.

Norm Building

Even if consensus is reached that the decision to kill must be made by a human being, the question arises of how such a norm is to be created. Given the history of technological development respecting armaments, it seems unlikely that we can prevent the wide availability of fully autonomous weapons. The more promising approach is indicated by the strategies used to create legal control of weapons that we already have. These include outright bans on certain types of weapons, restrictions on how and where certain weapons may be used, and limits on who may use certain weapons. Practitioners and scholars are aware of the need to engage the challenges autonomous systems pose to current law on weapons.[50] Much can be found in existing law to guide the development and use of autonomous systems. Nevertheless, the essential, core norm will be a new one: a ban on removing humans too far from the "kill chain."

In international law the most common way new norms are built is through multilateral treaty negotiation. Think only of the new norms that emerged in the course of the law of the sea or International Criminal Court negotiations. This form of norm development is also evident in the area of arms control. Some of the first multinational treaties concerned weapons bans, including the dum-dum bullet and asphyxiating gases. Some of the most recent successful treaties have also concerned weapons, including the Rome Statute of the International Criminal Court, the 1997 Ottawa Convention banning landmines, and the 2008 Dublin Convention banning cluster munitions.

On the other hand, the ICRC has employed a different approach to law development in recent decades, finding and compiling in written form principles of customary international law. In 2005, as mentioned above, the ICRC published its study of customary IHL rules; in 2009, it published its Interpretative Guidance on the Notion of Direct Participation in Hostilities. The Guidance includes the heightened necessity standard discussed in the section above on lawful and ethical killing. Developing a new express norm against autonomous killing is probably best achieved drawing on all categories of international legal norms: treaties and customary international law, as just described, but also general principles, and *jus cogens*.

International arms control agreements such as the conventions on land mines or cluster munitions indicate some of the difficulties that lie ahead in banning fully autonomous weapons. The requirement that a human being make the kill decision within a certain time of the killing is a novel form of arms control. Moreover, the technology of autonomous killing is under development, leaving the design of rules a matter of prediction. Richard Jackson, a civilian Pentagon lawyer, has asserted that it is not possible to develop a treaty on a technology that is not yet in use.[51] In fact, international law does prohibit certain future technologies. The ban on blinding laser weapons was developed before the technology came into use.[52] There are analogous bans, such as the ban on all forms of human cloning, despite the fact that the technology for human cloning did not exist when the ban was adopted.[53] Both of these bans rest on fundamental moral and ethical views of the technology.

Despite the challenge of regulating a developing weapons technology, there are advantages to acting before the technology is widely available. It may be possible, as in the case of blinding laser weapons, to get agreement in part because states do not yet have the technology. Additionally, the norm can be developed from existing principles used to regulate existing near-autonomous weapons systems. An incremental step in law development can enhance the perceived legitimacy of the new rule. Building on current rules overcomes the problem of regulating a hypothetical and may have a positive impact in developing new, more appropriate ethical/legal norms beyond the case of future robotic weapons. For example, land mines are a type of automatic weapon that operates by detonating under certain conditions rather than when a human being presses a button or pulls a trigger. A ban on fully autonomous killing could reinforce the existing legal and moral prohibition on land mines, even leading states not party to the land mines treaty to consider themselves bound. A ban on autonomous killing may also lead us to revisit other weapons technology, such as the intercontinental ballistic missile (ICBM), which apparently has no failsafe to interrupt it after launch. A norm against autonomous killing might necessitate retrofitting ICBMs and similar weapons. Currently a human being makes the decision to launch an ICBM, but after that the missile locks on its target, making it impossible to abort the strike. A norm against autonomous killing could drive the development of new technology that would allow a human being to abort an attack prior to the moment of impact that is now many minutes after the decision to launch.

Developing technologies do make the precise outlines of a new legal norm difficult to describe. This is where multilateral negotiation is helpful. Such negotiation proceeds after the development of position papers. As discussed in the section on autonomous weapons technology at the beginning of this chapter, consensus already exists that a human being should make the decision to deploy a lethal weapon. To keep a human conscience in any decision to kill, the temporal distance between the deployment of force and the lethal impact should be close. The nature of the target might indicate how much time should be required—more for a single enemy combatant, less for a squadron of attack drones. This concept is based on offensive weapon use. Defensive weapons, such as the Israeli Iron Dome, would arguably be permitted to have greater automaticity. Even then, however, the defensive system should operate on the state's own territory or on the high seas. Plainly, these sorts of issues need expert input and thoughtful discussion.

Florini argues that for preferred conduct to become a norm there must be three simultaneously favorable conditions: 1) initial prominence; 2) coherence;

and 3) and advantageous environment.⁵⁴ "Initial prominence" means that "someone is actively promoting the norm, or . . . the state where the norm first arose happens to be particularly conspicuous."⁵⁵ Drones have been a hot topic since the Bush administration began to use them regularly to kill in Pakistan, and the issue heated up when President Obama doubled, then quadrupled, his predecessor's efforts. In May 2010, the UN special rapporteur on extrajudicial, summary, or arbitrary executions discussed the use of drones in his study of targeted killing.⁵⁶ The discussion intensified in mid-2011, when the Obama administration tried to claim that "merely" deploying drones in the Libyan civil war did not involve the United States in a situation of armed-conflict hostilities. The debate escalated when the CIA used drone-launched missiles to kill two Americans and three Yemenis in Yemen on September 30, 2011, far from any hostilities involving the United States.

The type of prominence achieved respecting targeted killing with drones does not seem to be the precise condition to which Florini is pointing. She is indicating that the discussion should be about the norm, which, as identified here, would be a ban on autonomous robotic kill decisions. This norm has now been taken up by Human Rights Watch and other non-governmental organizations (NGOs). The UN special rapporteur on extrajudicial killing produced his first report on fully autonomous weapons in April 2013. Parties to the Convention on Certain Conventional Weapons are now studying autonomous weapons. In the space of three years, from 2010 to 2013, the situation changed quickly. The increasingly negative public opinion respecting combat drones is raising concerns about ever more distance and automaticity in killing.⁵⁷ Should a charismatic figure or group take up the cause of banning autonomous killing, the move toward a ban could become unstoppable.⁵⁸

Florini's reference to "coherence" relates to the new norm's legitimacy; that is, its ability to "fit coherently with other existing norms."⁵⁹ In other words, "emerging norms must make the case that they are logical extensions of that law—or necessary changes to it."⁶⁰ This condition does appear to be met in the case of a ban on fully autonomous kill decisions. The norm is the natural outgrowth or next step in the venerable and dynamic body of conventional and customary rules on arms control as well as in the *jus ad bellum, jus in bello*, and human rights law.

"Advantageous environment" refers to external environmental conditions confronting the new norm. If these are conducive to the development of the new norm, success is more likely. Conditions do not seem advantageous or advantageous enough as of mid-2013. States are busy acquiring drones and developing robots. Most indicate little interest in regulation beyond acknowledging that much current law already applies. Moreover, commercial interests are turning to drones and robots in the hope of continuing to win defense contracts at a time of tight government budgets. Manufacturers can argue that robots are a bargain in many respects and will deliver real military advantage to a state that is the first to have fully autonomous weapons systems.

The land mine ban developed as a result of multiple factors: media attention to the terrible injuries suffered by civilians as a result of land mines; charismatic figures, such as Princess Diana, taking up the cause; and ethical thinking of long standing about acceptable weapons. A grassroots movement ignited, and the Ottawa Convention was quickly adopted and has been widely ratified.[61] A ban on fully autonomous robotic weapons could develop along a similar trajectory. The ban would focus on the type of computer program loaded on a robotic weapon. This might seem like a quite intangible type of thing to ban with a treaty, certainly when compared with the land mine ban. Yet we humans are becoming very familiar with computer technology and should be able to address this challenge, just as people a generation earlier worked to end nuclear proliferation, chemical weapons, and biological weapons.

Conclusion

"Thou shalt not kill" is one of the most widely known and understood commandments. Humanity through its law has carved out exceptions to it, but those have been narrow: for self-defense and to respond collectively to threats to the peace. Resort to weapons has always been accompanied with some legal and moral restraint, including the complete ban on certain types of weapons. In the near future robotic weapons are expected to be available with programs able to select and destroy targets without a human operator in the loop. Such a development would conflict with the historical, legal, and moral understanding that killing should be based on a good-faith understanding of real necessity and carried out by someone who may be held accountable for a wrong decision. Even if a computer could be so programmed, it is imperative that human beings not give up sovereignty over these vital aspects of what it is to be human: to have a conscience and to be subject to accountability. Too much of our current system of community and personhood are based on these two factors to risk their elimination. This point is all the stronger when we realize the risk is being promoted for the sake of creating new means of killing. Before the weapons become widely available, a treaty should be negotiated to ban fully autonomous killing.

NOTES

Introduction

1. Dexter Filkins, "NATO Strike Cited in Afghan Civilian Deaths," *New York Times*, 15 August 2010.

2. See the chapter by Sahr Conway-Lanz in this volume. For a masterful study of the World War II air campaigns, from the perspective of civilian victims as well as military and political officials, see Richard Overy, *The Bombing War: Europe 1939–1945* (London: Allen Lane, 2013). Another excellent work focusing on the targeting of civilians is Pierre-Etienne Bourneuf, *Bombarder l'Allemagne. L'offensive alliée sur les villes pendant la Deuxième Guerre mondiale* (Paris: Presses Universitaires de France, 2014).

3. See the reports of the Eisenhower Research Project on the Costs of War, based at Brown University, at http://costsofwar.org/. On Afghanistan, see Susan G. Chesser, "Afghanistan Casualties: Military Forces and Civilians," Congressional Research Service Report 7-5700, 6 December 2012; and John Bohannon, "Counting the Dead in Afghanistan," *Science* 331 (11 March 2011): 1256–60.

4. Russia, for example, fought for years to regain control of the separatist republic of Chechnya, wreaking vast destruction by air and ground forces against cities and villages. The short but deadly incursions by Israel's forces into neighboring Lebanon and Gaza enacted a heavy toll on civilians there.

5. For an excellent overview, see Karl P. Mueller, "Air Power," in *The International Studies Encyclopedia*, vol. 1, ed. Robert A. Denemark (Oxford: Wiley-Blackwell, 2010), 47–65.

6. Colin Kahl, "In the Crossfire or the Crosshairs?" *International Security* 32, no. 1 (2007): 7–46.

7. Neta C. Crawford, *Argument and Change in World Politics: Ethics, Decolonization, and Humanitarian Intervention* (Cambridge: Cambridge University Press, 2002), 40–41.

8. William R. Slomanson, *Fundamental Perspectives on International Law*, 5th ed. (Belmont, CA: Thomson Wadsworth, 2007), 4.

9. Michael Byers, *Custom, Power and the Power of Rules: International Relations and Customary International Law* (New York: Cambridge University Press, 1999); Richard Price, "Emerging Norms and Anti-Personnel Landmines," in *The Politics of International Law*, ed. Christian Reus-Smit (Cambridge, UK: Cambridge University Press, 2004); Christiana Ochoa, "The Individual and Customary International Law Formation," *Virginia Journal of International Law* 48, no. 1 (Fall 2007): 119–86.

10. Jean-Marie Henckaerts and Louise Doswald-Beck, eds., *Customary International Humanitarian Law* (Cambridge, UK: Cambridge University Press, 2005). For a summary, see Jean-Marie Henckaerts, "Study on Customary International Humanitarian Law: A Contribution to the Understanding and Respect for the Rule of Law in Armed Conflict," *International Review of the Red Cross* 87, no. 857 (2005).

11. Milton Leitenberg, "Deaths in Wars and Conflicts in the 20th Century," 3rd ed., June 2006, Cornell University Peace Studies Program Occasional Paper 29, http://www.cissm.umd.edu/papers/files/deathswarsconflictsjune52006.pdf. In the twenty-first century, more civilians may be killed in wartime directly by other means, especially in Africa, although there are still many victims of bombing. See Jeffrey Gettleman, "Sudanese Struggle to Survive Endless Bombings Aimed to Quell Rebels," *New York Times*, 3 July 2011.

12. For a comparative study of the influence of other areas of international humanitarian law on state practice, see Matthew Evangelista and Nina Tannenwald, eds., *Do the Geneva Conventions Matter?* (New York: Oxford University Press, 2014).

13. Michael Byers, "Custom, Power, and the Power of Rules—Customary International Law from an Interdisciplinary Perspective," *Michigan Journal of International Law* 17 (1995–1996), 115. The literature on "hegemonic international law" is also relevant here. See, e.g., Detlev F. Vagts, "Hegemonic International Law," *American Journal of International Law* 95, no. 4 (2001): 843–48; and José E. Alvarez, "Hegemonic International Law Revisited," *American Journal of International Law* 97, no. 4 (October 2003): 873–88.

14. Henckaerts, "Study on Customary International Humanitarian Law," quoting the North Sea Continental Shelf cases of the International Court of Justice.

15. Nicholas J. Wheeler, *Saving Strangers: Humanitarian Intervention in International Society* (Oxford, UK: Oxford University Press, 2000).

16. Quoted in Thomas H. Greer, *The Development of Air Doctrine in the Army Air Arm, 1917–1941* (Maxwell Air Force Base, AL: USAF Historical Division, Research Studies Institute, 1955), 17. Thanks to Neta Crawford for this citation.

17. Giulio Douhet, *Il dominio dell'aria* (1921), revised version 1927, translated in 1942 by Dino Ferrari as *Command of the Air* (Washington, DC: Air Force History and Museums Program, 1998), 14. See also Paul K. Saint-Amour, "Air War Prophecy and Interwar Modernism," *Comparative Literature Studies* 42, no. 2 (2005): 130–61.

18. See the discussion in Michael S. Sherry, *The Rise of American Air Power: The Creation of Armageddon* (New Haven, CT: Yale University Press, 1987), chapter 2.

19. Donald Cameron Watt, "Restraints on War in the Air before 1945," in *Restraints on War: Studies in the Limitation of Armed Conflict*, ed. Michael Howard (Oxford, UK: Oxford University Press, 1979), 61; Marco Patricelli, *L'Italia sotto le bombe: Guerra aerea e vita civile, 1940–1945* (Rome: Laterza, 2007), 3–7.

20. Priya Satia, "The Defense of Inhumanity: Air Control and the British Idea of Arabia," *American Historical Review* 11, no. 1 (2006): 16–51.

21. With Churchill's blessing Trenchard sought to use the Iraqi campaign's results to boost the standing of his branch of the armed services, submitting a report to the cabinet on "The Development of Air Control in Iraq." Trenchard's report is available online from the British Public Records Office at http://www.nationalarchives.gov.uk/pathways/firstworldwar/transcripts/aftermath/air_power_iraq.htm. The quotations in this paragraph come from Geoff Simons, *Iraq: From Sumer to Saddam*, 3rd ed. (London: Palgrave Macmillan, 2004), 179. See also Yuki Tanaka, "British 'Humane' Bombing of Iraq during the Interwar Era," in *Bombing Civilians: A Twentieth Century History*, ed. Yuki Tanaka and Marilyn B. Young (New York: New Press, 2009). On the use of airpower in other colonies, see V. G. Kiernan, *Colonial Empires and Armies, 1815–1960* (Montreal: McGill/Queen's Press, 1998), 194–201.

22. The wording in this and following paragraphs draws on Matthew Evangelista, *Law, Ethics, and the War on Terror* (Cambridge, UK: Polity, 2008), 31–34.

23. Quotations from Robert C. Batchelder, *The Irreversible Decision, 1939–1950* (New York: Macmillan, 1961), 172–73. For more detail, see Tami Davis Biddle, *Rhetoric and Reality in Air Warfare: The Evolution of British and American Ideas about Strategic Bombing, 1914–1945* (Princeton, NJ: Princeton University Press, 2002).

24. Diary entry of F. Stevenson for 9 March 1934, quoted in Kiernan, *Colonial Empires and Armies*, 200. For a general argument about the distinction between civilized and uncivilized in the ethics of colonialism, see Crawford, *Argument and Change in World Politics*.

25. Batchelder, *The Irreversible Decision*, 174–75.

26. The atomic bombs were also distinctive in causing death by radiation poisoning and long-term genetic defects in the offspring of the survivors.

27. Quoted in A. C. Grayling, *Among the Dead Cities: The History and Moral Legacy of the WWII Bombing of Civilians in Germany and Japan* (New York: Walker and Company, 2006), 187.

28. Sherry, *The Rise of American Air Power*, 263.

29. Ronald Schaffer, *Wings of Judgment: American Bombing in World War II* (New York: Oxford University Press, 1985), 88–89.

30. Crawford, in this volume, cites Sahr Conway-Lanz, *Collateral Damage: Americans, Noncombatant Immunity, and Atrocity after World War II* (New York: Routledge, 2006), 13.

31. Alexander B. Downes, *Targeting Civilians in War* (Ithaca, NY: Cornell University Press, 2008).

32. Conway-Lanz, *Collateral Damage*. For a discussion of U.S. nuclear strategy, see David Alan Rosenberg, "'A Smoking Radiating Ruin at the End of Two Hours': Documents on American Plans for Nuclear War the Soviet Union," *International Security* 6, no. 3 (1981–82): 3–38.

33. Nina Tannenwald, *The Nuclear Taboo: The United States and the Nonuse of Nuclear Weapons since 1945* (Cambridge University Press, 2007).

34. Matthew Evangelista, "Nuclear Abolition or Nuclear Umbrella: Choices and Contradictions in US Proposals," in *Getting to Zero: The Path to Nuclear Disarmament*, ed. Catherine McArdle Kelleher and Judith Reppy (Stanford, CA: Stanford University Press, 2011).

35. This summary has benefited from discussions with both authors and borrows some language directly from e-mail correspondence with Conway-Lanz. The original distinction between fighting different wars and fighting differently comes from Henry Shue.

36. Quoted by Crawford, in this volume.

37. W. Hays Parks, "Rolling Thunder and the Law of War," *Air University Review* 33, no. 2 (1982): 2–23.

38. Protocol I, Article 52, para. 2.

39. US Air Force manuals, for example, quote the above passage verbatim in defining "military objective," as does the US Army Field Manual 27-10 of 1976, *The Law of Land Warfare*, http://www.afsc.army.mil/gc/files/fm27-10.pdf.

40. W. Hays Parks, "Air War and the Law of War," *Air Force Law Review* 32, no. 1 (1990): 33, 35.

41. Quoted in ibid., 33.

42. Parks's authority is M. W. Royce, a U.S. Marine aviator, who later received his PhD in international relations at Columbia University and participated in an ICRC conference of experts; ibid., 41. Parks is an important source for a similar argument in (Major) Jeanne M. Meyer, "Tearing down the Façade: A Critical Look at the Current Law on Targeting the Will of the Enemy and Air Force Doctrine," *Air Force Law Review* 51 (Spring 2001). She reproduces the summary of Royce (and Parks's misspelling of the title of his work), but mistakes it for a direct quotation rather than Parks's paraphrase; see 158n67.

43. This point is richly documented in Biddle, *Rhetoric and Reality.*

44. This is a central argument in Robert Pape, *Bombing to Win: Air Power and Coercion in War* (Ithaca, NY: Cornell University Press, 1996).

45. "Strategic Air Campaign against Iraq to Accomplish NCA Objectives," 16 August 1990, 3, quoted in Crawford in this volume.

46. Dunlap, in this volume. Regional specialists argue that the air war against Serbia actually delayed the popular effort to overthrow Slobodan Milošević, whose party had been consistently losing votes despite electoral manipulation. See V. P. Gagnon Jr., *The Myth of Ethnic War: Serbia and Croatia in the 1990s* (Ithaca, NY: Cornell University Press, 2004), 124–30. According to one authoritative account, the bombing "primed domestic support for Milošević in the short term at a time when the opposition was beginning to view the regime as increasingly vulnerable. . . . Even by the summer of 2000, there were good reasons to predict that the Milošević regime would continue in power." See Valerie J. Bunce and Sharon L. Wolchik, *Defeating Authoritarian Leaders in Postcommunist Countries* (New York: Cambridge University Press, 2011), 93, 105. For a representative response to the bombing from a member of Serbia's human rights movement, see Vojin Dimitrijević, "The Collateral Damage Is Democracy," *Institute for War and Peace Reporting*, 31 March 1999, http://iwpr.net/report-news/collateral-damage-democracy. For a review of the debate about the air strategy, see Stephen Biddle, "The New Way of War? Debating the Kosovo Model," *Foreign Affairs* 81, no. 3 (May–June 2002): 138–44. Worth mentioning is that some of the most anti-Milošević areas of Serbia, such as Nis and Novi Sad, were hit heavily by NATO bombs and sustained substantial civilian causalities. Thanks to Chip Gagnon for this point.

47. Dunlap, in this volume.

48. J. W. Crawford III, "The Law of Noncombatant Immunity and the Targeting of National Electrical Power Systems," *Fletcher Forum of World Affairs* 21, no. 2 (1997), 102.

49. I am grateful to Janina Dill for discussion of these points in an e-mail message, from which much of what follows is adapted.

50. Garraway, in this volume.

51. Even under that less stringent standard, however, the evidence of Russian war crimes, including those related to aerial bombardment, is compelling. See Matthew Evangelista, *The Chechen Wars: Will Russia Go the Way of the Soviet Union?* (Washington, DC: Brookings Institution, 2002), chapter 7.

52. Evangelista, *Law, Ethics, and the War on Terror*, chapter 3.

53. Quotation in Miller, in this volume.

54. Charles J. Dunlap Jr., *Shortchanging the Joint Fight? An Airman's Assessment of FM 3-24 and the Case for Developing Truly Joint COIN Doctrine*, Air University Monograph (Maxwell Air Force Base, AL: Airpower Research Institute, n.d.), 1, 12, http://carl.army.mil/docrepository/121007dunlap.pdf. Dunlap favors exploiting "the *psychological* dimension of today's airpower." By that he means not "the much-debated impact of airpower on civilian morale, but rather how today's precision capabilities influence the morale of *combatants*. It is about targeting the *insurgents*' hearts and minds." Ibid., 40, Dunlap's italics. He also favors creating "safe enclaves" or "gated communities," resembling somewhat the strategic hamlet program of the Vietnam War; ibid., 43–44.

55. For strong evidence of the impact of the sanctions program and the bombing of infrastructure on the Iraqi civilian population, see Joy Gordon, *Invisible War: The United States and the Iraq Sanctions* (Cambridge, MA: Harvard University Press, 2010).

56. For representative accounts, see Jack Goldsmith and Eric Posner, "Moral and Legal Rhetoric in International Relations: A Rational Choice Perspective," *Journal of Legal Studies* 31 (January 2002): 115–39; and Eric Posner, "International Law and the Disaggregated State," *Florida State University Law Review* 32 (2005): 797–842.

57. The most careful study of this phenomenon discusses not war but human rights: Beth A. Simmons, *Mobilizing for Human Rights: International Law in Domestic Politics* (New York: Cambridge University Press, 2009).

58. Christian Reus-Smit, ed., *The Politics of International Law* (Cambridge, UK: Cambridge University Press, 2004); Martha J. Finnemore and Stephen Toope, "Alternatives to 'Legalization': Richer Views of Law and Politics," *International Organization* 55, no. 3 (2001): 743–58.

59. Margaret Keck and Kathryn Sikkink, *Activists beyond Borders: Advocacy Networks in International Politics* (Ithaca, NY: Cornell University Press, 1998).

60. Petrova, in this volume.

61. Ryan, in this volume. For a similar argument, see Alessandro Colombo, *La guerra ineguale: Pace e violenza nel tramonto della societa internazionale* (Bologna: Il Mulino, 2006).

62. O'Connell, in this volume.

63. Some of the arguments are already evident in the campaigns of pacifist groups such as the "Hancock 38" drone resisters in central New York State. See Charles Ellis, "Update: Former U.S. Attorney General Ramsey Clark testifies at DeWitt trial of Hancock Field Drone Protesters," *The Post-Standard* (Syracuse, NY), 3 November 2011, http://www.syracuse.com/news/index.ssf/2011/11/former_us_attorney_general_ram.html; "'Hancock 38' Defendants Found Guilty for Bold Army Base Protest Against U.S. Drone Attacks Abroad," *Democracy Now* radio broadcast, 2 December 2011, transcript at http://www.democracynow.org/2011/12/2/hancock_38_defendants_found_guilty_for.

1. Strategic Bombardment

1. On the idea of "just war" and challenges to the principle of discrimination posed by the nature of political and scientific development in the nineteenth and early twentieth centuries, see Michael Howard, "Temperamenta Belli: Can War

Be Controlled?" in *Restraints on War: Studies in the Limitation of Armed Conflict*, ed. M. Howard (London: Oxford University Press, 1979), 1–15.

2. For a full development of this argument, see Tami Davis Biddle, *Rhetoric and Reality in Air Warfare: The Evolution of British and American Ideas about Strategic Bombing, 1914–1945* (Princeton, NJ: Princeton University Press, 2002).

3. The precise number of civilian deaths in World War II is impossible to know, but a number of sound estimates have been made. Air Force official historian Richard G. Davis argued that allied strategic bombing caused at least 250,000 deaths in Germany and that tactical bombing added another 55,000 deaths. See Richard G. Davis, *Carl A. Spaatz and the Air War in Europe* (Washington, DC: Center for Air Force History, 1993), 588. The British official historian placed the number of civilian deaths in Britain (from bombs, flying bombs, and rockets) at 60,447; Basil Collier, *The Defence of the United Kingdom* (London: HMSO, 1957), 528. The United States Strategic Bombing Survey report titled "The Effects of Bombing on Health and Medical Services in Japan" Medical Division (June 1947) estimated Japanese civilian deaths at 333,000. See page 156.

4. Hugo quoted in I. F. Clarke, *Voices Prophesying War 1763–1984* (Oxford: Oxford University Press, 1966), 3.

5. See David MacIsaac, "Voices from the Central Blue: The Air Power Theorists," in *Makers of Modern Strategy from Machiavelli to the Nuclear Age*, ed. Peter Paret (Princeton, NJ: Princeton University Press, 1986), 625–27.

6. Michael Howard, "The Influence of Clausewitz," in Carl von Clausewitz, *On War*, trans. M. Howard and P. Paret (Princeton, NJ: Princeton University Press, 1984), 39; M. Howard, *Clausewitz* (Oxford: Oxford University Press, 1983); Azar Gat, *The Development of Military Thought: The Nineteenth Century* (Oxford: Clarendon Press, 1992); Christopher Bassford, *Clausewitz in English* (Oxford: Oxford University Press, 1994), 104–12; Tim Travers, *The Killing Ground: The British Army, the Western Front and the Emergence of Modern Warfare, 1900–1918* (London: Unwin Hyman, 1987), 37–97; Antulio J. Echevarria II, "On the Brink of the Abyss: The Warrior Identity and German Military Thought before the Great War," *War and Society* 13, no. 2 (October 1995): 23–40.

7. See two lectures by T. Miller Maguire to the Royal United Services Institution, published as "Readiness or Ruin," *Journal of the Royal United Services Institution* 53, no. 382 (December 1909), 1579–1606, and "National Recuperation" *Journal of the Royal United Services Institution* 54, no. 385 (March 1910), 291–323. Many conservative writers associated urbanization with "decay" in the British national character.

8. There is an extensive literature on both the sociopolitical trends and the impact of these trends on thinking about air power. See, for instance, Samuel Hynes, *The Edwardian Turn of Mind* (Princeton, NJ: Princeton University Press, 1968); Alfred Gollin, *No Longer an Island: Britain and the Wright Brothers, 1902–1909* (Stanford, CA: Stanford University Press, 1984); David Edgerton, *England and the Aeroplane* (London: MacMillan, 1991); Michael Paris, *Winged Warfare* (New York: Manchester University Press, 1992); and Robert Wohl, *A Passion for Wings* (New Haven, CT: Yale University Press, 1994).

9. W. Hays Parks, "Air War and the Law of War," *Air Force Law Review* 32, no. 1 (1990): 16–17. The text of the 1899 and 1907 Hague Conventions can be viewed

in full online at the Avalon Project, Lillian Goldman Law Library, Yale Law School: http://avalon.law.yale.edu/default.asp

10. Tami Davis Biddle, "Air Power," in *The Laws of War: Constraints on Warfare in the Western World*, ed. M. Howard, G. Andreopoulos, and Mark Shulman (New Haven, CT: Yale University Press, 1994), 142–43; Parks, "Air War and the Law of War," 17–19.

11. In 1899, Lord Wolseley of the British Army argued that dropping bombs from balloons might be highly advantageous to Britain: "In war superior armament compensates for lack of numerical strength." See D. C. Watt, "Restraints on War in the Air before 1945," in *Restraints on War: Studies in the Limitation of Armed Conflict*, ed. M. Howard (London: Oxford University Press, 1979), 60.

12. See Lee Kennett, *A History of Strategic Bombing* (New York: Scribner's Sons, 1982), 13; Watt, "Restraints on War," 61.

13. For an excellent overview, see Richard J. Overy, "Strategic Bombardment before 1939: Doctrine, Planning and Operations," in *Case Studies in Strategic Bombardment*, ed. Cargill Hall (Washington, DC: Air Force History and Museums Program, 1998), 13.

14. For a useful introduction to World War I aviation, see John H. Morrow Jr., *The Great War in the Air* (Washington, DC: Smithsonian Institution Press, 1993).

15. Overy, "Strategic Bombing before 1939," 14, 17–18.

16. Behnke and Strasser quoted in ibid., 14.

17. On the development of British defenses, see John Ferris, "Airbandit: C3I and Strategic Air Defence during the First Battle of Britain, 1915–1918," in *Strategy and Intelligence: British Policy during the First World War*, ed. Michael Dockrill and David French (London: Hambleton, 1996). On Zeppelins and British responses, see, for instance, Douglas Robinson, *The Zeppelin in Combat, 1912–1918* (London: G. T. Foulis, 1962).

18. For a detailed analysis see Biddle, *Rhetoric and Reality in Air Warfare*, 29–40.

19. See "The Second Report of the Prime Minister's Committee on Air Organisation and Home Defence against Air Raids," 17 August 1917, AIR 1/515/1/3/83, The National Archives (hereafter TNA), Kew, London. Portions reprinted in Biddle, *Rhetoric and Reality in Air Warfare*, 33.

20. See Raymond H. Fredette, *The Sky on Fire: The First Battle of Britain, 1917–1918* (New York: Harcourt, Brace, Jovanovich, 1966). Responding sanctimoniously to the trends in London, the *New York Times* opined, "The bombing of undefended German cities promises no better result than the depriving of the Allies of a moral superiority they now possess. And that is not worthwhile"; *New York Times*, 4 October 1917, quoted in ibid., 154.

21. For detailed accounts of the events leading to a British bombing force in response to German attacks, see Malcolm Cooper, *The Birth of Independent Air Power* (London: Allen and Unwin, 1986); Fredette, *The Sky on Fire*, esp. 221–27.

22. See Overy, "Strategic Bombardment before 1939," 19, who quotes Memo for Supreme War Council, "Bombing Operations," January 1918, pp. 1–4, AIR 1/463/15/312/37, TNA, Kew, London.

23. See George K. Williams, "The Shank of the Drill: Americans and Strategical Aviation in the Great War," *Journal of Strategic Studies* 19, no. 3 (September 1996), 381–431.

24. For an extensive treatment, see George K. Williams, *Biplanes and Bombsights: British Bombing in World War I* (Maxwell Air Force Base, AL: Air University Press, 1999).

25. See M. Maurer, ed., *The US Air Service in World War I*, vols. 1, 2, and 4 (Washington, DC: USGPO, 1978).

26. See Overy, "Strategic Bombing before 1939," 24–25. In June of 1918, the chief of the RAF's air staff, Sir Frederick Sykes, produced a paper for the War Cabinet arguing that properly organized air power would afford in war "the best and most rapid return for the expenditure of national resources of man-power, material, and money." Quoted in Overy, 20.

27. See James Corum, "Airpower Thought in Continental Europe between the Wars," in *The Paths of Heaven: The Evolution of Air Power Theory*, ed. Phillip Meilinger (Maxwell Air Force Base, AL: Air University Press, 1997), esp. 168–75; Williamson Murray, "Strategic Bombing: The British, American and German Experiences," and Richard Muller, "Close Air Support: The German, British, and American Experiences, 1918–1941," both in *Military Innovation in the Interwar Period*, ed. Williamson Murray and Allan Millett (Cambridge: Cambridge University Press, 1996), 96–143(Murray) and 143–190 (Muller).

28. Overy, "Strategic Bombardment before 1939," 35–36.

29. On developments in the United States, see Peter Faber, "Interwar US Army Aviation and the Air Corps Tactical School: Incubators of American Air Power," in *The Paths of Heaven: The Evolution of Air Power Theory*, ed. Phillip Meilinger (Maxwell Air Force Base, AL: Air University Press, 1997), 183–238.

30. Biddle, *Rhetoric and Reality in Air Warfare*, 128–42.

31. Commandant's lecture, "Air Warfare," RAF Staff College, 1924, AIR 1/2385/228/10, TNA, Kew, London.

32. Biddle, *Rhetoric and Reality in Air Warfare*, 27–28, 37–38, 69–76.

33. "Why the Royal Air Force Should Be Maintained as Separate from the Navy and Army," Papers of Lord Trenchard, MFC 76/1/21, Royal Air Force Museum (hereafter RAF Museum), Hendon, London.

34. Fredette, *The Sky on Fire*, esp. 151–59 and 197–204.

35. For a detailed articulation of this argument, see Biddle, *Rhetoric and Reality in Air Warfare*, 74–81.

36. For the RAF argument on air control, see "Memorandum by the Air Staff on the Effects Likely to Be Produced by Intensive Aerial Bombing of Semi-Civilised People," n.d. [1922], Papers of Lord Trenchard, MFC 76/1/21, RAF Museum, Hendon, London. On the economics of air control, see Barry Powers, *Strategy without Slide Rule* (London: Croom Helm, 1976), 173. For analyses of the impact and implications of colonial air control, see Charles Townshend, "Civilisation and 'Frightfulness': Air Control in the Middle East Between the Wars," in *Warfare, Diplomacy, and Politics: Essays in Honor of A. J. P. Taylor*, ed. Chris Wrigley (London: H. Hamilton, 1986), 142–162; Toby Dodge, *Inventing Iraq: The Failure of Nation-Building and a History Denied* (New York: Columbia University Press, 2003); and Priya Satia, "The Defense of Inhumanity: Air Control and the British Idea of Arabia," *American Historical Review* 111, no. 1 (February 2006): 16–51.

37. I. F. Clarke, *Voices Prophesying War, 1763–1984* (Oxford: Oxford University Press, 1966), 169–70.

38. For overviews of Douhet in English from this period, see n.a., "The Air Doctrine of General Douhet," *Royal Air Force Quarterly* 4, no. 2 (April 1933): 164–67; "General Giulio Douhet—An Italian Apostle of Air Power," and "Air Warfare—The Principles of Air Warfare by General Giulio Douhet," *Royal Air Force Quarterly* 7, no. 2 (April 1936): 148–51 and 152–68, respectively. See also Azar Gat, "Futurism, Proto-Fascist Italian Culture and the Sources of Douhetism," *War and Society* 15, no. 1 (May 1997): 31–51.

39. Gat, "Futurism" 39.

40. Giulio Douhet, *Command of the Air* (Washington, DC: Office of Air Force History, 1983), 10, 22–23. This is a reprint of the 1942 edition translated by Dino Ferrari and published by Coward-McCann, Inc.

41. Douhet, *Command of the Air*, 61.

42. Michael S. Sherry, *The Rise of American Air Power* (New Haven, CT: Yale University Press, 1987), 27.

43. On the problems related to aircraft limitation, see Watt, "Restraints on War," 66.

44. See Parks, "Air War and the Law of War," 27–36. For the text, see the Avalon Project: http://avalon.law.yale.edu/default.asp.

45. Biddle, "Air Power," 150–51. In an articulate and perceptive essay on targeting and International Humanitarian Law (IHL), Timothy L.H. McCormack and Helen Dunham point out that, "International laws relating to targeting are highly complex and highly controversial." And they add, "Striking a balance between principles of humanity and military necessity is the underlying philosophical tension that continues to characterize the codification of IHL today." Tim McCormack and Helen Dunham, "Aerial Bombardment of Civilians: The Current International Legal Framework," in Yuki Tanaka and Marilyn B. Young, eds., *Bombing Civilians: A Twentieth Century History* (New York: The New Press, 2009), 215 and 218.

46. J. M. Spaight, "The Chaotic State of the International Law Governing Bombardment," *Royal Air Force Quarterly* 9, no. 1 (January 1938): 25; Mitchell quoted in Biddle, *Rhetoric and Reality in Air Warfare*, 153.

47. Quoted in Sir Charles Webster and Noble Frankland, *The Strategic Air Offensive against Germany, 1939–1945*, vol. 2, *Endeavor* (London: HMSO, 1961), 58. The authors were the official historians of the British strategic bombing offensive in World War II.

48. See Watt, "Restraints on War," 68–70; Uri Bialer, *The Shadow of the Bomber* (London: Royal Historical Society, 1980), 33–40; and Phillip Meilinger, "Clipping the Bomber's Wings," *War in History*, 6, no. 3 (1999), 306–330; Sir Charles Webster and Noble Frankland, *The Strategic Air Offensive, 1939–1945*, vol. 1, *Preparation* (London: HMSO, 1961), 58–60; Biddle, *Rhetoric and Reality in Air Warfare*, 150–53.

49. Watt, "Restraints on War," 69–70.

50. Webster and Frankland, *The Strategic Air Offensive*, 1:59–60.

51. Watt writes, "From 1933 to 1936 Britain and France devoted themselves to attempts to maneuver Germany back into the circle of international discussion"; "Restraints on War," 70.

52. There is an excellent account in Paul M. Kennedy, *The Realities behind Diplomacy* (London: Fontana, 1981), 226–301.

53. Quoted in John Terraine, *A Time for Courage* (New York: Macmillan, 1985), 82. Published in London as *The Right of the Line*.

54. Sherman, *Air Warfare*, 218.

55. On the shaping of American doctrine, see Biddle, *Rhetoric and Reality in Air Warfare*, 128–75.

56. See Sherry, *The Rise of American Air Power*, 78–79.

57. For the text of FDR's plea, see Department of State, *Foreign Relations of the United States, Diplomatic Papers 1939*, vol. 1 (Washington, DC: GPO, 1956), 541–42.

58. J. M. Spaight, *Air Power and War Rights*, 3rd ed. (London: Longman, Green and Co., 1947), 257–58.

59. Quoted in Parks, "Air War and the Law of War," 45.

60. "Air Ministry Instructions and Notes on the Rules to Be Observed by the Royal Air Force in War," 22 August 1939 and covering note, AIR14/249, TNA, Kew, London.

61. Webster and Frankland, *The Strategic Air Offensive*, 1:144–54, esp. 152; Terraine, *A Time for Courage*, 259–60. For a more recent account of British decisions in favor of bombing, see Richard Overy, *The Bombing War, 1939–1945* (London: Allen Lane, 2013), 237–54.

62. Webster and Frankland, *The Strategic Air Offensive*, 1:105–6; Brereton Greenhous, Stephen J. Harris, William C. Johnston, and William G. P. Rawling, *The Crucible of War, 1939–1945* (Toronto: University of Toronto Press, 1994), 530–34.

63. Webster and Frankland, *The Strategic Air Offensive*, 1:140, 204–11.

64. The analysis was called the "Butt Report" after the civil servant who led it. For the text, see Sir Charles Webster and Noble Frankland, *The Strategic Air Offensive, 1939–1945*, vol. 4, *Appendices* (London: HMSO, 1961), 205–13.

65. Shortly after the appearance of the Butt Report, Portal told a despondent Churchill that if he wished it, Britain could still shift to a grand strategic conception that would put the army in the lead. He warned that it would require a prompt restructuring of the RAF. Churchill, who wished no repeat of the WWI Western Front, turned the offer down. See Biddle, *Rhetoric and Reality in Air Warfare*, 1–2 and 195–99.

66. Webster and Frankland, *The Strategic Air Offensive*, 1:215. By mid-October of 1939 the chief of the air staff had indicated that Luftwaffe bombing in Poland had essentially lifted the RAF's obligations to adhere to strict limitations on bombing, including President Roosevelt's plea. See Parks, "Air War and the Law of War," 45.

67. Felix Brown, "Civilian Psychiatric Air Raid Casualties," *The Lancet* (31 May 1941): 691.

68. See the directive of 14 February 1942 in Webster and Frankland, *The Strategic Air Offensive*, 4:143–48.

69. In 1941, the British lost 4.3 million deadweight tons of matériel in the Battle of the Atlantic; by 1942 that total had soared to 7.3 million. In early 1943, British imports were one-third less than those in 1939, a fact that not only threatened the war effort but also risked malnutrition for the population. See Paul Kennedy, *Engineers of Victory* (New York: Random House, 3013), 5–73. See Michael Walzer's discussion of "supreme emergency" in *Just and Unjust Wars* (New York: Basic Books, 1977), 251–63.

70. On Harris, see Tami Davis Biddle, "Bombing by the Square Yard: Sir Arthur Harris at War, 1942–1945," *International History Review* 21, no. 3 (September 1999): 626–64.

71. Biddle, *Rhetoric and Reality in Air Warfare*, 197–200; Mark K. Wells, *Courage and Air Warfare: The Allied Aircrew Experience in the Second World War* (London: Frank Cass, 1995); and Terraine, *A Time for Courage*.

72. Adam Tooze, *The Wages of Destruction: The Making and Breaking of the Nazi Economy* (London: Allen Lane, 2006), 597–600.

73. Webster and Frankland, *The Strategic Air Offensive*, 2:108–211.

74. Webster and Frankland, *The Strategic Air Offensive*, 1:353–63; Biddle, *Rhetoric and Reality in Air Warfare*, 211–13.

75. Webster and Frankland, *The Strategic Air Offensive*, 2:39.

76. CG, Army Air Forces, "Combined Chiefs of Staff Air Plan for the Defeat of Germany," memorandum, 1 November 1943, Box 39, Henry Harley Arnold Papers, Manuscript Division, Library of Congress, Washington, DC. In April 1943, Arnold outlined the advantages of incendiary bombs and incendiary attacks. See Arnold to Asst. Chief of Air Staff, Materiel, Maintenance and Distribution, 26 April 1943, Box 38, Arnold Papers. See generally Ronald Schaffer, *Wings of Judgment: American Bombing in World War II* (Oxford: Oxford University Press, 1985).

77. Biddle, *Rhetoric and Reality in Air Warfare*, 228–29, 243–44; Richard G. Davis, "German Railyards and Cities: US Bombing Policy, 1944–45," *Air Power History* 42, no. 2 (1995): 48–49. See also generally W. Hays Parks, "Precision and Area Bombing: Who Did Which and When?" *Journal of Strategic Studies* 18, no. 1 (March 1995): 145–74.

78. See, for instance, Richard G. Davis, *Carl A. Spaatz and the Air War in Europe* (Washington, DC: Center for Air Force History, 1993), 287–418; W. W. Rostow, *Pre-Invasion Bombing Strategy* (Austin: University of Texas Press, 1981).

79. On Allied concerns about secret weapons, see Hanson Baldwin, "How Long Will the War Last?" *New York Times Magazine*, 23 July 1944, 3, 38–39.

80. Significantly, the report stated that "a heavy flow of refugees from Berlin in the depth of winter coinciding with the trekking westwards of a population fleeing from Eastern Germany would be bound to *create great confusion, interfere with the orderly movement of troops to the front, and hamper the German military and administrative machine*" (my italics). Joint Intelligence Sub-Committee, "Strategic Bombing in Relation to the Present Russian Offensive," 25 January 1945, 1–2, CAB 81/127, TNA, Kew, London.

81. Churchill quoted in Sir Charles Webster and Noble Frankland, *The Strategic Air Offensive, 1939–1945*, vol. 3, *Victory* (London: HMSO, 1961), 101. Churchill's role is made clear on 98–104. See also Sebastian Cox, "The Dresden Raids: Why and How," in *Firestorm: The Bombing of Dresden, 1945*, ed. Paul Addison and Jeremy Crang (London: Pimlico, 2006), 22–23.

82. Webster and Frankland, *The Strategic Air Offensive*, 3:103.

83. Quoted in Webster and Frankland, *The Strategic Air Offensive*, 3:104. See also the summary of the plan, included in "Notes of the Allied Air Commanders' Conference held at S.H.A.E.F. on 1st February 1945, at 1130 Hours," decimal file K239.046-38, Air Force Historical Research Agency, Maxwell Air Force Base, Alabama. For a full discussion of these events, and their crescendo, see Tami Davis Biddle, "Dresden 1945: Reality, History, and Memory," *Journal of Military History* 72, no. 2 (April 2008), 413–50.

84. See Cox, "The Dresden Raids," 29–48.

85. On the history of the death toll at Dresden, see Biddle, "Dresden 1945," 423–24. An extensive examination appears in Richard Evans, *Telling Lies about Hitler* (London: Verso, 2002). On the wartime reactions to the raid, see Tami Davis Biddle, "Wartime Reactions," in *Firestorm: The Bombing of Dresden, 1945*, ed. Paul Addison and Jeremy Crang (London: Pimlico, 2006), 96–122. While the tactics and techniques used in the late war raids on eastern German cities were no different than those the United States and Britain had used earlier, the bombing directive of January 1945, which ordered attacks designed to exploit the movement of refugees, placed these attacks in a different moral category—one that spoke to the heightened fears and fully brutalized atmosphere of the moment.

86. Sherry, *The Rise of American Air Power*, 264–300; Conrad C. Crane, *Bombs, Cities, and Civilians: American Airpower Strategy in World War II* (Lawrence: University Press of Kansas, 1993), 120–42; Kenneth Werrell, *Blankets of Fire* (Washington, DC: Smithsonian, 1998), 98, 112.

87. See Haywood Hansell, *The Strategic Air War against Germany and Japan* (Washington, DC: Office of Air Force History, 1986), esp. 135–204.

88. See "Analysis of Incendiary Phase of Operations, 9–19 March 1945," Headquarters, XXI Bomber Command, volume 7, Narrative History, Headquarters 20th Air Force Air Force Historical Research Center (AFHRC), Maxwell Air Force Base, Alabama, dec. file no. 760.01; "Staff Presentation, 10 April 1945," Binder XIII, Operations Div. Reports, Docs. 103–109, volume 14 Narrative History, 20th Air Force in AFHRC dec. file no. 760.01.

89. For an excellent analysis that also examines the abandonment of responsibility on the part of civilian leaders in Washington, see William Ralph, "Improvised Destruction: Arnold, LeMay, and the Firebombing of Japan," *War in History* 13, no. 4 (November 2006): 495–522.

90. See "Analysis of Incendiary Phase of Operations"; and Crane, *Bombs, Cities, and Civilians*, 127–29.

91. Some of the earliest published critiques of World War II bombing include Sir Gerald Dickens, *Bombing and Strategy: The Fallacy of Total War* (London: S. Low, Marston, 1947); P.M. S. Blackett, *Fear, War, and the Bomb: Military and Political Consequences of Atomic Energy* (New York: Whittlesey House, 1949); Marshall Andrews, *Disaster through Air Power* (New York: Reinhart, 1950); and David Divine, *The Broken Wing: A Study in the British Exercise of Air Power* (London: Hutchinson, 1949). On the impact of atomic bombs, see Paul Boyer, *By the Bomb's Early Light* (New York: Pantheon, 1985). John Hersey's essay on Hiroshima for the 31 August 1946 issue of the *New Yorker* had an important impact on the American conscience.

92. Overy, *The Bombing War, Europe 1939-1945* (London: Allen Lane, 2013) 630.

2. Bombing Civilians after World War II

1. For such arguments, see George E. Hopkins, "Bombing and the American Conscience during World War II," *Historian* 28, no. 3 (1966): 451–73; Richard Shelly Hartigan, *The Forgotten Victim: A History of the Civilian* (Chicago: Precedent, 1982), 1–10; Ronald Schaffer, *Wings of Judgment: American Bombing in World War II* (New York: Oxford University Press, 1985), 3, 217–18; H. Bruce Franklin, *War Stars: The Superweapon and the American Imagination* (New York: Oxford University Press, 1988),

105; Paul Boyer, *Fallout: A Historian Reflects on America's Half-Century Encounter with Nuclear Weapons* (Columbus: Ohio State University Press, 1998), 12; John W. Dower, *Cultures of War: Pearl Harbor/Hiroshima/9-11/Iraq* (New York: W.W. Norton and New Press, 2010), 161, 166–70, 192–96. For a contrary view, see Biddle, this volume.

2. *Congressional Record*, 75th Cong., 3rd sess., vol. 83, pt. 8: 9524–26, 9545; *Public Papers and Addresses of Franklin D. Roosevelt*, vol. 8 (New York: Macmillan, 1941), 454.

3. See, for example, *New York Times*, April 29 and May 10, 1940.

4. Schaffer, *Wings of Judgment*, 70; Conrad C. Crane, *Bombs, Cities, and Civilians: American Airpower Strategy in World War II* (Lawrence: University of Kansas Press, 1993), 31.

5. *New York Times*, February 25, 1945; Michael S. Sherry, *The Rise of American Air Power: The Creation of Armageddon* (New Haven, CT: Yale University Press, 1987), 289.

6. *Public Papers of the Presidents of the United States: Harry S. Truman, 1945* (Washington: U.S. Government Printing Office, 1961), 197, 212.

7. Crane, *Bombs, Cities, and Civilians*, 29–30.

8. For an early example of this, see Conference Minutes, July 7, 1949, Box 2389, 514.2, Central Decimal Files 1945–1949, Record Group (hereafter RG) 59, National Archives and Records Administration (hereafter NARA), College Park, MD.

9. Message of the Joint Chiefs of Staff to MacArthur, 29 June 1950, in *Foreign Relations of the United States 1950* (hereafter *FRUS*), 7 (Washington, DC: US Government Printing Office, 1976), 240–41.

10. O'Donnell to LeMay, July 11, 1950, Box 65, series B, Papers of Curtis E. LeMay (hereafter LeMay Papers), Library of Congress (hereafter LC).

11. Stratemeyer to O'Donnell, July 11, 1950, Box 103, Series B, LeMay Papers; HQ USAF, *An Evaluation of the Effectiveness of the United States Air Force in the Korean Campaign* (Barcus Report), vol. 5, 2, Box 906, Project Decimal Files 1942–1954, Directorate of Plans, Office of the Deputy Chief of Staff for Operations, RG 341, NARA, College Park, MD.

12. *New York Times*, September 3, 1950. See also "Report of the United Nations Command Operations in Korea," *US Department of State Bulletin*, October 2, 1950, 534–40; "Fifth Report of the U.N. Command Operations in Korea," *US Department of State Bulletin*, October 16, 1950, 603–6.

13. Message, London Embassy to Secretary of State, July 1, 1950, Box 4264, 795.00, Central Decimal Files 1950–1954, RG 59, NARA; *New York Times*, July 4, 11, 12, 14, 18, and 26, 1950; message, Moscow Embassy to Secretary of State, July 14, 1950, Box 4265, 795.00, Central Decimal Files 1950–1954, RG 59, NARA; message, Moscow Embassy to Secretary of State, July 17, 1950, Box 4265, 795.00, Central Decimal Files 1950–1954, RG 59, NARA; *Daily Worker*, July 4–6, 10, 12, 14, 17, 18, 20, 24–28, and 31, 1950; *United Nations Security Council Official Records*, August 8, 1950, 5th year, 484th mtg., S/PV.484, 20.

14. "North Korea Slanders U.N. Forces to Hide Guilt of Aggression," *US Department of State Bulletin*, September 18, 1950, 454.

15. I want to thank Alexander B. Downes and his book *Targeting Civilians in War* (Ithaca, NY: Cornell University Press, 2008) for helping me understand the larger significance of this dynamic of escalation.

16. Memorandum of conversation, John Muccio, November 17, 1950, *FRUS 1950*, 7:1175.

17. William T. Y'Blood, ed., *The Three Wars of Lt. Gen. George E. Stratemeyer: His Korean War Diary* (Washington, DC: Air Force History and Museums Program, 1999), 236–37.

18. Ibid., 253–55.

19. Douglas MacArthur, *Reminiscences* (New York: McGraw-Hill, 1964), 366; Conrad C. Crane, *American Airpower Strategy in Korea, 1950–1953* (Lawrence: University of Kansas Press, 2000), 46; Y'Blood, *The Three Wars of Lt. Gen. George E. Stratemeyer*, 258–61; message, Stratemeyer to Vandenberg, November 5, 1950, Box 86, Vandenberg Papers, LC.

20. Robert Futrell, *The United States Air Force in Korea, 1950–1953*, rev. ed. (Washington: U.S. Government Printing Office, 1983), 221–23, 226; *New York Times*, November 9, 1950; Y'Blood, *The Three Wars of Lt. Gen. George E. Stratemeyer*, 269, 371–72; interview transcript from 98th Bomb Group, November 30, 1950, Box 905, Project Decimal File 1942–1954, Directorate of Plans, Office of the Deputy Chief of Staff for Operations, RG 341, NARA; Crane, *American Airpower Strategy in Korea*, 63, 168.

21. Wiley D. Ganey to LeMay, September 7, 1952, series B, Box 65, LeMay Papers, LC. For a fuller discussion of targeting "buildings," see Sahr Conway-Lanz, *Collateral Damage: Americans, Noncombatant Immunity, and Atrocity after World War II* (New York: Routledge, 2006), 107.

22. See the press releases printed daily in the *New York Times*, starting with "Korean Release, No. 627," November 9, 1950. By spring 1951, supply centers or areas were frequently mentioned as targets for UN air attacks in the releases. Releases December 1950–December 1951 are also in boxes 2–3, Korean War Communiques and Press Releases 1950–1951, Office of the Chief of Information, RG 319, NARA. Terms such as "supply center" were not just used by the military for public consumption. Similar terms were used in internal documents by American officers. Message, G-2, Department of the Army to USCINCEUR et al., November 24, 1952, Box 756, Chronological File 1949–June 1954, Office of Security Review, Office of the Assistant Secretary of Defense for Legislative and Public Affairs, RG 330, NARA.

23. Memorandum, Office of the Chief of Information, HQ FEC to Public Information Office, FEAF, 1 August 1951, Box 36, Office of the Chief of Information, Office of the Chief of Staff, Supreme Commander for the Allied Powers, RG 331, NARA.

24. *Chicago Tribune*, November 8, 1950; *St. Louis Post-Dispatch*, November 8, 1950; *Detroit News*, November 8, 1950; *Philadelphia Bulletin*, November 8 and 9, 1950; *Los Angeles Times*, November 8 and 9, 1950; *San Francisco Examiner*, November 8 and 9, 1950; *Houston Chronicle*, November 8 and 9, 1950; *Washington Post*, November 8–10, 1950; *Baltimore Sun*, November 8–10, 1950; *Boston Post*, November 8–11, 1950; *New York Times*, November 9, 1950; *Cleveland Press*, November 9, 1950. Of the twelve daily newspapers surveyed, only the *Detroit News* and *Cleveland Press* did not call Sinuiju a supply base or use a similar term. For additional evidence of the wider public embrace of this persisting vision of a war fought with discrimination, see Conway-Lanz, *Collateral Damage*, 114–19.

25. Futrell, *The United States Air Force in Korea*.

26. For example, Max Hastings, *The Korean War* (New York: Simon and Schuster, 1987); Crane, *Bombs, Cities, and Civilians*, 147–50.

27. Bruce Cumings, *The Roaring of the Cataract, 1947–1950*, vol. 2 of *The Origins of the Korean War* (Princeton, NJ: Princeton University Press, 1990); Crane, *American Airpower Strategy in Korea*; Steven Hugh Lee, *The Korean War* (New York: Longman, 2001).

28. First Radio Broadcast and Leaflet Group, "Plan for Psychological Warfare Operations Designed to Support the United Nations Air Force," June 12, 1952, Box 20, General Correspondence 1952, Psychological Warfare Section, General Headquarters, Far East Command, RG 338, NARA; "Plan for Psychological Warfare Operations in Support of Air Attack Program," July 7, 1952, Box 7, General Correspondence 1952, Psychological Warfare Section, General Headquarters, Far East Command, RG 338, NARA; "Monthly Report for August 1952," Box 14, General Correspondence 1952, Psychological Warfare Section, General Headquarters, Far East Command, RG 338, NARA; "Report of the U.N. Command Operations in Korea," *US Department of State Bulletin*, January 26, 1951, 155–59; "Psychological Warfare Weekly Bulletin," n.d., Box 20, General Correspondence 1952, Psychological Warfare Section, General Headquarters, Far East Command, RG 338, NARA; Crane, *American Airpower Strategy in Korea*, 122–25; message, CINCFE to PsyWar, October 9, 1952, Box 759, Chronological File 1949–June 1954, Office of Security Review, Office of the Assistant Secretary of Defense for Legislative and Public Affairs, RG 330, NARA; "Reports of U.N. Command Operations in Korea: Sixty-Fifth Report for the Period March 1–15, 1953," *US Department of State Bulletin*, July 13, 1953, 52–53.

29. "The Right Track," *Time*, July 21, 1952, 32; "Will Bombing End Korean War?" *US News and World Report*, September 12, 1952, 13–15; "Truth about the Air War," *US News and World Report*, November 7, 1952, 20–21; Carl Spaatz, "Stepped-Up Bombing in Korea," *Newsweek*, August 18, 1952, 27; *New York Times*, August 5, 6, 8–10, 19, 21, and 29–30, September 14 and 20, and October 3 and 5, 1952.

30. James A. Field, *The History of United States Naval Operations: Korea* (Washington: U.S. Government Printing Office, 1962), 304; Futrell, *The United States Air Force in Korea*, 269.

31. *New York Times*, December 25, 1950, January 19, February 11, June 16, and July 30, 1951, November 4, 1952, and January 14 and May 25, 1953; *San Francisco Examiner*, December 1, 1950; Nora Waln, "Our Softhearted Warriors in Korea," *Saturday Evening Post*, December 23, 1950, 28–29, 66–67; "Waifs of War," *Time*, January 1, 1951, 16; "The Greatest Tragedy," *Time*, January 15, 1951, 23–24; "Helping the Hopeless," *Time*, January 29, 1951, 31; Bill Stapleton, "Little Orphan Island," *Collier's*, July 14, 1951, 51; Michael Rougier, "The Little Boy Who Wouldn't Smile," *Life*, July 23, 1951, 91–98; James Finan, "Voyage from Hungnam," *Reader's Digest*, November 1951, 111–12; "Christian Soldiers," *Time*, June 15, 1953, 75–76.

32. Department of the Army, *Department of the Army Field Manual: The Law of Land Warfare* (Washington, DC: Department of the Army, 1956), 16.

33. For an example of the challenge in assessing individual officers' principled commitments to protecting civilians, see Conway-Lanz, *Collateral Damage*, 52–55.

34. For a more extensive examination of this argument, see Conway-Lanz, *Collateral Damage*.

35. Geoffrey Best, *War and Law Since 1945* (New York: Clarendon), 115–16, 204–5; conference minutes, July 7, 1949, 514.2, Central Decimal Files 1945–1949,

RG 59, NARA; Raymund T. Yingling and Robert W. Ginnane, "The Geneva Conventions of 1949," *American Journal of International Law* 46, no. 3 (1951): 427.

36. Paul Ruegger, "Press Conference Statement," April 9, 1951, Box 4380, 800.571, Central Decimal Files 1950–1954, RG 59, NARA; *New York Times*, July 23 and September 27, 1952; K.R. Kreps to Secretary of State, April 20, 1951, Box 879, 014, Project Decimal File 1942–1954, Directorate of Plans, Office of Deputy Chief of Staff for Operations, RG 341, NARA.

37. U.N. General Assembly, "Respect for Human Rights in Armed Conflicts," Resolution 2444, December 19, 1968; Adam Roberts and Richard Guelff, eds., *Documents on the Laws of War* (Clarendon: Oxford, 1989), 415, 455.

38. For an additional example from a prominent booster of air power, see Alexander De Seversky, *Air Power: Key to Survival* (New York: Simon and Schuster, 1950), 184–85.

39. Department of the Army, *Department of the Army Field Manual*, 19.

40. The chapter by Janina Dill in this volume suggests that such temptations remain in current U.S. Air Force practice.

41. For examples from the Vietnam War, see Westmoreland to Commander, All Subordinate Units, July 7, 1965, History Files, in *The War in Vietnam: Papers of William C. Westmoreland*, microfilm collection (Bethesda, MD: University Publications of America, 1993); "Combat Operations Minimizing Non-Combatant Battle Casualties," MACV Directive 525–3, September 7, 1965, History Files, in ibid.; memorandum, George M. Gallagher, September 15, 1965, History Files, in ibid.; "Tactics and Techniques for Employment of U.S. Forces in the Republic of Vietnam," MACV Directive 525–4, September 17, 1965, History Files, in ibid.; "Combat Operations Control, Disposition, and Safeguarding of Vietnamese Property, Captured Materiel and Food Supplies," MACV Directive 525–9, April 10, 1967, 2021 (MACJ4-Logistics), MACV Historical Office, *Records of the Military Assistance Command Vietnam*, microfilm collection (Bethesda, MD: University Publications of America, 1988); Division Order 003330.2, August 9, 1967, attachment to August 1967 Command History of the 1st Marine Division, *Records of the US Marine Corps in the Vietnam War*, microfilm collection (Bethesda, MD: University Publications of America, 1990); Appendix 10 to Annex A to 9th Infantry Division Field SOP, attachment to Major General George G. O'Connor, US Army Senior Officer Debriefing Report, February 23, 1968, in *U.S. Armed Forces in Vietnam 1954–1975*, ed. Paul Kesaris, microfilm collection (Frederick, MD: University Publications of America, 1983).

3. Targeting Civilians and U.S. Strategic Bombing Norms

I thank participants at the 2011 Cornell University Workshop on "Bombing: How Legal and Ethical Norms Change" for comments on the original draft of this paper and Matt Evangelista and Henry Shue for suggestions on shortening it. I expand these arguments in a book-length manuscript, *"To Make Heaven Weep": Civilians and the American Way of War*. Also see Neta C. Crawford, *Accountability for Killing: Moral Responsibility for Collateral Damage in America's Post-9/11 Wars* (Oxford: Oxford University Press, 2013).

1. While nuclear weapons would kill many more civilians than conventional bombing, norms about targeting civilians follow a more complex trajectory than norms and practices about conventional bombing. The normative beliefs of the U.S. public about the morality of targeting civilians with nuclear weapons now diverge from the beliefs of nuclear planners. Public support for using nuclear weapons against enemy civilians was high immediately following their first use against Japanese cities in World War II, but public support has dramatically declined since then. Although greatly reduced in number, nuclear weapons still remain in the strategic arsenals of the United States, and deterrence—threatening nuclear strikes in retaliation for an attack in the hope that this will prevent the other from striking—was and remains the stated nuclear weapons policy. The difference between the tolerance of civilian casualties for conventional and nuclear war is puzzling.

2. I focus here on why the norms changed. If space allowed, I would also more closely attend to the process of doctrinal change—the arguments won and lost within the U.S. military and the leadership and organizational learning of civilians.

3. Sahr Conway-Lanz, *Collateral Damage: Americans, Noncombatant Immunity, and Atrocity after World War II* (New York: Routledge, 2006), 19; and Conway-Lanz's chapter in this volume.

4. Conway-Lanz, *Collateral Damage*, 19–20.

5. Ibid., 230.

6. Ward Thomas, *The Ethics of Destruction: Norms and Force in International Relations* (Ithaca, NY: Cornell University Press, 2001), 150–51.

7. See ibid.; Robert Pape, *Bombing to Win: Airpower and Coercion in War* (Ithaca, NY: Cornell University Press, 1996).

8. Mark Clodfelter, *The Limits of Air Power: The American Bombing of North Vietnam* (New York: Free Press, 1989), 127.

9. On the politics of Rolling Thunder, see W. Hays Parks, "Rolling Thunder and the Law of War," *Air University Review* 33, no. 2 (1982): 2–21; and Earl H. Tilford Jr., *Crosswinds: The Air Force's Setup in Vietnam* (College Station: Texas A&M University Press, 1993), 59–103.

10. Russell F. Weigley, *The American Way of War: A History of United States Military Strategy and Policy* (Bloomington: Indiana University Press, 1973), 464; Pape, *Bombing to Win*, 177–81.

11. Quoted in Thomas, *The Ethics of Destruction*, 156.

12. Quoted in ibid., 157–58.

13. Parks, "Rolling Thunder," 13–14; Tilford, *Crosswinds*.

14. H. R. McMaster, *Dereliction of Duty: Lyndon Johnson, Robert McNamara, the Joint Chiefs of Staff and the Lies that Led to Vietnam* (New York: Harper Perennial, 1997), 329.

15. Ibid., 327.

16. Parks, "Rolling Thunder," 9–12; Thomas, *The Ethics of Destruction*, 152–53 and 190; William W. Momyer, *Air Power in Three Wars* (Washington: Air Force, 1979), 134–35.

17. Pape, *Bombing to Win*, 188.

18. United States National Security Study Memorandum estimate quoted in Pape, *Bombing to Win*, 190. Clodfelter notes that the CIA estimated that by 1967,

the 200,000 tons of bombs dropped had caused 29,600 deaths. Clodfelter, *Limits of Air Power*, 136.

19. Momyer, *Air Power in Three Wars*, 179–80.

20. Quoted in John Morrocco, *Rain of Fire: Air War, 1969–1973* (Boston: Boston Publishing Company, 1985), 46.

21. Paul Wiseman, "30-Year-Old Bombs Still Very Deadly in Laos," *USA Today*, 11 December 2003, http://www.usatoday.com/news/world/2003-12-11-laos-bombs_x.htm, accessed 20 July 2009.

22. Nicole Barrett, "Holding Individual Leaders Responsible for Violations of Customary International Law: The U.S. Bombardment of Cambodia and Laos," *Columbia Human Rights Law Review* 32 (2001): 429–76, 456.

23. William Shawcross, *Sideshow: Kissinger, Nixon and the Destruction of Cambodia* (New York: Simon and Schuster, 1979), 29; Marilyn B. Young, *The Vietnam Wars 1945–1990* (New York: HarperCollins, 1991), 238.

24. The Finnish Inquiry Commission report on Kampuchea of 1984 is quoted in Barrett, "Holding Individual Leaders Responsible," 437–38.

25. Ibid., 438n and 456.

26. Estimate from Taylor Owen and Ben Kiernan, "Bombs over Cambodia," *The Walrus*, October 2006, 62–63, http://www.walrusmagazine.com/articles/2006.10-history-bombing-cambodia/, accessed 29 July 2009.

27. During the first day alone, for instance, B-52s made more than 400 sorties. The United States made 41,653 total sorties from April to October 23, an average 184 sorties per day over the 226 days. Guenter Lewy, *America in Vietnam* (New York: Oxford University Press, 1978), 411.

28. W. Hays Parks, "Linebacker and the Law of War," *Air University Review* 34, no. 2 (1983): 2–30.

29. Lewy, *America in Vietnam*, 451. Lewy's estimate is based on pro-rating the estimated 52,000 North Vietnamese killed by U.S. bombing during Rolling Thunder, when the United States dropped 643,000 tons of bombs. This estimate is also consistent with estimates for deaths per ton during Linebacker II.

30. The United States dropped approximately 15,287 tons of bombs from B-52s, F-111s and F-4s, and the Navy and Marines dropped an additional five tons in Linebacker II. The United States did not bomb for 36 hours over Christmas itself. Calculated from Clodfelter, *The Limits of Air Power*, 194.

31. Joint Chiefs of Staff instructions quoted in Parks, "Linebacker and the Law of War," 13.

32. Clodfelter, *Limits of Air Power*, 191.

33. Momyer, *Air Power in Three Wars*, 188.

34. Morocco, *Rain of Fire*, 158.

35. Pape, *Bombing to Win*, 208.

36. Sebastian Kaempf, "Double Standards in US Warfare: Exploring the Historical Legacy of Civilian Protection and the Complex Nature of the Moral-Legal Nexus," *Review of International Studies* 35 (2009): 663.

37. Parks, "Linebacker and the Law of War."

38. Momyer, *Air Power in Three Wars*, 339.

39. Parks, "Linebacker and the Law of War."

40. Momyer, *Air Power in Three Wars*, 338.
41. Tilford, *Crosswinds*, xv.
42. Ibid., 75.
43. Ibid.
44. See Conway-Lanz, *Collateral Damage*, 214–20; and Nick Turse, *Kill Anything That Moves: The Real American War in Vietnam* (New York: Henry Holt, 2013).
45. United States Department of the Air Force, *International Law—The Conduct of Armed Conflict and Air Operations*, Air Forces Pamphlet 110-31 (Washington, DC: U.S. Government Printing Office, 1976), paragraph 1–3a (1).
46. Ibid., 1–6.
47. Ibid., 5–9. Also see Parks, "Linebacker and the Law of War"; and George N. Walne, "AFP-110-1: *International Law—The Conduct of Armed Conflict and Air Operations* and the Linebacker Bombing Campaigns of the Vietnam War," Center for Naval Analysis Professional Paper 487 (November 1987), 5 and 12.
48. United States Department of the Air Force, *International Law*, paragraph 5-3(c)(2)(b).
49. Thomas, *The Ethics of Destruction*, 169. These deaths were part of an estimated 5,000 excess deaths. For an estimate of excess deaths, see Frank Hobbs, "Population Estimates for Iraq," Population Studies Branch, Center for International Research, U.S. Bureau of the Census, January 1992, cited in Pape, *Bombing to Win*, 357. Beth Osborne Daponte estimates 3,500 civilian deaths from direct war effects before the cease-fire. Beth Osborne Daponte, "A Case Study in Estimating Casualties from War and its Aftermath: The 1991 Persian Gulf War," *PSR Quarterly* 3, no. 2 (1993): 57–66.
50. Michael W. Lewis, "The Law of Aerial Bombardment in the 1991 Gulf War," *American Journal of International Law* 97 (2003): 488. Also see Thomas Keaney and Eliot A. Cohen, *Gulf War Air Power Survey Summary Report* (Washington, DC: U.S. Government Printing Office, 1993). The entire survey was five volumes.
51. Thomas, *The Ethics of Destruction*, 158.
52. Keaney and Cohen, *Gulf War Air Power Survey Summary Report*, 69.
53. "Strategic Air Campaign against Iraq to Accomplish NCA Objectives," 16 August 1990, 3, as quoted in Pape, *Bombing to Win*, 222.
54. Pape, *Bombing to Win*, 222.
55. Quoted in Lewis, "The Law of Aerial Bombardment in the 1991 Gulf War," 487.
56. Crane, *Bombs, Cities and Civilians*, 156. Of the total ordnance dropped (88,500 tons), precision guided munitions accounted for 6,250 tons.
57. William M. Arkin, "Baghdad: The Urban Sanctuary in Desert Storm? *Airpower Journal* 11, no. 1 (1997): 5 and 7.
58. Thomas, *The Ethics of Destruction*, 88.
59. Lewis, "The Law of Aerial Bombardment in the 1991 Gulf War," 503.
60. Human Rights Watch, "Ticking Time Bombs: NATO's Use of Cluster Munitions in Yugoslavia," June 1999, http://www.hrw.org/legacy/reports/1999/nato2/, accessed 24 November 2002; Human Rights Watch Briefing Paper, "Cluster Munitions a Foreseeable Hazard in Iraq," March 2003.
61. Lewis, "The Law of Aerial Bombardment in the 1991 Gulf War," 504–7. See Henry Shue's chapter in this volume.

62. The "no-fly zones," which were implemented by Coalition forces in 1992, were maintained until 2003 to prevent Iraq from attacking its own civilians. The BBC reported in February 2001 that Iraq claimed that 300 civilians had been killed in the zones, which were maintained in the northern and southern areas of Iraq. "No-fly Zones: The Legal Position," *BBC News*, 19 February 2001, http://news.bbc.co.uk/2/hi/middle_east/1175950.stm; Susan Taylor Martin, "'No-Fly' Zone Perils Were for Iraqis, Not Allied Pilots," *St. Petersburg Times*, 29 October 2004, http://www.sptimes.com/2004/10/29/Columns/_No_fly__zone_perils_.shtml, accessed 23 November 2009. Jeremy Scahill reported in 2002 that "Baghdad says over the last decade more than 1,400 civilians have been killed in the US and British attacks in the no fly zones. While this cannot be independently verified, UN statistics say that more than 300 civilians have been killed in the raids since December 1998"; Scahill, "No-Fly Zones over Iraq," *CounterPunch*, 4 December 2004, http://www.counterpunch.org/2002/12/04/no-fly-zones-over-iraq.

63. Colonel Charles Wald quoted in Patrick M. Shaw, "Collateral Damage and the United States Air Force" (Thesis, School of Advanced Airpower Studies, Air University Maxwell Air Force Base, June 1997), 92.

64. Richard L. Sargent, "Weapons Used in Deliberate Force," in *Deliberate Force: A Case Study in Effective Air Campaigning*, ed. Robert C. Owen (Maxwell Air Force Base, Air University Press, 2000), 265.

65. Of the 622 precision guided weapons used by the United States, 567 were laser-guided bombs.

66. NATO dropped a total of 1,026 bombs: 708 were precision guided, and 318 were "dumb" or nonprecision weapons. Richard P. Hallion, *Precision Guided Munitions and the New Era of Warfare*, Air Power Studies Centre Paper Number 53 (RAAF Australia: Air Power Studies Centre, 1995); GlobalSecurity.org, "Operation Deliberate Force," http://globalsecurity.org/military/ops/deliberate_force.htm, accessed 28 July 2007. Shaw, "Collateral Damage and the United States Air Force," 92.

67. Scott A. Cooper, "The Politics of Air Strikes," in *Immaculate Warfare: Participants Reflect on the Air Campaigns over Kosovo, Afghanistan and Iraq*, ed. Stephen D. Wrage (Westport, CT: Praeger, 2003), 71–83.

68. Short, quoted in William Drozdiak, "NATO General Predicts Victory in Two Months," *The Washington Post*, 24 May 1999, http://www.washingtonpost.com/wp-srv/inatl/longterm/balkans/stories/airwar052499.htm.

69. Thomas, *The Ethics of Destruction*, 162.

70. The United States had not made public a detailed assessment of the causes of collateral damage. Human Rights Watch reported: "Human Rights Watch was able to determine the weapons involved in the cause of the civilian deaths in only twenty-eight of the ninety incidents. Of these, twenty-one are incidents about which it can be confirmed that precision-guided munitions (PGMs) were used (though there could be others). This includes all of the attacks on bridges or targets in and around the Belgrade area. Cluster bomb use can be positively determined in seven incidents (another five are possible but unconfirmed). In almost all of the other instances, it is impossible to establish the weapon used." Human Rights Watch, "Civilian Deaths in the NATO Air Campaign." Human Rights Watch Reports, February 2000, vol. 12, no. 1, http://www.hrw.org/reports/2000/nato/.

71. Derek S. Reveron, "Coalition Warfare: The Commander's Role," in *Immaculate Warfare: Participants Reflect on the Air Campaigns over Kosovo, Afghanistan, and Iraq*, ed. Stephen D. Wrage (Westport: Praeger, 2003), 56; Stephen D. Wrage, "The Ethics of Precision Air Power," in Wrage, *Immaculate Warfare*, 91.

72. Michael C. Short, quoted in Derek S. Reveron, "Coalition Warfare: The Commander's Role," in Wrage, *Immaculate Warfare*, 57.

73. Marcus Tanner, "Up to 100 Feared Dead as NATO Bombers Strike Kosovo Village," *The Independent*, 15 May 1999, http://www.independent.co.uk/news/up-to-100-feared-dead-as-nato-bombers-strike-kosovo-village-1093530.html; Ivo H. Daalder and Michael E. O'Hanlon, *Winning Ugly: NATO's War to Save Kosovo* (Washington: Brookings Institution, 2000), 241, 123, and 144.

74. See Human Rights Watch, "Ticking Time Bombs."

75. Clark quoted in Linda Kozaryn, "Air Chief's Kosovo Lesson: Go for Snake's Head First," *Defense News*, 26 October 1999, http://www.defenselink.mil/news/newsarticle.aspx?id=42877.

76. William S. Cohen and Henry H. Shelton, Joint Statement, "Kosovo after Action Review," Hearings of Senate Armed Services Committee, 14 October 1999, 1. Quoted in Nicholas J. Wheeler, "The Kosovo Bombing Campaign," in *The Politics of International Law*, ed. Christian Reus-Smit (Cambridge: Cambridge University Press, 2004), 197.

77. Human Rights Watch, "Civilian Deaths in the NATO Air Campaign."

78. DoD News Briefing—Secretary Rumsfeld and Gen. Myers, Monday, October 29, 2001, http://www.defense.gov/transcripts/transcript.aspx?transcriptid=2226.

79. Eric Schmitt, "Rumsfeld Says Dozens of Important Targets Have Been Avoided," *New York Times*, 24 March 2003, 12.

80. "Joint Methodology for Estimating Collateral Damage and Casualties for Conventional Weapons: Precision, Unguided and Cluster (U)," in Chairman of the Joint Chiefs of Staff Manual 3160.01, 20 September 2002.

81. Michael R. Gordon, "After the War: Preliminaries; U.S. Air Raids in '02 Prepared for War in Iraq," *New York Times*, 20 July 2003; Sharon Otterman, "The Calculus of Civilian Death," *New York Times*, 6 January 2009.

82. The details of the contents, values, and methods of the algorithms used to estimate potential harm to noncombatants are classified. See Douglas D. Martin and Steven C. Gordon, "Collateral Damage Estimation: Transforming Time-Sensitive Command and Control," Interservice/Industry Training, Simulation, and Education Conference, Paper no. 1768 (2004).

83. Senior Defense Official, "Background Briefing on Targeting," 5 March 2003, http://www.defense.gov/transcripts/transcript.aspx?transcriptid=2007.

84. Bradley Graham, "'Bugsplat' Computer Program Aims to Limit Civilian Deaths at Targets," *Washington Post*, 22 February 2003.

85. Eric V. Larson and Bogdan Savych, *Misfortunes of War: Press and Public Reactions to Civilian Deaths in Wartime* (Santa Monica: RAND Corporation, 2007), 209.

86. Ibid., 216.

87. Ibid.

88. U.S. Department of the Army, *U.S. Army/Marine Corps Counterinsurgency Field Manual*, Field Manual no. 3-24 (Chicago: University of Chicago Press, 2007), 45.

89. Ibid., 48.

90. Luke N. Condra, Joseph H. Felter, Radha K Iyengar, and Jacob N. Shapiro, "The Effect of Civilian Casualties in Afghanistan and Iraq," *NBER Working Paper Series*, National Bureau of Economic Research, July 2010.

91. U.S. Senate Committee on Armed Services, *Hearing to Consider Nominations*, 111th Cong., 1st sess., 2 June 2009, 11.

92. Quoted in Dexter Filkins, "U.S. Toughens Airstrike Policy in Afghanistan," *New York Times*, 22 June 2009.

93. Declassified excerpt from NATO Tactical Directive, 2 July 2009, released by NATO ISAF Headquarters 6 July 2009, http://www.nato.int/isaf/docu/offi cial_texts/Tactical_Directive_090706.pdf.

94. John Bohannon, "Counting the Dead in Afghanistan," *Science* 331 (11 March 2011): 1256–60, http://www.sciencemag.org/content/331/6022/1256.summary, accessed 11 March 2011.

95. Quoted in Ken Dilanian, "U.S. Counter-Terrorism Strategy to Rely on Surgical Strikes, Unmanned Drones," *Los Angeles Times*, 29 June 2011, http://articles.latimes.com/2011/jun/29/news/la-pn-al-qaeda-strategy-20110629; "Remarks of John O. Brennan, Assistant to the President for Homeland Security and Counterterrorism, on Ensuring al-Qa-ida's Demise," 29 June 2011, http://www.whitehouse.gov/the-press-office/2011/06/29/remarks-john-o-brennan-assistant-president-homeland-security-and-counter.

96. See, for instance, "The Covert Drone War," Bureau of Investigative Journalism web site, http://www.thebureauinvestigates.com/category/projects/drones/; Bill Roggio and Alexander Mayer, "Charting the Data for US Airstrikes in Pakistan, 2004–2013," The Long War Journal web site, http://www.longwarjournal.org/pakistan-strikes.php; "Drone War Pakistan: Analysis," The New America Foundation web site, http://counterterrorism.newamerica.net/drones/; Pakistan Body Count web site, http://www.pakistanbodycount.org/dattacks.php.

97. Bob Woodward, *Obama's Wars* (New York: Simon and Schuster, 2010), 26.

98. Barack Obama, "Remarks by the President at the National Defense University," 23 May 2013, http://www.whitehouse.gov/the-press-office/2013/05/23/remarks-president-national-defense-university.

99. Ibid. Also see "Fact Sheet: U.S. Policy Standards and Procedures for the Use of Force in Counterterrorism Operations Outside the United States and Areas of Active Hostilities," 20 May 2013, http://www.whitehouse.gov/the-press-office/2013/05/23/fact-sheet-us-policy-standards-and-procedures-use-force-counter terrorism; and "Background Briefing by Senior Administration Officials on the President's Speech on Counterterrorism," 23 May 2013, http://www.whitehouse.gov/the-press-office/2013/05/23/background-briefing-senior-administration-officials-presidents-speech-co.

4. The Law Applies, But Which Law?

1. Hersch Lauterpacht, "The Problem of Revision of the Law of War," *British Yearbook of International Law* 29 (1952): 382.

2. Convention for the Amelioration of the Condition of the Wounded in Armies in the Field, 22 August 1864, in *The Laws of Armed Conflicts: A Collection of*

Conventions, Resolutions, and Other Documents, 4th ed., ed. Dietrich Schindler and Jiří Toman (Leiden: Martinus Nijhoff, 2004), 3.

3. St. Petersburg Declaration Renouncing the Use, in Time of War, of Explosive Projectiles Under 400 Grammes Weight, 29 November 1868, in *Documents on the Laws of War*, 3rd ed., ed. Adam Roberts and Richard Guelff (New York: Oxford University Press, 2000), 54.

4. Ibid., 55.

5. See Prefatory Note, St. Petersburg Declaration, in Roberts and Guelff, *Documents on the Laws of War*, 53.

6. Hague Declaration 2 Concerning Asphyxiating Gases, 29 July 1899, and Hague Declaration 3 Concerning Expanding Bullets, 29 July 1899, both in Roberts and Guelff, *Documents on the Laws of War*, 60 and 64, respectively.

7. Hague Convention IX Concerning Bombardment by Naval Forces in Time of War, 18 October 1907, in Roberts and Guelff, *Documents on the Laws of War*, 112.

8. Article 25, Regulations Respecting the Laws and Customs of War on Land, Annexed to Hague Convention IV Respecting the Laws and Customs of War on Land, 18 October 1907, in Roberts and Guelff, *Documents on the Laws of War*, 73, 78.

9. Roberts and Guelff, *Documents on the Laws of War*, 78–79.

10. Declaration (IV, 1) to Prohibit, for the Term of Five Years, the Launching of Projectiles and Explosives from Balloons, and Other Methods of Similar Nature, The Hague, 29 July 1899, http://www.icrc.org/ihl.nsf/FULL/160?OpenDocument.

11. Declaration (XIV), Prohibiting the Discharge of Projectiles and Explosives from Balloons, The Hague, 18 October 1907, http://www.icrc.org/ihl.nsf/FULL/245?OpenDocument.

12. See Richard Gabriel and Karen Metz, "The Emergence of Modern War, Naval and Air Weaponry," in Gabriel and Metz, *A Short History of War: The Evolution of Warfare and Weapons* (Carlisle Barracks, PA: Strategic Studies Institute, US Army War College, 1992).

13. See Prefatory Note, 1923 Hague Rules of Aerial Warfare, in Roberts and Guelff, *Documents on the Laws of War*, 139.

14. The first air raid from ships is reputed to have taken place on Christmas Day 1914.

15. Draft Rules of Aerial Warfare, without Commentary, in Roberts and Guelff, *Documents on the Laws of War*, 141.

16. Article 18, in Roberts and Guelff, *Documents on the Laws of War*, 144.

17. Articles 22–26, in *Documents on the Laws of War*, 144–46.

18. *Parliamentary Debates*, Commons, 5th Series, 21 June 1938, vol. 337, cols. 937–38.

19. Protection of the Civilian Population Against Bombing from the Air in Time of War, League of Nations Assembly Resolution adopted on 30 September 1938, *Official Journal/League of Nations*, Special Supplement No. 182, Records of the XIXth Ordinary Session of the Assembly, 15–17.

20. Convention for the Amelioration of the Condition of the Wounded and Sick in Armed Forces in the Field, 12 August 1949; Convention for the Amelioration of the Condition of Wounded, Sick and Shipwrecked Members of Armed Forces at Sea, 12 August, 1949; Convention Relative to the Treatment of Prisoners of War, 12 August, 1949; Convention Relative to the Protection of Civilian Persons in Time

of War, 12 August, 1949, all in Roberts and Guelff, *Documents on the Laws of War*, 197, 222, 244, and 301, respectively.

21. See extract from the Judgment of the International Military Tribunal at Nuremberg, November 1948, in Roberts and Guelff, *Documents on the Laws of War*, 178.

22. Charter of the United Nations, 24 October 1945, in *Basic Documents in International Law*, 5th ed., ed. Ian Brownlie (Oxford: Oxford University Press, 2002), 2.

23. Universal Declaration of Human Rights, 10 December 1948, in Brownlie, *Basic Documents in International Law*, 192.

24. International Covenant on Economic, Social and Cultural Rights, 16 December 1966, in Brownlie, *Basic Documents in International Law*, 197.

25. International Covenant on Civil and Political Rights, 16 December 1966, in Brownlie, *Basic Documents in International Law*, 205.

26. European Convention for the Protection of Human Rights and Fundamental Freedoms, 4 November 1950, in Brownlie, *Basic Documents in International Law*, 245. It entered into force on 3 September 1953.

27. Article 15, in Brownlie, *Basic Documents in International Law*, 249.

28. Common Article 3 to the four Geneva Conventions of 12 August 1949, in Roberts and Guelff, *Documents on the Laws of War*, 198, 223, 245, and 302, respectively.

29. Protocol Additional to the Geneva Conventions of 12 August 1949, and Relating to the Protection of Victims of International Armed Conflicts, 8 June 1977 (hereinafter AP I), and Protocol Additional to the Geneva Conventions of 12 August 1949, and Relating to the Protection of Victims of Non-International Armed Conflicts, 8 June 1977 (hereinafter AP II, both in *Documents on the Laws of War*, 422 and 483, respectively.

30. Common Article 3.

31. Article 1(1), AP II, *Documents on the Laws of War*, 484.

32. Article 52(2), AP I, *Documents on the Laws of War*, 450.

33. Article 51(5)(b), AP I, *Documents on the Laws of War*, 449.

34. Article 43(2), AP I, *Documents on the Laws of War*, 444.

35. Article 50(1), AP I, *Documents on the Laws of War*, 448.

36. Article 57, AP I, *Documents on the Laws of War*, 452.

37. Article 49(3), AP I, *Documents on the Laws of War*, 448.

38. Article 58, AP I, *Documents on the Laws of War*, 448.

39. Part IV, AP II, *Documents on the Laws of War*, 490.

40. Prosecutor v Tadi , Decision on the Defence Motion for Interlocutory Appeal on Jurisdiction, Appeals Chamber, 2 October 1995, Case No. IT-94-1-AR72, 35 I.L.M. 32.

41. Ibid., 67.

42. Article 8(2)(c) and (e), Rome Statute of the International Criminal Court, in Roberts and Guelff, *Documents on the Laws of War*, 678–79.

43. Jean-Marie Henckaerts and Louise Doswald-Beck, eds., *Customary International Humanitarian Law*, 2 volumes (Cambridge: Cambridge University Press, 2005).

44. See Jean-Marie Henckaerts, "The ICRC Study on Customary International Humanitarian Law—An Assessment," in *Custom as a Source of International Humanitarian Law*, ed. Larry Maybee and Benarji Chakka (New Delhi: ICRC, 2006), 50.

45. See, for example, *McCann and Others v. United Kingdom*, 27 September 1995, 21 Eur. Ct. H. R. 97.

46. See *Cyprus v. Turkey*, 10 May 2001, 35 Eur. Ct. H. R. 30; *Varnava and Others v. Turkey*, 18 September 2009 (2010) 49 ILM 358.

47. *Bankovic and Others v. Belgium and Others*, 19 December 2001, 41 ILM 517.

48. *Isayeva and Others v. Russia*, 24 February 2005, 41 Eur. Ct. H. R. 39.

49. In two later cases, *Khamzayev and Others v. Russia*, no. 1503/02, 3 May 2011 and *Kerimova and Others v. Russia*, no. 17170/04, 20792/04, 22448/04, 23360/04, 5681/05, and 5684/05, 4 May 2011, both decided on 3 May 2011, there is no explicit reference to the law of armed conflict at all. See http://hudoc.echr.coe.int/sites/eng/Pages/search.aspx#{"documentcollectionid2":["GRANDCHAMBER", "CHAMBER"]}. For a more detailed analysis of the jurisprudence of the European Court of Human Rights, see Andrea Gioia, "The Role of the European Court of Human Rights," in *International Humanitarian Law and International Human Rights Law: Pas de Deux*, ed. Orna Ben-Naftal (Oxford: Oxford University Press, 2011), 201 and esp. 223–34.

50. See Andrew Clapham, "Human Rights Obligations of Non-State Actors in Conflict Situations," *International Review of the Red Cross* 88, no. 863 (September 2006): 491.

51. *Isayeva v. Russia*, 24 February 2005, 41 Eur. Ct. H. R. 847, para. 191.

52. Article 6, International Covenant on Civil and Political Rights.

53. Article 2, European Convention for the Protection of Human Rights.

54. Article 52(2), A P I, 450, in Roberts and Guelff, *Documents on the Laws of War*.

55. Art. 51(5)(b), A P I.

56. See UK Statement (c), made on ratification of the AP I, in Roberts and Guelff, *Documents on the Laws of War*, 510.

57. See Executive Summary—Main Findings, the Independent International Commission on Kosovo, in *The Kosovo Report. Conflict, International Response and Lessons Learned* (Oxford: Oxford University Press: 2000).

58. *HPCR Manual on International Law Applicable to Air and Missile Warfare, Bern, 15 May 2009, with Commentary* (Cambridge, MA: Program on Humanitarian Policy and Conflict Research, Harvard University School of Public Health, 2013).

59. Ibid., Commentary, 6.

60. Ibid.

61. International Court of Justice, Legal Consequences of the Construction of a Wall in the Occupied Palestinian Territory, Advisory Opinion, 2004 I.C.J. 136, para. 106 (9 July).

62. Article 1(1), AP II, *Documents on the Laws of War*, 484.

63. Convention with Respect to the Laws and Customs of War on Land, 29 July 1899, 1 *American Journal of International Law* 129 (1907).

64. Article 1(2), AP I, *Documents on the Laws of War*, 423.

5. Clever or Clueless?

1. "Airstrikes Clear Way for Libyan Rebels' First Major Advance," *New York Times*, March 27, 2011.

2. Stuart Patrick, "A New Lease on Life for Humanitarianism," *Foreign Affairs* (March 24, 2011), http://www.foreignaffairs.com/articles/67674/stewart-patrick/a-new-lease-on-life-for-humanitarianism?page=show.

3. Security Council Res. 1973, UN Doc S/RES/1973 (March 17, 2011), para. 6.

4. Phillip S. Meilinger, "Airpower: Myths and Facts" (Maxwell Air Force Base, AL: Air University Press, 2003), http://www.dtic.mil/cgi-bin/GetTRDoc?AD=ADA421894.

5. Rebecca Grant, "Nine Myths about Kosovo," *Air Force Magazine*, June 2000, 50, http://www.airforce-magazine.com/MagazineArchive/Documents/2000/June%202000/0600myths.pdf.

6. Frederick Taylor, *Dresden, Tuesday, February 13, 1945* (New York: HarperCollins, 2004), 416–17.

7. Richard Overy, "The Air War in Europe, 1939–1945," in *A History of Air Warfare*, ed. John Andreas Olsen (Washington, DC: Potomac Books, 2010), 49.

8. Ibid., 50.

9. Phillip S. Meilinger, "The USSBS' Eye on Europe," *Air Force Magazine*, October 2011, 74.

10. Daniel J. Hughes, "Professors in the Colonels' World," in *Military Culture and Education*, ed. Douglas Higbee (Burlington, VT: Ashgate, 2010), 153.

11. See Stephen B. Johnson, *The United States Air Force and the Culture of Innovation, 1945–1965* (Washington, DC: Air Force History and Museums Program, 2002). "Since its inception the United States Air Force has depended on advanced technologies to maintain an edge over its actual and potential enemies. Continuous innovation became a way of life," 221.

12. See, generally, "How Precision Weapons Revolutionized Warfare," Strategy Page web site, September 27, 2007, http://www.strategypage.com/htmw/htairw/articles/20070924.aspx.

13. John T. Correll, "The Emergence of Smart Bombs," *Air Force Magazine*, March 2010, 64, http://www.airforcemag.com/MagazineArchive/Pages/2010/March%202010/0310bombs.aspx.

14. Rebecca Grant, "Airpower in Afghanistan: How a Faraway War Is Remaking the Air Force," Mitchel Institute special report, February 2009, 4, http://www.afa.org/Mitchell/reports/0209airpowerinafghan.pdf.

15. See, generally, Michael C. Sirak, "ISR Revolution," *Air Force Magazine*, June 2010, 36.

16. Tom Vanden Brook, "Air Force to Train More on Drones," *USA Today*, June 16, 2009.

17. Mark Benjamin, "Killing 'Bubba' from the Skies," Salon.com, February 15, 2008, http://www.salon.com/news/feature/2008/02/15/air_war/.

18. Anna Mulrine, "A Look inside the Air Force's Control Center for Iraq and Afghanistan," *US News & World Report*, May 29, 2008, my italics.

19. See, e.g., Daniel Baltrusaitis, "Airpower: The Flip Side of COIN," *Conflict & Security* (Summer/Fall 2008): 89.

20. Thom Shanker, "Civilian Risks Curb Strikes in Afghan War," *New York Times*, July 23, 2008.

21. Ibid.

22. "Targeting and Weaponeering," in *Air Force Operations & the Law* (Maxwell Air Force Base, AL: Judge Advocate General's School, 2009), 247–64, http://www.afjag.af.mil/shared/media/document/AFD-100510-059.pdf.

23. Shanker, "Civilian Risks Curb Strikes."

24. What exactly constitutes "customary international law" is often hotly disputed. See, e.g., Hays Parks, "The ICRC Customary Law Study: A Preliminary Assessment," *Proceedings of the Annual Meeting of the American Society of International Law* 99 (2005): 208–12.

25. For a list, see Hays Parks, "National Security Law in Practice: The Department of Defense Law of War Manual," ABA Standing Committee on Law and National Security Breakfast Series, November 18, 2010, http://www.americanbar.org/content/dam/aba/migrated/natsecurity/hays_parks_speech11082010.authcheckdam.pdf.

26. See, e.g., Protocol Additional to the Geneva Conventions of 12 August 1949, and relating to the Protection of Victims of International Armed Conflicts (Protocol I), 8 June 1977, Part IV. Although the United States is not a party to Protocol I, it accepts that most of it comprises customary international law that binds all nations.

27. See Robert Chesney, "Hays Parks on the Demise of the DoD Law of War Manual," Lawfare Web site, December 8, 2012, http://www.lawfareblog.com/2012/12/hays-parks-on-the-demise-of-the-dod-law-of-war-manual/.

28. *HPCR Manual on International Law Applicable to Air and Missile Warfare, Bern, 15 May 2009, with Commentary* (Cambridge, MA: Program on Humanitarian Policy and Conflict Research, 2009), http://www.ihlresearch.org/amw/manual/.

29. James E. Baker, "LBJ's Ghost: A Contextual Approach to Targeting Decisions and the Commander in Chief," 4 *Chicago Journal of International Law* 407 (2003): 417.

30. Declaration Renouncing the Use, in Time of War, of Explosive Projectiles under 400 Grammes Weight, Saint Petersburg, 29 November / 11 December 1868, http://www.icrc.org/ihl.nsf/FULL/130?OpenDocument.

31. Ibid.

32. Carl von Clausewitz, *On War*, Book I, Ch. 1, my italics.

33. Sun Tzu, *The Art of War*, trans. Lionel Giles (London: Luzac & Co, 1910), chapter 3.

34. See Stephen T. Hosmer, *The Conflict over Kosovo: Why Milosevic Decided to Settle the Conflict over Kosovo When He Did*, RAND Research Brief RB-71 (RAND, 2001).

35. See, generally, Michael N. Schmitt, "The Interpretive Guidance on the Notion of Direct Participation in Hostilities: A Critical Analysis," *Harvard Law School National Security Journal* 1, no. 5 (2010).

36. Daniel J. Boorstin, "Myths of Popular Innocence," *US News & World Report*, March 4, 1991, 41.

37. Ibid.

38. *HPCR Manual on International Law Applicable to Air and Missile Warfare*, Sec. D(II), para. 18.

39. Dwight D. Eisenhower, Speech before the American Society of Newspaper Editors, April 16, 1953, http://www.informationclearinghouse.info/article9743.htm.

40. *Commentary on the HPCR Manual on International Law Applicable to Air and Missile Warfare* (Cambridge, MA: Program on Humanitarian Policy and Conflict Research, 2010), 91, Commentary on Rule 14, para. 2, http://ihlresearch.org/amw/Commentary%20on%20the%20HPCR%20Manual.pdf.

41. Convention on the Prohibition of the Use, Stockpiling, Production and Transfer of Anti-Personnel Mines and on their Destruction, September 18, 1997, http://www.un.org/Depts/mine/UNDocs/ban_trty.htm.

42. "CBU-89 Gator," Deagel.com web site, March 10, 2011, http://www.deagel.com/Bombs-and-Guidance-Kits/CBU-89-Gator_a000943001.aspx.

43. See, e.g., John B. Alexander, "Optional Lethality: Evolving Attitudes towards Nonlethal Weaponry," *Harvard International Business Review*, May 7, 2006.

44. Convention on the Prohibition of the Development, Production, Stockpiling and Use of Chemical Weapons and on Their Destruction, 1993, http://www.opcw.org/chemical-weapons-convention/.

45. Gabriella Blum, "The Laws of War and the 'Lesser Evil'," 35 *Yale Journal of International Law* 1 (2010): 68.

46. Department of the Air Force, "US Air Force Transformation Flight Plan 2004," http://www.af.mil/shared/media/document/AFD-060328-005.pdf.

47. For a discussion of the legality of fuel-air (thermobaric) and other blast weapons, see William H. Boothby, *Weapons and the Law of Armed Conflict* (New York: Oxford University Press, 2009), 233–35.

48. Noah Shachtman, "When a Gun Is More Than a Gun," *Wired*, March 20, 2003, http://www.wired.com/politics/law/news/2003/03/58094.

49. For a discussion of an approach to treaty law that advocates a focus on the broader principles underlying technologies rather than on the specific technologies themselves, see Benjamin Kastan, "The Chemical Weapons Convention and Riot Control Agents: Limitations of Technology-Specific Arms Control Treaties," December 5, 2010, unpublished paper in author's possession.

50. See, generally, Andrew Feickert and Paul K. Kerr, "Cluster Munitions: Background and Issues for Congress," Congressional Research Service, January 11, 2011, http://www.fas.org/sgp/crs/weapons/RS22907.pdf.

51. Submunitions that fail to detonate as designed can unexpectedly explode, putting at risk civilians who enter a battle area after the fighting is over.

52. Feickert and Kerr, "Cluster Munitions."

53. Ibid., 1.

54. For example, torpedoes that used magnetic devices to home in on their target's metal composition were widely used in World War II. See Theodore Roscoe, *United States Submarine Operations in World War II* (Annapolis: United States Naval Institute, 1949), 253–58.

55. See, e.g., Ryan J. Vogel, "Drone Warfare and the Law of Armed Conflict," 39 *Denver Journal of International Law and Policy* 1 (2010).

56. This writer has said previously: "In fact, such weaponry is basic to warmaking. David slew Goliath with a missile weapon before the giant could bring his weapons to bear; the sixteen-foot pikes of Alexander the Great's phalanxes reached their targets well ahead of the twelve-foot pikes wielded by their opponents; English longbowmen destroyed the flower of French knighthood at Agincourt from afar when they rained arrows down upon the horsemen; and, more recently, US and British tanks destroyed the heart of Saddam's armor forces during 1991's Battle of 73 Easting much because their guns outranged those of Iraq's T-72 tanks. There is nothing really new about killing from a distance." Charles J. Dunlap Jr., "Does Lawfare Need an Apologia?" 43 *Case Western Reserve Journal of International Law* (2010): 132.

57. Geoffrey Best, *War and Law since 1945* (Oxford: Clarendon Press, 1994), 289.

58. Richard Falk, "Think Again: Human Rights," *Foreign Policy* 141 (March–April 2004): 18.

59. Manual for Courts-Martial, United States, 2012 ed., Part IV, para. 14c(2)(a)(iv), http://www.loc.gov/rr/frd/Military_Law/pdf/MCM-2012.pdf.

60. Department of the Army, *Counterinsurgency*, Field Manual no. 3-24, D-8, 15 December 2006, http://www.scribd.com/doc/9137276/US-Army-Field-Manual-FM-324-Counterinsurgency.

61. Dan Ephron, "The Book on Iraq," *Newsweek* (web exclusive), December 14, 2007.

62. See Gian P. Gentile, "The Selective Use of History in the Development of American Counterinsurgency Doctrine," *Army History* 72 (Summer 2009): 21. Gentile maintains that the army's *Counterinsurgency* field manual "draws narrowly on a body of writing from the French Revolutionary War school of counterinsurgency theory and practice of the early 1960s."

63. Alan J. Vick, Adam Grissom, William Rosenau, Beth Grill, and Karl P. Mueller, *Air Power in the New Counterinsurgency Era: The Strategic Importance of USAF Advisory and Assistance Missions* (Santa Monica, CA: RAND, 2006), 111.

64. One of the few book-length treatments of airpower in COIN operations concludes that "the support role of airpower . . . is usually the most important and effective mission in guerrilla war." See James S. Corum and Wray R. Johnson, *Airpower in Small Wars* (Lawrence: University of Kansas Press, 2003), 427.

65. See, e.g., Department of the Army, *Counterinsurgency*, para. 4-1: "Ultimately, the long-term objective for both sides in that struggle remains acceptance by the people of the state or region of the legitimacy of one side's claim to political power."

66. See, e.g., Department of the Army, *Counterinsurgency*, para. A-5.

67. "Link Hard, Soft Power," *DefenseNews*, September 8, 2008, 28 (editorial).

68. Sarah Sewall, "Modernizing US Counterinsurgency Practice: Rethinking Risk and Developing a National Strategy," *Military Review* (September–October 2006): 103.

69. Department of the Army, *Counterinsurgency*, para. 2-42.

70. Steve Coll, "The General's Dilemma," *New Yorker*, September 8, 2008.

71. Department of the Army, *Counterinsurgency*, app. E, para. E-5 admonishes ground commanders to: "exercise exceptional care when using airpower in the strike role. Bombing, even with the most precise weapons, can cause unintended civilian casualties. Effective leaders weigh the benefits of every air strike against its risks. An air strike can cause collateral damage that turns people against the host-nation (HN) government and provides insurgents with a major propaganda victory. Even when justified under the law of war, bombings that result in civilian casualties can bring media coverage that works to the insurgents' benefit."

72. Department of the Air Force, "MQ-1 Predator Unmanned Aerial Vehicle," July 20, 2010, http://www.af.mil/information/factsheets/factsheet.asp?fsID=122.

73. Department of the Air Force, "MQ9 Reaper," August 18, 2010, http://www.af.mil/AboutUs/FactSheets/Display/tabid/224/Article/104470/mq-9-reaper.aspx.

74. Department of the Air Force, "RQ-4 Global Hawk," November 19, 2009, http://www.af.mil/AboutUs/FactSheets/Display/tabid/224/Article/104516/rq-4-global-hawk.aspx.

75. Department of the Air Force, "Joint Direct Attack Munitions," May 10, 2006, http://www.minot.af.mil/library/factsheets/factsheet.asp?id=3900.

76. General Barry R. McCaffrey, "Memorandum for Colonel Mike Meese, United States Military Academy, Subject: After Action Report," October 15, 2007, 5, my italics, http://www.mccaffreyassociates.com/pages/documents/AirForceAAR-101207.pdf.

77. A single center once held as many as 26,000 detainees in 2007. See Jason Keyser, "Camp Bucca: Military Closes Largest Detention Camp in Iraq," *Huffington Post*, September 16, 2009, http://www.huffingtonpost.com/2009/09/16/camp-bucca-military-close_n_289285.html.

78. In September 2007, *USA Today* reported that insurgent deaths already exceeded the total for all of 2006 by 25 percent. See Jim Michaels, "19,000 Insurgents Killed in Iraq Since '03," *USA Today*, September 26, 2007.

79. See, e.g., Anthony H. Cordesman, "US Airpower in Iraq and Afghanistan: 2004–2007," Center for Strategic and International Studies, December 13, 2007, http://www.csis.org/media/csis/pubs/071213_oif-oef_airpower.pdf.

80. Michael M. Dunn, "The Pile-On Effect," Air Force Association, July 9, 2008, http://secure.afa.org/EdOp/2008/PileOnEffect.pdf.

81. "Off-Target: The Conduct of the War and Civilian Casualties in Iraq," Human Rights Watch, New York, 2003, 20.

82. Madelyn Hsiao-Rei, Hamit Dardagan, Gabriela Guerrero Serdán, Peter M. Bagnall, John A. Sloboda, and Michael Spagat, "The Weapons That Kill Civilians—Deaths of Children and Noncombatants in Iraq, 2003–2008," *New England Journal of Medicine* 360 (April 16, 2009): 1585.

83. "Security Gains Reverse Iraq's Spiral Though Serious Problems Remain," ABC/BBC/ARD/NHK Poll, March 17, 2008, http://www.abcnews.go.com/images/PollingUnit/1060a1IraqWhereThingsStand.pdf.

84. See, generally, Benjamin Lambeth, "Operation Enduring Freedom, 2001," in *A History of Air Warfare,* ed. John Andreas Olsen (Washington, DC: Potomac Books, 2010), 256.

85. Major John Thomas, spokesman for NATO's International Security Assistance Force, quoted in Associated Press, "US Coalition Airstrikes Kill, Wound Civilians in Southern Afghanistan, Official Says," *International Herald Tribune,* June 30, 2007.

86. Brigadier General Richard Blanchette, chief spokesman for NATO forces, quoted in Pamela Constable, "NATO Hopes to Undercut Taliban with Surge of Projects," *Washington Post,* September 27, 2008, A12.

87. United Nations Assistance Mission in Afghanistan (UNAMA), "Afghan Civilian Casualties Rise 31 Per Cent in First Six Months of 2010," August 10, 2010, Relief Web, http://reliefweb.int/report/afghanistan/afghan-civilian-casualties-rise%C2%A031%C2%A0-cent-first-six-months-2010.

88. See, e.g., David Zuchinno, "As US Deaths in Afghanistan Rise, Military Families Grow Critical," *L.A. Times,* September 2, 2010.

89. Luke N. Condra, Joseph H. Felter, Radha K. Iyengar, and Jacob N. Shapiro, "The Effect of Civilian Casualities in Afghanistan and Iraq," NBER Working Paper no. 16152, July 2010, 34, http://www.nber.org/papers/w16152.

90. Ibid.

91. Noah Shachtman, "Petreaus Launches Afghan Air Assault; Strikes Up 172 Percent," *Wired,* October 12, 2010, http://www.wired.com/dangerroom/2010/10/gloves-come-off-afghan-air-war-strikes-spike-172/.

92. UNAMA, "Afghan Civilian Casualties."

93. United Nations Assistance Mission in Afghanistan and the Afghanistan Independent Human Rights Commission, "Afghanistan: Annual Report 2010, Protection of Civilians in Armed Conflict" (March 2011), 24, http://unama.unmissions.org/Portals/UNAMA/human%20rights/March%20PoC%20Annual%20Report%20Final.pdf.

94. Ibid., i.

95. Ben Arnoldy, "History Sides with Taliban, for Now," *Christian Science Monitor,* May 10, 2010, 9. See also Condra, Felter, Iyengar, and Shapiro, "The Effect of Civilian Casualties," 28–29.

96. See transcript of Jeremy Shapiro's participation in "Afghanistan and Pakistan Index and Assessments," Brookings Institution panel, October 5, 2009, http://www.brookings.edu/~/media/Files/events/2009/1005_afghanistan_pakistan/20091005_afghanistan_pakistan.pdf.

97. Christopher Swift, "The Drone Blowback Fallacy," *Foreign Affairs* (July 1, 2012).

98. C. Christine Fair, Karl C. Kaltenthaler, and William J. Miller, "You Say Pakistanis All Hate the Drone War? Prove It," *The Atlantic,* January 23, 2013, http://www.theatlantic.com/international/archive/2013/01/you-say-pakistanis-all-hate-the-drone-war-prove-it/267447/.

99. Ibid.

100. William R. Polk, *Violent Politics: A History of Insurgency, Terrorism & Guerrilla War, From the American Revolution to Iraq* (New York: Harper, 2007), xvi–xv.

101. See Zbigniew Brzezinski, "The Smart Way Out of a Foolish War," *Washington Post,* March 30, 2008, B03. Brzezinski wrote, "It is also important to recognize that most of the anti-US insurgency in Iraq has *not* been inspired by al-Qaeda. Locally based jihadist groups have gained strength only insofar as they have been able to identify themselves with the fight against a hated foreign occupier."

102. See also Maria Costigan, "Interview with Jacqueline (Jill) Hazelton: Does Counterinsurgency as State-Building Work?" Belfer Center, December 3, 2010, http://belfercenter.ksg.harvard.edu/publication/20860/interview_with_jacqueline_jill_hazelton.html?breadcrumb=%2Fexperts%2F2085%2Fjacqueline_l_hazelton.

103. Jacqueline L. Hazelton, "The Hearts-and-Minds Approach versus the High-Force Low Accommodation Approach," in "Compellence and Accommodation in Counterinsurgency Warfare, September 2010," unpublished manuscript in author's possession.

104. Hazelton, "The Hearts-and-Minds Approach."

105. Michael Few, "The Wrong War: An Interview with Bing West," *Small Wars Journal,* February 21, 2011, my italics.

106. Ibid., my italics.

107. For a general discussion of just war theory, see Alexander Moseley, "Just War Theory," *Internet Encyclopedia of Philosophy,* February 10, 2009, http://www.iep.utm.edu/justwar/.

108. Caroline Frost, "Colin Powell: A Profile," BBC, August 5, 2003, http://www.bbc.co.uk/bbcfour/documentaries/profile/colin-powell.shtml.

109. Ibid.

110. Special Rapporteur on Extrajudicial, Summary or Arbitrary Executions, *Report of the Special Rapporteur on Extrajudicial, Summary or Arbitrary Executions: Study on Targeted Killings*, delivered to the Human Rights Council, UN General Assembly Doc. A/HRC/14/24/Add.6 (May 28, 2010), http://www2.ohchr.org/english/bodies/hrcouncil/docs/14session/A.HRC.14.24.Add6.pdf.

111. Ibid., 25.

112. Marc Pitzke, "Interview with Defense Expert P. W. Singer," *Spiegel Online International*, March 12, 2010, http://www.spiegel.de/international/world/0,1518,682852,00.html.

113. Ibid.

114. Aaron Retica, "Drone-Pilot Burnout," *New York Times Magazine*, December 12, 2008.

115. Jeff Schogol and Markeshia Ricks, "Demand Grows for UAV Pilots, Sensor Operators," *Air Force Times*, April 21, 2012.

116. James Dao, "Drone Pilots Are Found to Get Stress Disorders as Much as Those in Combat Do," *New York Times*, February 23, 2013, A9.

117. Michael Spagat, "The Iraq Sanctions Myth," *Pacific Standard*, April 26, 2013, http://www.psmag.com/politics/the-iraq-sanctions-myth-56433/.

118. David Deptula, "Retired Lt. Gen. Deptula: Drones the Best Weapons We've Got for Accuracy, Control, Oversight: Critics Don't Get It," *Breaking Defense*, February 15, 2013, http://breakingdefense.com/2013/02/15/retired-gen-deputula-drones-best-weapons-weve-got-for-accurac/.

119. Joshua Foust, "Targeted Killing, Pro and Con: What to Make of US Drone Strikes in Pakistan," *Atlantic*, September 26, 2012, http://www.theatlantic.com/international/archive/2012/09/targeted-killing-pro-and-con-what-to-make-of-us-drone-strikes-in-pakistan/262862/, referencing the joint study by Stanford University and New York University, "Living Under Drones."

120. Patrick B. Johnston and Anoop K. Sarbahi, "The Impact of US Drone Strikes on Terrorism in Pakistan and Afghanistan," January 3, 2013, http://patrickjohnston.info/materials/drones.pdf.

121. New America Foundation, "The Drone War in Pakistan," http://natsec.newamerica.net/drones/pakistan/analysis, citing statistics current through June 8, 2013.

122. Daniel Byman, "Why Drones Work," *Foreign Affairs*, July/August 2013, 32; but see also Audrey Kurth Cronin, "Why Drones Fail," *Foreign Affairs*, July/August 2013, 44.

123. Elizabeth Mendes and Joy Wilke, "Americans' Confidence in Congress Falls to Lowest on Record," *Gallup^R Politics*, June 13, 2013, http://www.gallup.com/poll/163052/americans-confidence-congress-falls-lowest-record.aspx.

124. "The public continues to hold the military and military leaders in high regard, despite the recent scandal involving former CIA director Gen. David Petraeus"; Pew Research Center for the People & Press, "As Fiscal Cliff Nears, Democrats Have Public Opinion on Their Side," December 13, 2012, http://www.people-press.org/2012/12/13/section-1-views-of-obama-congress-the-parties. See also Gallup, "Honesty/Ethics in Professions," November 19–21, 2010, http://www.gallup.com/poll/1654/honesty-ethics-professions.aspx.

125. Francisco Martin, Stephen J. Schnably, Richard Wilson, Jonathan Simon, and Mark Tushnet, *International Human Rights & Humanitarian Law: Treaties, Cases & Analysis* (New York: Cambridge University Press, 2006), 536.

126. *Gilligan v. Morgan,* 413 US 1, 10 (1973).

127. "China's Hu Tells Sarkozy Dialogue Way out of Libya Crisis," Reuters, March 30, 2011, http://www.reuters.com/article/2011/03/30/china-france-libya-idUSB9E7EN00T20110330, accessed October 20, 2013.

128. William F. Owen, "Seek and Destroy: The Forgotten Strategy for Countering Armed Rebellion," *Infinity Journal* (Spring 2011): 14.

129. Taylor Dinerman, "The Politics of Airpower," Gatestone Institute, March 30, 2011, http://www.hudson-ny.org/1995/the-politics-of-air-power.

130. Julian E. Barnes and Adam Entous, "US Military Proposal to Arm Rebels Includes No-Fly Zone in Syria,"*Wall Street Journal,* June 13, 2013, http://online.wsj.com/article/SB10001424127887323734304578543761501124132.html.

131. John Reed, "Here's a Map of the 23 Places the US Will Bomb If There Is a Syria No-Fly Zone," ForeignPolicy.com, June 14, 2013, http://killerapps.foreignpolicy.com/posts/2013/06/14/heres_a_map_of_the_places_the_us_will_bomb_if_theres_a_syria_no_fly_zone?wp_login_redirect=0.

132. Lewis Mackenzie, "Why This Strategy Won't Fly in Syria," *Globe and Mail,* June 25, 2013, http://www.theglobeandmail.com/commentary/why-this-strategy-wont-fly-in-syria/article12786362/.

133. The full quote is: "War is an ugly thing, but not the ugliest of things. The decayed and degraded state of moral and patriotic feeling which thinks that nothing is worth war is much worse. The person who has nothing for which he is willing to fight, nothing which is more important than his own personal safety, is a miserable creature and has no chance of being free unless made and kept so by the exertions of better men than himself." John Stuart Mill, "The Quotation Page," http://www.quotationspage.com/quote/27169.html (last accessed June 26, 2013).

134. Carl von Clausewitz, *On War,* Book II, Ch. 7.

135. George Friedman and Meredith Friedman,*The Future of War: Power, Technology and American World Dominance in the 21st Century* (New York: Crown, 1996), xi.

136. Ibid.

6. The American Way of Bombing and International Law

1. Tami Davis Biddle's chapter in this volume demonstrates that the definition of military objectives proved contentious even before precision technology provided the ability to reach selected targets with any degree of certainty. For further discussion of the flexibility of the concept of military objective, see also Sahr Conway-Lanz's chapter in this volume.

2. The exception is close air support of troops in contact, which follows the pace of ground combat. So-called time-sensitive targeting is, contrary to what the name may suggest, often "pre-planned."

3. Geoffrey Best, *Humanity in Warfare: The Modern History of the International Law of Armed Conflict* (London: Weidenfeld and Nicolson, 1983), 265.

4. Protocol Additional to the Geneva Conventions of 12 August 1949, and relating to the Protection of Victims of International Armed Conflicts (Protocol I), 3 June 1977, http://www.icrc.org/ihl/INTRO/470.

5. The increase in targeted killings of suspected terrorists with unmanned aerial vehicles during the Obama administration has drawn attention to the question of how distinction applies to persons. Without entering into the debate about whether the "battlefield" in the war on terror extends to Pakistan, Yemen, and Somalia, it is safe to say that what is colloquially referred to as drone strikes differs from regular all-out war from the air. During the latter, specifically in the course of the initial air assault on a country, targeting of individual combatants, mostly in the form of decapitation strikes, is dwarfed by comparison to attacks on objects considered to be military, which means distinction is first and foremost a matter of categorizing objects. Of course, the designation of persons as either combatants or civilians arises during the application of the principle of proportionality, when decision makers calculate the so-called collateral damage to be expected from an air strike. This complicated issue is, however, beyond the scope of this chapter. For a discussion of the increasing use of unmanned aerial vehicles during the Obama administration, see the chapters by Hugh Gusterson and Klem Ryan in this volume.

6. Most authors simply assume that the two criteria logically presuppose each other. See for instance, Yoram Dinstein, "Legitimate Military Objectives under the Current Jus In Bello," in *Legal and Ethical Lessons of NATO's Kosovo Campaign*, ed. Anru E. Wall (Newport, RI: Naval War College, 2002), 4; and Marco Sassòli, *Bedeutung einer Kodifikation für das allgemeine Völkerrecht mit besonderer Betrachtung der Regeln zum Schutz der Zivilbevölkerung vor den Auswirkungen von Feindseligkeiten* (Basel: Helbig & Lichtenhahn, 1990), 363. For an opposing view, see Timothy L. H. McCormack and Helen Durham, "Aerial Bombardment of Civilians: The Current International Legal Framework," in *Bombing Civilians: A Twentieth-Century History*, ed. Y. Tanaka and M. B. Young (New York: New Press, 2009), 222.

7. Not every military objective is at all times a legitimate target of attack. The latter must also conform to the principle of proportionality and the attack must be subject to precautions. Military objectives are hence only presumptive legitimate targets, which is what I mean when using the term "legitimate target" throughout this chapter.

8. Dinstein considers a truck carrying foodstuffs for the military a legitimate target, but the driver of the vehicle does not lose her status as a civilian. Yoram Dinstein, *The Conduct of Hostilities under the Law of International Armed Conflict*, 2nd ed. (Cambridge: Cambridge University Press, 2010), 150. In a slight variation on the example described, the *Joint Fires and Targeting Handbook* lists "basic processing and equipment production" industry as well as its "end products," even though they are described as "chiefly civilian," as part of a target set for legitimate attack. U.S. Joint Forces Command, *Joint Fires and Targeting Handbook* (Suffolk, VA: U.S. Joint Forces Command, 2007), III-49.

9. See, for instance, A. P. V. Rogers, *Law on the Battlefield* (Manchester: Manchester University Press, 2004); Michael N. Schmitt, "21st Century Conflict: Can the Law Survive?" *Melbourne Journal of International Law* 8 (2007): 443; and Michael N. Schmitt, "Fault Lines in the Law of Attack," in *Testing the Boundaries of International Humanitarian Law*, ed. Susan Breau and Agnieszka Jachec-Neale (London: British Institute of International and Comparative Law, 2006).

10. Michael Bothe, Karl J. Partsch, and Waldemar A. Solf, *New Rules for Victims of Armed Conflicts: Commentary on the Two 1977 Protocols Additional to the Geneva*

Conventions of 1949 (Geneva: Martinus Nijhoff, 1982), 326; Igor Primoratz, *Civilian Immunity in War* (Oxford: Oxford University Press, 2007). Another Commentary requires the advantage to be "substantial and relatively close"; see Yves Sandoz and Bruno Zimmermann, *Commentary on the Additional Protocols of 8 June 1977 to the Geneva Conventions of 12 August 1949* (Geneva: Martinus Nijhoff, 1987), 2209.

11. The following recent *Operational Law Handbooks* contain the formulation of Article 52 (2) API: U.S. Army Judge Advocate General's Legal Center and School, International and Operational Law Department, *Operational Law Handbook 2002*, 16; *Operational Law Handbook 2003*, 8; *Operational Law Handbook 2004*, 12, 20; *Operational Law Handbook 2007*, 21, 446; *Operational Law Handbook 2008*, 19, 614; *The Military Commander and the Law 2008*, 19ff.; *Operational Law Handbook 2010*, 12; *Operational Law Handbook 2011*, 20; *Operational Law Handbook 2012*, 22.

12. Air Land Sea Application Center, *The Joint Targeting Process and Procedures for Targeting Time-Critical Targets*, Field Manual 90-36 (Hampton, Independent City, VA: Air Land Sea Application Center, Langley Air Force Base, 1997), i–10.

13. "Civilian objects consist of all civilian property and activities other than those used to support or sustain the adversary's war fighting capability"; U.S. Joint Chiefs of Staff, *Joint Doctrine for Targeting*, Joint Publication 3-60, 17 January 2002, A-2.

14. U.S. Department of Defense, Military Commission Instruction No. 2, 30 April 2003, Article 5d, http://www.defense.gov/news/May2003/d20030430milcominstno2.pdf, my italics. The 2007 *Joint Doctrine for Targeting* brings back the link between war-sustaining and war-*fighting*. U.S. Joint Chiefs of Staff, *Joint Doctrine for Targeting*, Joint Publication 3-60, 13 April 2007, last changed 28 July 2011, 91.

15. U.S. Army Judge Advocate General's Legal Center and School, International and Operational Law Department, *Operational Law Handbook 2007*, 22; *Operational Law Handbook 2010*, 19ff.; *Operational Law Handbook 2011*, 21; *Operational Law Handbook 2012*, 23.

16. U.S. Army Judge Advocate General's Legal Center and School, International and Operational Law Department, *Operational Law Handbook 2006*, 20; *Operational Law Handbook 2007*, 21; *Operational Law Handbook 2008*, 19.

17. U.S. Army Judge Advocate General's Legal Center and School, International and Operational Law Department, *Operational Law Handbook 2010*, 20; *Operational Law Handbook 2011*, 20; *Operational Law Handbook 2012*, 22. Previous handbooks provided the slightly more specific formulation "military objectives that enable an opposing state and its military forces to wage war"; see, among others, *Operational Law Handbook 2006*, 20.

18. See, for instance, Dinstein, *Legitimate Military Objectives*, 6; Francoise J. Hampson, "Means and Methods of Warfare in the Conflict in the Gulf," in *The Gulf War 1990–1991 in International and English Law*, ed. Peter Rowe (London: Routledge, 1993), 94; and Primoratz, *Civilian Immunity in War*.

19. U.S. Army Judge Advocate General's Legal Center and School, International and Operational Law Department, *Operational Law Handbook 2010*, 12.

20. Sandoz and Zimmermann, *Commentary on the Additional Protocols*, 2209.

21. U.S. Department of Defense, Department of the Air Force, *Targeting,* Doctrine Document 3-60 of 8 June 2006, last changed 28 July 2011, vii; United States Air Force, *Air Force Basic Doctrine*, Doctrine Document 1, 17 November 2003, 18,

http://www.globalsecurity.org/military/library/policy/usaf/afdd/1/afdd1-2003.pdf, italics in original; United States Air Force, *Targeting*, Air Force Doctrine Document 3-60, 8 June 2006, last changed 28 July 2011 (Maxwell Air Force Base, AL: LeMay Center for Doctrine Development and Education, 2011), vii.

22. United States Air Force, *Air Force Basic Doctrine*, Doctrine Document 1, 17 November 2003, 18, http://www.globalsecurity.org/military/library/policy/usaf/afdd/1/afdd1-2003.pdf, italics in original; United States Air Force, *Targeting*, Air Force Doctrine Document 3-60, 8 June 2006, last changed 28 July 2011 (Maxwell Air Force Base, AL: LeMay Center for Doctrine Development and Education, 2011), vii.

23. United States Air Force, *Strategic Attack*, Doctrine Document 2-1.2, 12 June 2007, 2, http://www.fas.org/irp/doddir/usaf/afdd3-70.pdf.

24. U.S. Joint Forces Command, *Joint Fires and Targeting Handbook*, III-20.

25. Israeli Defence Forces, for instance, systematically targeted police stations in the Gaza Strip; see Amnesty International, *Israel/Gaza: Operation "Cast Lead": 22 Days of Death and Destruction* (London: Amnesty International Publications, 2009), 6.

26. See for instance, Nicholas Wheeler, "The Kosovo Bombing Campaign," in *The Politics of International Law*, ed. Christian Reus-Smit (Cambridge: Cambridge University Press, 2004).

27. United States Air Force, *Targeting*, 14.

28. Even though these efficiency-seeking strategies are meant to keep casualty rates down by ending wars sooner and sparing friendly forces and some enemy combatants, I argue elsewhere that at least for civilians they make war more, not less, dangerous. Specifically, the calculation that wars waged in this way lead to quick political victory is often not borne out by reality. For this argument, see Janina Dill, *Legitimate Targets? The Partial Effectiveness of International Law in US Air Warfare* (Cambridge: Cambridge University Press, forthcoming).

29. This explanation is not at all meant to suggest that the strategies employed during operations Allied Force, Iraqi Freedom, and Cast Lead were legally or morally less problematic because they closely followed the respective political goals of the three wars. The legality and legitimacy of the resort to force, some methods of war employed, and some targets chosen are in doubt in all three cases.

30. "Sharp wars are brief" is one of the most famous propositions contained in the Lieber Code. Francis Lieber, *Instructions for the Government of Armies of the United States in the Field* (Washington, DC: Government Printing Office, 1898), Article XXIX, http://www.loc.gov/rr/frd/Military_Law/pdf/Instructions-gov-armies.pdf. The Lieber Code was originally issued as General Orders No. 100, Adjutant General's Office, 1863.

31. Giulio Douhet, *Command of the Air: USAF Warrior Studies* (Washington, DC: United States Government Printing Office, 1983); Mark Clodfelter, "Molding Air Power Convictions: Development and Legacy of William Mitchell's Strategic Thought," in *The Paths of Heaven: The Evolution of Air Power Theory*, ed. Phillip S. Melinger (Maxwell Air Force Base, Montgomery, AL: Air University Press, 1997). It is hence not surprising that the U.S. Air Force is the greatest advocate of EBO. While the army and the Joint Forces Command have eschewed the terminology of EBO since 2007 and 2008, respectively, the air force has in the same time expanded its application. See Commander U.S. Joint Forces Command, Assessment of Effects

Based Operations, memorandum, 14 August 2008, 20; and United States Air Force, *Air Force Basic Doctrine*, 19.

32. Elliot Cohen, "The Mystique of US Air Power," *Foreign Affairs* 73 (1994): 109.

33. Ibid.

34. Ibid. See also Elliot Cohen, "Change and Transformation in Military Affairs," *Journal of Strategic Studies* 27 (2004): 3; and Elliot Cohen, "The Meaning and Future of Air Power," *Orbis* (Spring 1995): 189.

35. The targeting of civilian morale, of course, also challenges Article 51 (2) sentence 2 API. It reads "acts or threats of violence the primary purpose of which is to spread terror among the civilian population are prohibited." A separate discussion of this provision is beyond the scope of this chapter.

36. Charles J. Dunlap, "Law and Military Interventions, Preserving Humanitarian Values in 21st Century Conflicts," 14, paper presented at the conference Humanitarian Challenges in Military Intervention, Washington, DC, 2001, http://www.duke.edu/~pfeaver/dunlap.pdf. The words "concrete and direct" describe the military advantage required for the proportionality calculus in accordance with Article 51(5b), not the advantage defining a military objective according to Article 52(2).

7. Force Protection, Military Advantage, and "Constant Care" for Civilians

This chapter has benefited greatly from comments by Charles Garraway at the Cornell workshop, two reviewers for the press, and members of the Oxford War Workshop, especially Janina Dill, Cécile Fabre, Seth Lazar, and Cheyney Ryan. Final judgments are of course entirely my responsibility, even though some of the best ideas are not.

1. U.S. Department of Defense, *Conduct of the Persian Gulf War: Final Report to Congress Pursuant to Title V* (Washington, DC: US Government Printing Office, 1992), Appendix O, 615 (hereafter *Final Report*). Washington insiders often refer to this document as a Title V Report. Also see claims about protection of civilians on 611, 612, and 614 of Appendix O.

2. I have struggled with a different case in Henry Shue, "Bombing to Rescue?: NATO's 1999 Bombing of Serbia," in *Ethics and Foreign Intervention*, ed. Deen K. Chatterjee and Don E. Scheid (Cambridge: Cambridge University Press, 2003), 97–117; and with issues common to both cases in Henry Shue and David Wippman, "Limiting Attacks on Dual-Use Facilities Performing Indispensable Civilian Functions," 35 *Cornell International Law Journal* (2002): 559–79. The latter is, thanks to Wippman, considerably more sophisticated about alternative interpretations of international law, while the former unimaginatively treats revision of treaties as the only option to law as currently interpreted.

3. Joy Gordon, *Invisible War: The United States and the Iraq Sanctions* (Cambridge: Harvard University Press, 2010), 90. Her source for the two quotations is World Health Organization/UNICEF, "Special Mission to Iraq," February 1991, 12 and 2, respectively.

4. Gordon, *Invisible War*, 34–35. Her source is United Nations, "Report to the Secretary-General on Humanitarian Needs in Kuwait and Iraq in the Immediate

Post-crisis Environment by a Mission to the Area Led by Mr. Martti Ahtisaari, Under-Secretary-General for Administration and Management, Dated 20 March 1991," S/22366, annex (1991) (Ahtisaari Report), para. 8.

5. Gordon, *Invisible War*, 37.

6. Joy Gordon (Fairfield University), personal communication, 17 June 2013. In her original multipage note regarding the analyses that yielded the figure of 500,000 deaths of children Gordon relied on the calculations of respected scholar Tim Dyson of the London School of Economics. Dyson reportedly no longer stands by those calculations and himself believes that the figure of 500,000 is inflated. See Gordon, *Invisible War*, 255–57n82.

7. Gordon documents meticulously the methods by which and the extent to which the United States dominated important decisions about the UN sanctions through their entire life.

8. I am immensely grateful to Janina Dill for suggesting this way of summarizing the essential legal structure and for insightful suggestions about the difficult question of the status of Rule 2.

9. First Additional Protocol, Article 35(1).

10. Adam Roberts and Richard Guelff, eds., *Documents on the Laws of War*, 3rd ed. (Oxford: Oxford University Press, 2000), 9. This basic rule of the customary law for the conduct of war appears widely, for example, in International Declaration Concerning the Laws and Customs of War, Brussels, 27 August 1874, Article 12; in The Laws of War on Land, Oxford, 9 September 1880, Article 4; Convention (II) with Respect to the Laws and Customs of War on Land and Its Annex: Regulations concerning the Laws and Customs of War on Land, The Hague, 29 July 1899, Regulation 22; and Convention (IV) Respecting the Laws and Customs of War on Land and Its Annex: Regulations concerning the Laws and Customs of War on Land, The Hague, 18 October 1907. "The principle which states that the Parties to the conflict do not have an unlimited right was contested in the name of total war. The Protocol answered by declaring that the negation of this principle is incompatible with the preservation of civilisation and humanity." Claude Pilloud, Yves Sandoz, Christophe Swinarski, Bruno Zimmermann, *Commentary on the Additional Protocols of 8 June 1977 to the Geneva Conventions of 12 August 1949* (Geneva: Martinus Nijhoff, 1987), 392.

11. First Additional Protocol, Article 48.

12. First Additional Protocol, Article 51(5)(b). Compare customary Rule 14 in Jean-Marie Henckaerts and Louise Doswald-Beck, *Customary International Humanitarian Law*, vol. 1, *Rules* (Cambridge: Cambridge University Press for the ICRC, 2005), 46. "Which may be expected" means not "is actually expected" but "could reasonably be expected." Failure to make reasonable judgments about what to expect is not exculpating.

13. I have analyzed some of the many complications here in Henry Shue, "Proportionality in War," in *The Encyclopedia of War*, ed. Gordon Martel (Oxford: Blackwell, 2011).

14. As everyone knows, the United States chose not to ratify the First Additional Protocol as a whole because it finds some Articles objectionable, but the United States has raised no objections to the Articles under discussion here, and it claims generally to abide by the provisions of the First Protocol.

15. Article 57 of the First Protocol is expressive of what is presented as Rule 15 of customary international humanitarian law in Henckaerts and Doswald-Beck, *Customary International Humanitarian Law*, 1:51.

16. First Additional Protocol, Article 57(3), my italics. This is expressive of what is presented as Rule 21 of customary international humanitarian law in ibid., 1:65. See final section below for discussion.

17. We shall see later that this applies directly to the plans for the bombing of Iraq in 1991.

18. First Additional Protocol, Article 57(2)(a)(ii), my italics. This is expressive of what is presented as Rule 17 of customary international humanitarian law in Henckaerts and Doswald-Beck, *Customary International Humanitarian Law*, 1:56.

19. The final clause, "except at the extreme," does not directly reflect the language of the law, and I will explain my reasoning for it later.

20. Christopher Greenwood, "The Law of Weaponry at the Start of the New Millennium," in *The Law of Armed Conflict: Into the Next Millennium*, ed. Michael N. Schmitt and Leslie C. Green (Newport, RI: Naval War College, 1998), 201. Christopher Greenwood is now a judge on the International Court of Justice; this was written prior to his appointment. Judge Greenwood went on immediately to write: "Where the proportionality principle differs from the unnecessary suffering principle is that it is clearly established that it does not stop at the prohibition of unnecessary collateral injury and damage, but also requires a belligerent to abstain from an attack altogether, even if that means losing a military advantage which cannot be obtained by other means, if the military advantage would not be worth the expected civilian casualties and damage" (ibid.). Clearly civilian losses that are not minimal are excessive, according to Article 57. Michael Walzer has argued on moral grounds for the same thesis that the running of additional risk is required. See his distinctive and valuable conceptions of "double intention" and "due care" in *Just and Unjust Wars: A Moral Argument with Historical Illustrations* [1977], 4th ed. (New York: Basic Books, 2006), 151–56. Also see David Luban, "Risk Taking and Force Protection," in *Reading Walzer*, ed. Yitzhak Benbaji and Naomi Sussman (London: Routledge, 2013), 277–301.

21. I have briefly explored this clash in Henry Shue, "Civilian Protection and Force Protection," in *Ethics, Law and Military Operations*, ed. David Whetham (Basingstoke: Palgrave Macmillan, 2010), 135–47.

22. Whether through, say, nonlethal weapons this could be changed to any degree is a long story that I cannot enter upon here. The most imaginative discussion is in Elaine Scarry, *The Body in Pain: The Making and Unmaking of the World* (New York: Oxford University Press, 1985), 60–157.

23. "Force protection" means protection of the combatants on one's own side. Protections in the Geneva Conventions for wounded and captured combatants from opposing forces are discussed in Charles Garraway's chapter in this volume.

24. Here I oppose what I take to be the spirit of the dominant trend in the discussion of the morality of war in contemporary analytic philosophy. This too of course is a very long story that cannot be taken up again here. See, for example, Henry Shue, "Do We Need a 'Morality of War?'" in *Just and Unjust Warriors: The Moral and Legal Status of Soldiers*, ed. David Rodin and Henry Shue (Oxford: Oxford University Press, 2008), 87–111; Henry Shue, "Laws of War," in *The Philosophy of*

International Law, ed. Samantha Besson and John Tasioulas (Oxford: Oxford University Press, 2010), 511–27; Janina Dill and Henry Shue, "Limiting the Killing in War: Military Necessity and the St. Petersburg Assumption," *Ethics and International Affairs* 26 (2012): 311–33; and Henry Shue, "Laws of War, Morality, and International Politics: Compliance, Stringency, and Limits," *Leiden Journal of International Law* 26, no. 2 (2013): 271–92. In this volume, for a powerful plea for taking seriously casualties on the other side of the war in Afghanistan, see the chapter by Richard Miller.

25. Charles J. Dunlap Jr., "Kosovo, Casualty Aversion, and the American Military Ethos: A Perspective," 10 *Air Force Journal of Legal Studies* (2000): 99. Thus, I agree entirely with the statement in Gen. Dunlap's chapter in this volume that "the life of the civilian—'innocent' or not—is not intrinsically more worthy than that of the combatant" but see no basis for any inference to the claim that "the notion that combatants are legally or morally obliged to take more risk than those holding civilian status is . . . deeply flawed." Neta Crawford suggests in her chapter that the increasing unacceptability since World War II of "wasting" one's own soldiers is underwritten by the spreading consensus on universal human rights.

26. Again, I am indebted to Janina Dill for legal perspicacity here.

27. Sometimes as soldiers approach the end of their tour of combat duty, they understandably switch into the mode of giving almost no weight to anything except "making it home." And of course in wars in which the soldiers genuinely cannot see the point, as Vietnam, Iraq (2003), and Afghanistan became for many U.S. soldiers, the only real goal becomes surviving the tour of duty and making it "back to the world." But I take these to be facts about wars that are and/or seem to the soldiers fighting them to be pointless, not facts about war and soldiers in general, who are remarkably willing to run risks.

28. The image of balancing itself tells us nothing and often serves, as Jeremy Waldron has perceptively observed, as a cover for the failure to spell out one's reasons in sufficient detail. See Jeremy Waldron, *Torture, Terror, and Trade-Offs: Philosophy for the White House* (Oxford: Oxford University Press, 2010), 22.

29. So is the "strategy" of suicide bombings among civilians. And, one might well think, so obviously is the strategy of targeting masses of civilians with nuclear weapons. Nuclear deterrence, however, is more complicated than it looks. For discussion, see Henry Shue, "Having It Both Ways: The Gradual Wrong Turn in American Strategy" and Steven Lee, "Morality, the SDI, and Limited Nuclear War," both in *Nuclear Deterrence and Moral Restraint: Critical Choices for American Strategy*, ed. Henry Shue (New York: Cambridge University Press, 1989).

30. Gordon, *Invisible War*.

31. U.S. Department of Defense, *Conduct of the Persian Gulf War*, 96.

32. See the chapter in this volume by Janina Dill on the logic of efficiency versus the logic of sufficiency.

33. Although it is little known, the United States employed the same "wholesale" approach to its dominant management of the sanctions imposed on Iraq for the next decade and beyond, preventing the import of spare parts for infrastructure repairs. The story is told in detail in Gordon, *Invisible War*. Thus when U.S. troops arrived in Baghdad in 2003, many of the repairs desperately needed since 1991 were still waiting to be made. Funds heartlessly wasted on his lavish life-style by the unlamented

dictator, Saddam Hussein, made a relatively minor difference in the absence of authorization to import spare parts, as Gordon shows. It is hardly surprising that the arrival of U.S. troops was greeted not with the welcoming flowers predicted by the U.S. politicians who wanted the war but with years of insurrection.

34. See the central argument in the chapter by Janina Dill.

35. It seems to me not out of the question that the 1991 bombing was in addition a violation of Article 51(5)(a) of the First Protocol, which prohibits treating "as a single objective [e.g., the grid as a whole] a number of clearly separated and distinct military objectives" [e.g., individual radar sites, communications centers, etc.]. This Article expresses the content of customary Rule 13; see Henckaerts and Doswald-Beck, *Customary International Humanitarian Law*, 1:43. I will not pursue that possibility here, however.

36. Many of them were in fact predicted: "Prior to the war, the US Defense Department anticipated that the damage from such a bombing campaign would be devastating to critical public services, which were already precarious. In January 1991, shortly before the bombing began, a memo from the Defense Intelligence Agency (DIA) to Central Command described the vulnerability of Iraq's water system. The memo noted that Iraq depended on imported and specialized equipment, as well as imported chemicals for water treatment. Iraq had no domestic means of producing the equipment and materials for water and sewage treatment, and no significant alternative sources"; Gordon, *Invisible War*, 89–90. Gordon's source is U.S. Defense Intelligence Agency, "Iraq Water Treatment Vulnerabilities as of 18 Jan. 91—Key Judgments," January 18, 1991, para. 5 (Gordon, *Invisible War*, 271n3).

37. Henckaerts and Doswald-Beck, *Customary International Humanitarian Law*, 1:65. This study is subjected to vigorous critique in Elizabeth Wilmshurst and Susan Breau, eds., *Perspectives on the ICRC Study on Customary International Humanitarian Law* (Cambridge: Cambridge University Press for British Institute of International and Comparative Law and Chatham House, 2007), which includes a chapter specifically on "The Law of Targeting" by Michael N. Schmitt.

38. First Additional Protocol, Article 57(2)(a)(iii).

39. J. W. Crawford III, "The Law of Noncombatant Immunity and the Targeting of National Electrical Power Systems," *Fletcher Forum of World Affairs* 21, no. 2 (1997): 114.

40. Ibid., 102.

41. Michael N. Schmitt, "The Principle of Discrimination in 21st Century Warfare," 2 *Yale Human Rights and Development Law Journal* (1999): 162. Schmitt was obviously echoing the content of Article 57(2)(a)(ii) of the First Additional Protocol and Customary Rule 17, discussed above. Schmitt is currently the chair of the Department of International Law of the U.S. Naval War College.

42. Ibid., 168.

43. Michael N. Schmitt, "Wired Warfare: Computer Network Attack and *Jus in Bello*," 846 *International Review of the Red Cross* (2002): 392. Neta Crawford also discusses long-term effects of targeting infrastructure in her chapter in this volume.

44. U.S. Joint Forces Command [USJFCOM], *Joint Fires and Targeting Handbook*, 19 October 2007, I-20–I-21.

45. Ibid., I–22.

46. U.S. Army Judge Advocates General's Legal Center & School, *Operational Law Handbook* (Charlottesville, VA: International and Operational Law Department, 2010), 148, my italics.

47. I do not have space here to trace U.S. practices forward in time. But neither the unnecessary bombing of 1991 nor the heartless sanctions of 1991–2003 have so far been repeated elsewhere by the United States.

48. Any adversary whose installations each had their own electricity generation capacity would have to be disabled one installation at a time.

49. See note 36.

8. Civilian Deaths and American Power

1. I further define the goal of power and offer evidence of its overriding force and moral consequences in Richard W. Miller, *Globalizing Justice* (Oxford: Oxford University Press, 2010), especially chapters 5 and 6.

2. See Evan Wright, *Generation Kill* (New York: Penguin, 2004), especially 151–52, 159–60, 166–72, 178–80, 190–91.

3. George W. Ball, *The Past Has Another Pattern* (New York: Norton, 1982), 409.

4. Ibid., 433.

5. From *Public Papers of the Presidents of the United States: Harry S. Truman* (Washington: U.S. Government Printing Office, 1961), 197, 212, cited in Sahr Conway-Lanz's chapter in this book.

6. They have not been entirely absent, either, as in the leveling of Fallujah in 2003 (see below).

7. *The U.S. Army–Marine Corps Counterinsurgency Field Manual* (Chicago: University of Chicago Press, 2007), 45.

8. Armed Forces Medical Intelligence Center, "Disease Information," www.gulflink.osd.mil/declassdocs/dia/19950901/950901_0504rept_91.html.

9. See Andrew Cockburn and Patrick Cockburn, *Out of the Ashes* (New York: Harper Collins, 1999), 4, 131.

10. Descriptions of the objective by a senior air force officer "who played a central role in the air campaign but declined to be named" and an unnamed "Air Force planner," reported in Barton Gellman, "Allied Air War Struck Broadly in Iraq," *Washington Post*, June 23, 1991, A1. Gellman attributes similar characterizations of this goal of "long-term leverage," in less pithy formulations, to Colonel John A. Warden III, deputy director of strategy, doctrine and plans for the air force.

11. G. Jones, "Iraq—Under-Five Mortality" (UNICEF, 1999), www.unicef.org/reseval/pdfs/irqu5est/pdf; UNICEF, "Child and Maternal Mortality Survey" (1999), www.unicef.org/reseval/iraqr/html (accessed June 7, 2002).

12. Joy Gordon, "Cool War: Economic Sanctions as a Weapon of Mass Destruction," *Harper's Magazine*, November 2002, 4, 2, 8, http://harpers.org/archive/2002/11/cool-war/.

13. See Madeleine Albright, *Madam Secretary* (New York: Miramax, 2003), 274–75.

14. Tabulations of deaths reported by at least two major news sources in Iraq Body Count, "A Dossier of Civilian Casualties 2003–2005" include 5,232 to 6,882 deaths of civilians killed by US forces during the invasion. On June 11, 2003, the Associated Press Baghdad bureau reported a "fragmentary" count of civilian

deaths during the month of war against Saddam, based solely on deaths recorded by sixty of Iraq's 124 hospitals. They further excluded records that did not distinguish between civilian and military deaths, a precaution they took to exclude "hundreds, possibly thousands" of civilian victims. This death tally was 3,240. See Niko Price, "AP Tallies 3,240 Civilian Deaths in Iraq," http://news.google.com/newspapers?nid=1980&dat=20030611&id=XyoyAAAAIBAJ&sjid=Ma8FAAAAIBAJ&pg=5931,1407496.

15. While the records that would have provided a death toll of Iraqi soldiers have been destroyed, the U.S. military estimated that at least 2,320 Iraqis were killed in one operation, the attack on troops near Baghdad preliminary to the taking of the city. See "Special Analysis: Iraq Has Fallen," *Independent* (London), April 16, 2003, 7. Reuters reports "unofficial think-tank estimates" of 4,895 to 6,370 Iraqi military deaths in the invasion; "Table of Military Deaths in Iraq," April 7, 2004, www.reuters.com.

16. See U.S. Department of Defense, "Global War on Terrorism—Operation Iraqi Freedom," siadapp.dmdc.osd.mil/personnel/CASUALTY/oif-total-by-month.pdf.

17. Wright, *Generation Kill*, 56, 68.

18. Ibid., 191.

19. See Human Rights Watch, *Off-Target: The Conduct of the War and Civilian Casualties in Iraq* (New York: Human Rights Watch, 2003), 94–97.

20. See Jonathan Steele, "Iraq War Logs: Apache Helicopters Kill 14 Civilians in Hunt for Insurgents," *Guardian*, October 22, 2010, with accompanying video.

21. See Gilbert Burnham, Riyadh Lafta, Shannon Doocy, and Les Roberts, "Mortality after the 2003 Invasion of Iraq," *The Lancet* 368 (2006): 1421, 1425. After the onset of insurgency, Iraq Body Count's tabulations of civilian deaths specifically reported by at least two major news sources were highly vulnerable to the inadequacy of reports from the areas of most intense conflict. Still, among the reported 9,270 Coalition-caused violent civilian deaths in Iraq Body Count, "A Dossier of Civilian Casualties 2003–2005," 26 percent occurred after the end of the invasion, and 88 percent of these occurred in the second year after the start of the invasion. See the section "Fact Sheet—When Did They Die?"

22. Scott Peterson, "Behind Fallujah Strategy," *Christian Science Monitor*, October 29, 2004.

23. See Patrick Cockburn, "Ambulance Torn Apart in Fallujah as US Launches 'Precision' Strikes," *Independent*, September 14, 2004; Andrew Buncombe, "'Take Them Out, Dude': Pilots Toast Hits on Iraqi 'Civilians,'" *Independent*, October 6, 2004.

24. See Fadhil Bafdrani, "Inside Besieged Falluja," BBC News, October 18, 2004.

25. See Fadel al-Barani, "U.S. Bombs Rain on Falluja, Rebels Attack in Samarra," Reuters, November 6, 2004.

26. See Richard Oppel and Robert Worth, "G.I.'s Open Attack to Take Falluja from Iraq Rebels," *New York Times*, November 8, 2009.

27. See Kim Sengupta, "Battle for Fallujah Rages," *Independent*, November 9, 2004.

28. See Miles Schuman, "Falluja's Health Damage," *The Nation*, November 24, 2004.

29. See Kim Sengupta and Raymond Whitaker, "When the Smoke Has Cleared around Fallujah, What Horrors Will Be Revealed?" *Independent*, November 14, 2004.

30. See Rory McCarthy, "Civilian Cost of Battle for Falluja Emerges," *Observer*, November 14, 2004; CNN, "Iraqi Forces Find Chemical Weapons in Lab," November 26, 2004.

31. Jim Krane, "Layers of Aircraft Press War on Fallujah," Associated Press, November 12, 2004.

32. Michael Janofsky, "Rights Experts See Possibility of a War Crime," *New York Times*, November 14, 2004.

33. See Sam Cage, "U.N.: War 'Wreaking Havoc' on Iraq Young," Associated Press, November 23, 2004.

34. "Aid Reaches Falluja's Citizens," BBC News, November 27, 2004.

35. See "Iraq: Death Toll in Fallujah Rising, Doctors Say," IRIN News, UN Office for the Coordination of Humanitarian Affairs, January 4, 2005.

36. The tally by the United Nations Assistance Mission in Afghanistan of civilians killed by military action of pro-government forces was 440 in 2010, 518 in 2011, and 316 in 2012, the vast majority due to actions of U.S.-NATO forces. This is the sum of findings that UNAMA regards as clearly and reliably established by the work of their Kabul-based investigators. The exclusion of inconclusive results of these investigations, which are burdened by terrible insecurity, virtually guarantees that this is an underestimate. See *Annual Report on Protection of Civilians in Armed Conflict 2010* (Kabul: UNAMA, 2011), 21; and *Annual Report 2012* (Kabul: UNAMA, 2013), 2.

37. *The U.S. Army–Marine Corps Counterinsurgency Field Manual*, 365.

38. According to United Nations Development Program estimates, life expectancy at birth in Afghanistan in 2012 was 49.1, one year more than the lowest, Sierra Leone. The chance of dying before the age of five was 15 percent, close to the highest (Mali, 18 percent). By comparison, the chance of death before five was 9 percent in neighboring Pakistan and 5 percent in Nepal and Bangladesh. See UNDP, *International Human Development Indicators* (2012), http://hdr.undp.org/en/statistics/.

39. ABC News, "Afghanistan Poll: Where Things Stand 2010," December 6, 2010, http://abcnews.go.com/Politics/Afghanistan/afghanistan-poll-things-stand-2010/story?id=12277743; ABC News/BBC/ARD/Washington Post, "Afghanistan: Where Things Stand: Afghan Views Worsen as Setbacks Counter U.S. Progress in Helmand," December 6, 2010, http://www.langerresearch.com/uploads/1116a1Afghanistan.pdf,question 29. For the Helmand and Kandahar responses, I am grateful for information provided by Gary Langer, director of Langer Research Associates, which conducted the poll.

40. U.S. Senate Committee on Armed Services, "Hearing to Receive Testimony on Afghanistan," December 2, 2009, http://www.docstoc.com/docs/48525776/HEARING-TO-RECEIVE-TESTIMONY-ON-AFGHANISTAN, 6–7.

41. I present a detailed moral argument for pursuit of a political settlement with the Taliban, conceding substantial territorial authority, together with an end to the offensive in the Pashtun south, in Richard W. Miller, "The Ethics of America's Afghan War," *Ethics and International Affairs* 25 (2011): 103–31.

42. *The U.S. Army–Marine Corps Counterinsurgency Field Manual*, ii.

43. Ibid., 45.

44. UNAMA, *Mid-Year Report 2010: Protection of Civilians in Armed Conflict* (Kabul: UNAMA, 2010), Glossary. The definition varies. In *Annual Report 2012*, it is almost entirely borrowed verbatim from a definition supplied by the U.S.-NATO command, including the specification that a civilian is not "part of a *levée en masse* (mass uprising)" (*Annual Report 2012*, Glossary, 31). This specification is a good

setting for UNAMA's decision not to count deaths and injuries caused by Afghan police firing into demonstrations as civilian casualties because "demonstrations . . . [are] a rule of law and policing issue" (67–68). Very broad construals of "civilian" are always implicit in the detailed descriptions of representative incidents in which civilian casualties are attributed to anti-government forces.

45. Steve Mufson, *Fighting Years: Black Resistance and the Struggle for a New South Africa* (Boston: Beacon Press, 1990), 200.

46. Afghanistan NGO Safety Office, "ANSO Quarterly Data Report," fourth quarter 2010, "Summary and Assessment," http://reliefweb.int/sites/reliefweb.int/files/resources/F7EE02609B7F7A0F4925782200200E4D-Full_Report.pdf.

47. *The U.S. Army–Marine Corps Counterinsurgency Field Manual*, xxxv–xxxvi, xxxiii.

48. John Stuart Mill, "A Few Words on Non-Intervention" (1867) in *Collected Works of John Stuart Mill,* John Robson, ed. (Toronto: University of Toronto Press, 1984), vol. 21, 119.

9. Proportionality and Restraint on the Use of Force

1. Jean-Marie Henckaerts and Louise Doswald-Beck, *Customary International Humanitarian Law,* vol. 1, *Rules* (Cambridge: Cambridge University Press, 2005).

2. Judith Gail Gardam, "Proportionality and Force in International Law," 87 *American Journal of International Law* (1993): 391–413; Dale Stephens and Michael W. Lewis, "The Law of Armed Conflict—A Contemporary Critique," 6 *Melbourne Journal of International Law* (2005): 55–85; Janina Dill, "Applying the Principle of Proportionality in Combat Operations," Policy Briefing, Oxford Institute for Ethics, Law, and Armed Conflict, December 2010, http://www.elac.ox.ac.uk/downloads/proportionality_policybrief_%20dec_2010.pdf.

3. *Final Report to the Prosecutor by the Committee Established to Review the NATO Bombing Campaign against the Federal Republic of Yugoslavia* (2000), para. 48, http://www.un.org/icty/pressreal/nato061300.htm.

4. Gardam, "Proportionality and Force in International Law," 405, 409; Stephens and Lewis, "The Law of Armed Conflict," 76.

5. Michael Walzer, "Responsibility and Proportionality in State and Nonstate Wars," *Parameters* (Spring 2009): 40–52.

6. Jonathan F. Keiler, "The End of Proportionality," *Parameters* (Spring 2009): 53–64; Jefferson D. Reynolds, "Collateral Damage on the 21st Century Battlefield: Enemy Exploitation of the Law of Armed Conflict, and the Struggle for a Moral High Ground," *Air Force Law Review* 56 (2005): 1–108.

7. "Collective expectations for the proper behavior of actors with a given identity"; Peter J. Katzenstein, "Introduction: Alternative Perspectives on National Security" in *The Culture of National Security,* ed. Peter J. Katzenstein (New York: Columbia University Press, 1996), 5.

8. Wayne Sandholtz, "Explaining International Norm Change," in *International Norms and Cycles of Change,* ed. Wayne Sandholtz and Kendall Stiles (Oxford: Oxford University Press, 2008), 1–26; Neta Crawford, *Argument and Change in World Politics: Ethics, Decolonization, and Humanitarian Intervention* (Cambridge: Cambridge University Press, 2002); Richard Price, "Reversing the Gun Sights: Transnational Civil Society Targets Landmines," *International Organization* 52, no. 3 (1998): 613–44.

9. Robert D. Benford and David A. Snow, "Framing Processes and Social Movements: An Overview and Assessment," *Annual Review of Sociology* 26 (2000): 613–14.

10. David A. Snow and Robert D. Benford, "Ideology, Frame Resonance, and Participant Mobilization," *International Social Movement Research* 1 (1998): 198.

11. "Frame amplification" involves "highlighting some issues, events, or beliefs as being more salient than others"; ibid.

12. "Counterframing" is defined as attempts "to rebut, undermine, or neutralize a person's or group's myths, versions of reality, or interpretive framework"; quoted in ibid.

13. Christopher Greenwood, "Legal Issues Regarding Explosive Remnants of War," UN document CCW/GGE/I/WP.10 (2002).

14. William H. Boothby, "Cluster Bombs: Is There a Case for New Law?" HPCR Occasional Paper Series no. 5 (2005), 36, 40, http://reliefweb.int/report/world/cluster-bombs-there-case-new-law.

15. Virgil Wiebe, "Footprints of Death: Cluster Bombs as Indiscriminate Weapons under International Humanitarian Law," *Michigan Journal of International Law* 22 (2000): 85–167.

16. ICRC, "Explosive Remnants of War: An Examination of Legal Issues Raised in the ERW Discussions," UN document CCW/GGE/II/WP.8 (2002), 2, emphasis in original.

17. Charles Garraway, "How Does Existing International Law Address the Issue of Explosive Remnants of War?" UN document CCW/GGE/XII/WG.1/WP.15 (2005).

18. Reynolds, "Collateral Damage on the 21st Century Battlefield," 96.

19. Anthony H. Cordesman, "The Lessons and Non-Lessons of the Air and Missile Campaign in Kosovo" (2000), 250, http://csis.org/files/media/csis/pubs/kosovolessons-full.pdf.

20. Author's interview with NGO representative, 10 December 2003, Washington, DC; author's interview with government official, 30 March 2006, Brussels.

21. Quoted in Michael Krepon, "Weapons Potentially Inhumane: The Case of Cluster Bombs," *Foreign Affairs* (April 1974): 599.

22. Quoted in "M26 Multiple Launch Rocket System," http://www.globalsecurity.org/military/systems/munitions/m26.htm.

23. U.S. Department of Defense, "Kosovo/Operation Allied Force After-Action Report: Report to Congress," 31 January 2000, 90, http://www.globalsecurity.org/military/library/report/2000/kaar02072000.pdf.

24. United Kingdom, "Military Utility of Cluster Munitions," Working Paper Prepared by the UK Group of Governmental Experts of the Parties to the CCW, UN document CCW/GGE/X/WG.1/WP.1 (21 February 2005).

25. Statement of Edward Cummings, Head of the U.S. Delegation to the Second Preparatory Committee of the 2001 CCW Review Conference, 5 April 2001, http://www.state.gov/documents/organization/16694.pdf; Defense Science Board, "Munitions System Reliability," Report, September 2005, http://www.acq.osd.mil/dsb/reports/ADA441959.pdf

26. France, "Technical Improvements to Submunitions," UN document CCW/GGE/II/WP.6, Geneva, 15–26 July 2002.

27. Gal Scellos, "Colloque Handicap International—Intervention du Gal Scellos (en tant que représentant du ministère de la Défense)," 6 October 2005, http://www.sousmunitions.org/dyn/actus/InterventionGBSCELLOS.pdf. Last accessed 5 January 2006; my translation.

28. Colin King, "Explosive Remnants of War: A Study on Submunitions and Other Unexploded Ordnance," Report commissioned by the International Committee of the Red Cross (August 2000), 37. On file with author.

29. Quoted in Pax Christi Netherlands, "Cluster Weapons: Necessity or Convenience?" Report (June 2005), 20, http://www.stopclustermunitions.org/wp/wp-content/uploads/2008/07/cluster-munitions-pxc.pdf.

30. Rae McGrath, "Cluster Bombs: The Military Effectiveness and Impact on Civilians of Cluster Munitions," (September 2000), 52, http://www.stopclustermunitions.org/wp/wp-content/uploads/2008/07/cluster_bombs-lma.pdf.

31. Pax Christi Netherlands, "Cluster Weapons: Necessity or Convenience?" 22, 25.

32. Rosy Cave, "Disarmament as Humanitarian Action? Comparing Negotiations on Anti-Personnel Mines and Explosive Remnants of War," in *Disarmament as Humanitarian Action,* ed. John Borrie and V. Martin Randin (Geneva: UNDIR, 2006), 62. However, at the time the landmine campaign was launched, the perceived military utility of the weapons was much higher.

33. Human Rights Watch, *Ticking Time Bombs: NATO's Use of Cluster Munitions in Yugoslavia* (June 1999), http://www.hrw.org/reports/1999/nato2/; Human Rights Watch, *Fatally Flawed: Cluster Bombs and Their Use by the United States in Afghanistan* (December 2002), http://www.hrw.org/reports/2002/us-afghanistan/; Human Rights Watch, *Off Target: The Conduct of the War and Civilian Casualties in Iraq* (2003), http://www.hrw.org/sites/default/files/reports/usa1203_sumrecs.pdf; McGrath, "Cluster Bombs"; ICRC, "Explosive Remnants of War: Cluster Bombs and Landmines in Kosovo" (August 2000), http://www.icrc.org/eng/assets/files/other/icrc_002_0780.pdf; Brian Rappert, "Out of Balance: The UK Government's Efforts to Understand Cluster Munitions and International Humanitarian Law," *Landmine Action Report* (November 2005), http://www.landmineaction.org/resources/Out%20of%20Balance.pdf; Handicap International, *Circle of Impact: The Fatal Footprint of Cluster Munitions on People and Communities* (May 2007), http://www.stopclustermunitions.org/wp/wp-content/uploads/2009/02/circle-of-impact-may-07.pdf; Norwegian People's Aid, "Yellow Killers: The Impact of Cluster Munitions in Serbia and Montenegro" (2007), http://www.stopclustermunitions.org/wp/wp-content/uploads/2008/07/yellowkillers-npa.pdf.

34. See, for example, Human Rights Watch, "Convention on Conventional Weapons (CCW): Time to Begin a New International Instrument on Cluster Munitions," statement at the CCW 3rd Review Conference, 9 November 2006, http://www.hrw.org/news/2006/11/08/convention-conventional-weapons-ccw-time-begin-new-international-instrument-cluster-.

35. John Borrie, "Unacceptable Harm: A History of How the Treaty to Ban Cluster Munitions Was Won," UN document UNIDIR/2009/8 (2009), 180–81.

36. Human Rights Watch, *Fatally Flawed*, 30.

37. Human Rights Watch, "Cluster Munitions and the Proportionality Test," Memorandum to Delegates of the Convention on Conventional Weapons (April 2008), 3,

http://www.hrw.org/news/2008/04/07/cluster-munitions-and-proportionality-test.

38. McGrath, "Cluster Bombs," 49; Pax Christi Netherlands, "Cluster Weapons: Necessity or Convenience?" 4; Handicap International, *Circle of Impact*, 70.

39. ICRC, "Existing Principles and Rules of International Humanitarian Law Applicable to Munitions That May Become Explosive Remnants of War," UN document CCW/GGE/XI/WG.1/W P.7 (2005), 4–5.

40. Timothy L. H. McCormack and Paramdeep B. Mtharu, "Expected Civilian Damage and the Proportionality Equation," 3rd Review Conference of the States Parties to the CCW, UN document CCW/CONF.III/W P.9 (2006), 7, 10.

41. CCW, "Final Document, Part II, Final Declaration," Third CCW Review Conference, UN document CCW/CONF.III/11 (Part II) (2006), 4.

42. Human Rights Watch, "Human Rights Watch Memorandum to Delegates to the Convention on Conventional Weapons Group of Governmental Experts on Explosive Remnants of War," Geneva, March 2003, http://www.hrw.org/backgrounder/arms/cluster031003.htm; Human Rights Watch, "Non-Paper to CCW Delegates: Draft Protocol and Annotations on Prohibitions or Restrictions on the Use of Cluster Munitions (Protocol VI) Annexed to the Convention on Prohibitions or Restrictions on the Use of Certain Conventional Weapons which May Be Deemed to be Excessively Injurious or to Have Indiscriminate Effects", 8-19 November 2004, on file with author; ICRC, Explosive Remnants of War.

43. Author's interview with NGO representative, 17 March 2006, Brussels.

44. Rappert, "Out of Balance," 2.

45. Ibid., 32.

46. Human Rights Watch, "Convention on Conventional Weapons (CCW)."

47. Jonas Gahr Støre, "Opening Statement by Minister of Foreign Affairs," Oslo Conference on Cluster Munitions, Oslo, 22 February 2007, http://www.regjeringen.no/upload/UD/Vedlegg/NorwayOpening%20Statement.pdf.

48. Colin King, Ove Dullum, and Grethe Østern, "M85: An Analysis of Reliability," report for Norwegian People's Aid (December 2007), http://www.landmineaction.org/resources/M85%20report.pdf.

49. For example, Price, "Reversing the Gun Sights"; Rappert, "Out of Balance;" Brian Rappert and Richard Moyes, "The Prohibition of Cluster Munitions: Setting International Precedents for Defining Inhumanity," *Nonproliferation Review* 16, no. 2 (2009): 237–56.

50. Rappert and Moyes, "The Prohibition of Cluster Munitions."

51. Yves Sandoz, Christophe Swinarski, and Bruno Zimmermann, eds., *Commentary on the Additional Protocols of 8 June 1977 to the Geneva Conventions of 12 August 1949* (Geneva: Martinus Nijhoff Publishers, 1987), 626.

52. Eitan Barak, "None to Be Trusted: Israel's Use of Cluster Munitions in the Second Lebanon War and the Case for the Convention on Cluster Munitions," *American University International Law Review* 25 (2010): 423–83.

53. ICBL (International Campaign to Ban Landmines), *Second NGO Conference on Landmines: Report of Proceedings, Geneva, 9–11 May 1994* (Upland, PA: Diane Pub, 1994), 82.

54. For example, Human Rights Watch, *Off Target*.

55. Rappert, "Out of Balance."
56. McGrath, "Cluster Bombs," 52.
57. Director of Landmine Action, quoted in ICRC, "Humanitarian, Military, Technical, and Legal Challenges of Cluster Munitions," Report, Montreux Expert Meeting, 18–20 April 2007, 17, http://www.icrc.org/eng/resources/documents/publication/p0915.htm.
58. Human Rights Watch, *Off Target*.
59. Price, "Reversing the Gun Sights," 632.
60. *Parliamentary Debates*, House of Lords, "Cluster Munitions," 17 May 2007.
61. "Cluster Bombs Don't Work and Must Be Banned," *The Times,* 19 May 2008.
62. Landmine Action, "UK Signs Off for Final Stage of Cluster Bomb Ban," 22 February 2008, http://www.landmineaction.org/resources/resource.asp?resID1088 (accessed 1 June 2008); "UK Ready to Scrap Killer Cluster Bombs: Ministers Overrule Opposition from Military over Controversial Weapons," *Guardian,* 28 May 2008.
63. CMC (Cluster Munition Coalition), "Prohibiting Cluster Munitions: Summary of Key Issues," (2007) http://www.minesactioncanada.org/tool_kit/more%20info/Cluster%20Munition%20Coaition/en/CMC%20key%20issues%20summary.pdf.
64. Mines Action Canada director, quoted in Julie Burtinshaw, "Stop Cluster Bombs," 22 March 2007, http://disarmament.suite101.com/article.cfm/stop_cluster_bombs, accessed 9 February 2009.
65. CMC co-chair, quoted in "Cluster Bomb Treaty Takes Shape after Successful Ban Talks," ICBL press release, 25 May 2007, http://www.icbl.org/index.php//Library/News/The-Treaties/limafinalPR.
66. ICRC, "Humanitarian, Military, Technical, and Legal Challenges of Cluster Munitions."
67. Ove Dullum, "Cluster Munitions: Military Utility and Alternatives," executive summary of FFI-report 2007/02345 (Norwegian Defense Research Establishment, 2008), 3 and 154, italics added, http://www.clusterconvention.org/files/2011/01/CM-alternatives-FFI-report.pdf.
68. Ibid., 143.
69. Michael N. Schmitt, "The Conduct of Hostilities during Operation Iraqi Freedom," *Yearbook of International Humanitarian Law* 6 (2003): 73–109.
70. Explosive weapons "cause injury, death or damage by projecting explosive blast, and often fragmentation, from the detonation of an explosive device"; Landmine Action, "Explosive Violence: The Problem of Explosive Weapons," August 2009, http://www.landmineaction.org/resources/Explosive%20violence.pdf.
71. Human Rights Watch and Harvard Law School International Human Rights Clinic, "Memorandum to CCW Delegates: The Need to Re-Visit Protocol III on Incendiary Weapons," November 2010, 9, http://www.hrw.org/news/2010/11/22/memorandum-ccw-delegates.
72. Richard Moyes, "Explosive Violence in Areas of Civilian Concentration," presentation at the meeting "Cities Are Not Targets!" 28 October 2008, 4, http://www.landmineaction.org/resources/Explosive%20violence%20in%20areas%20of%20civilian%20concentration%20-%20presentation%2028%20Nov%2008.pdf.

73. Landmine Action, "Explosive Violence."

74. Moyes, "Explosive Violence in Areas of Civilian Concentration," 4.

75. UNIDIR (UN Institute for Disarmament Research), "Explosive Weapons: Framing the Problem," Background Paper, May 2010, http://www.inew.org/site/wp-content/uploads/2011/07/DEW-paper-No-1.pdf.

76. For example, UNIDIR, "The Humanitarian Challenge of Explosive Weapons Use in Populated Areas," Report, January 2011, 3, http://explosiveweapons.info/wp-content/uploads/2011/01/UNIDIR-Report-of-the-Glion-symposium-Jan-2011.pdf.

77. Twenty-one out of thirty-three state or UN actors in statements related to explosive weapons refer to IHL; see INEW (International Network on Explosive Weapons), "Acknowledging the Harm," http://www.inew.org/acknowledgements.

78. Author's interview with NGO representative, 26 November 2012; author's interview with senior researcher, 27 November 2012, Geneva.

79. Esther Cann and Katherine Harrison, "100 Incidents of Humanitarian Harm: Explosive Weapons in Populated Areas, 2009–2010," Action on Armed Violence report, March 2011, 6, 8, http://www.inew.org/site/wp-content/uploads/2011/06/100-Incidents-of-Humanitarian-Harm.pdf.

80. AOAV researcher and INEW communications coordinator, quoted in Deutsche Welle report, available at http://www.inew.org/site/wp-content/uploads/2011/07/9E7A0F2A_1.mp3, 30 March 2011.

81. Author's interview with researcher, 23 November 2012, Geneva.

82. Author's interview with NGO representative, 26 November 2012.

83. David Kennedy, *The Dark Sides of Virtue: Reassessing International Humanitarianism* (Princeton, NJ: Princeton University Press, 2004), 294, Kennedy's italics.

84. Rappert and Moyes, "The Prohibition of Cluster Munitions."

10. Toward an Anthropology of Drones

My thanks to Matthew Evangelista and Henry Shue for their finely judged feedback and encouragement and to the other participants in this project for their comments.

1. Jane Mayer, "The Predator War," *New Yorker*, October 26, 2009, http://www.newyorker.com/reporting/2009/10/26/091026fa_fact_mayer.

2. Michael Hastings, "The Rise of the Killer Drones: How America Goes to War in Secret," *Rolling Stone*, April 16, 2012.

3. Scott Shane, "Coming Soon: The Drone Arms Race," *New York Times*, October 24, 2011.

4. Cor Oudes and Wim Zwijnenburg, "Does Unmanned Make Unacceptable? Exploring the Debate on Using Drones and Robots in Warfare," report for IKV Pax Christi, March 2011, 12, http://www.ikvpaxchristi.nl/media/files/does-u-make-ulowspreads_0.pdf; Peter Singer, *Wired for War: The Robotics Revolution and Conflict in the Twenty-First Century* (New York: Penguin, 2009), 263–65.

5. Shane, "Coming Soon"; *Remote Control War*, television documentary, 24 February 2011, written by Leif Kaldor and Leslea Mair, directed by Leif Kaldor, distributed

by Zoot Pictures. See also Friends Committee on National Legislation, "Understanding Drones," http://fcnl.org/issues/foreign_policy/understanding_drones/.

6. Michael Evans, "US Soldiers to Carry Miniature Drones to Launch from Backpacks," *The Australian*, October 19, 2011, http://www.theaustralian.com.au/national-affairs/defence/us-soldiers-to-pack-miniature-drones-to-launch-from-backpacks/story-e6frg8yo-1226170741400; Glenn Greenwald, "Domestic Drones and Their Unique Dangers," *The Guardian* March 29, 2013.

7. Mayer, "The Predator War." See also Singer, *Wired for War*, 116–18.

8. "Drones Today, Drones Tomorrow," Privacy SOS web site, 10 April 2012, http://www.privacysos.org/node/574; Michael Kelley, "Homeland Security Wants to Double Its Domestic Drone Fleet," *Yahoo News*, November 20, 2012, http://finance.yahoo.com/news/homeland-security-wants-double-domestic-225614596.html.

9. Singer, *Wired for War*.

10. Mayer, "The Predator War."

11. Hastings, "The Rise of the Killer Drones"; see also New America Foundation, "Drone Wars Pakistan: Analysis," http://natsec.newamerica.net/drones/pakistan/analysis.

12. Daniel Klaidman, "Drones: How Obama Learned to Kill," *Newsweek*, May 28, 2012, http://www.thedailybeast.com/newsweek/2012/05/27/drones-the-silent-killers.html.

13. "Washington Post: CIA Wants Expansion of Drone Strike Authority in Yemen," AlterNet web site, http://www.alternet.org/newsandviews/article/907007/washington_post%3A_cia_wants_expansion_of_drone_strike_authority_in_yemen/.

14. New America Foundation, "Drone Wars Pakistan." For a critique of the New America Foundation's counting methods, see Gareth Porter, "Cover-Up of Civilian Drone Deaths Revealed by New Evidence," Truthout Web site, August 17, 2012, http://truth-out.org/news/item/10907-cover-up-of-civilian-drone-deaths-revealed-by-new-evidence. A report jointly issued by the law schools at Stanford and NYU, is similarly critical of low estimates of civilian casualties; see "Living under Drones: Death, Injury and Trauma to Civilians from US Drone Practices in Pakistan," http://www.livingunderdrones.org/report/. For yet another set of numbers, and a spectacular technique of visual representation, see Pitch Interactive, "Out of Sight, Out of Mind," http://drones.pitchinteractive.com/.

15. Jo Becker and Scott Shane, "Obama's Secret Kill List," *New York Times*, May 29, 2012.

16. Peter Beaumont, "Former CIA Legal Chief Wanted for Murder by Drone," *The Guardian*, July 16, 2011. See also Tara McKelvey, "Arrest Warrant Sought for CIA Lawyer," *The Daily Beast*, July 28, 2011, http://readersupportednews.org/off-site-news-section/122-122/6857-arrest-warrant-sought-for-cia-lawyer-who-authorized-drone-attacks. For a lively debate on allegations by human rights activists that drone attacks in Pakistan have killed large numbers of innocent civilians, including 168 children, see the April 30 Al Jazeera show "The Stream" at http://stream.aljazeera.com/. See also Scott Shane, "CIA Is Disputed on Civilian Toll in Drone Strikes," *New York Times*, August 11, 2011.

17. David Kilcullen and Andrew McDonald Exum, "Death from Above, Outrage Down Below," *New York Times*, May 16, 2009.

18. Saeed Shah and Peter Beaumont, "US Drone Strikes Killing and Harming Many Civilians," *The Guardian*, July 18, 2011.

19. Mayer, "The Predator War."

20. Hastings, "The Rise of the Killer Drones."

21. See Christopher Coker, *Humane Warfare* (New York: Routledge, 2001); Michael Ignatieff, *Virtual War: Kosovo and Beyond* (New York: Picador, 2001).

22. See http://www.af.mil/AboutUs/FactSheets/Display/tabid/224/Article/104469/mq-1b-predator.aspx.

23. Reuters, "Factbox: What does Lockheed's F-35 fighter jet really cost?" *Reuters* March 15, 2013 http://www.reuters.com/article/2013/03/15/us-lockheed-fighter-cost-idUSBRE92E12E20130315

24. Mayer, "The Predator War."

25. Sheila Jasanoff, *States of Knowledge: Co-Production of Science and the Social Order* (New York: Routledge, 2006).

26. The lyrics of Waters's song "The Bravery of Being out of Range" can be found at http://www.metrolyrics.com/the-bravery-of-being-out-of-range-lyrics-roger-waters.html. It has been set to an arresting set of images at http://www.youtube.com/watch?v=ybHhq48b33Q.

27. Quoted in Mair and Kaldor, *Remote Control War*.

28. In "The Predator War," Mayer says "The drones, which make a buzzing noise, are nicknamed *machay* ("wasps") by the Pashtun natives, and can sometimes be seen and heard, depending on weather conditions."

29. To complicate this narrative a little, there is a long history in war of using booby traps (of which the IED is the latest example), and booby traps also disarticulate in space the body and the weapon of the attacker. However, booby traps and land mines have been largely passive technologies activated fortuitously by the enemy, whereas drones enable the attacker to act while being remote from the scene of the attack.

30. Mair and Kaldor, *Remote Control War*.

31. Megan McCloskey, "The War Room: Daily Transition between Battle, Home Takes Toll on Drone Operators," *Stars and Stripes,* October 27, 2009, http://www.stripes.com/news/the-war-room-daily-transition-between-battle-home-takes-a-toll-on-drone-operators-1.95949.

32. McCloskey, "The War Room."

33. Klem Ryan, this volume.

34. Glenn Greenwald, "Bravery and Drone Pilots," *Salon.com*, July 10, 2012, http://www.salon.com/2012/07/10/bravery_and_drone_pilots/.

35. For a similar argument, see D. Munoz-Rojas and J. J. Fresard, *The Roots of Behavior: Understanding and Preventing IHL Violations* (Geneva: International Committee of the Red Cross, 2004).

36. Mary Kaldor, *New and Old Wars: Organized Violence in a Global Era* (Stanford, CA: Stanford University Press, 2007).

37. Beaumont, "Former CIA Legal Chief Wanted for Murder by Drone."

38. Carl von Clausewitz, *On War* (London, Penguin, 9182); Paul Virilio, *War and Cinema: The Logistics of Perception* (London: Verso, 2009); Paul Virilio and Sylvère Lotringer, *Pure War* (New York: Semiotext(e), 2008).

39. McCloskey, "The War Room."

40. Thom Shanker, "A New Medal Honors Drone Pilots and Computer Experts," *New York Times,* February 14, 2013, A16.

41. Ernesto Londono, "Pentagon Cancels Divisive Distinguished Warfare Medal for Cyber Ops, Drone Strikes," *Washington Post*, April 15, 2013.

42. Mayer, "The Predator War."

43. Ibid.

44. Greenwald, "Bravery and Drone Pilots."

45. Jonathan Schell, "Attacking Libya—and the Dictionary," *Los Angeles Times*, June 21, 2011.

46. Eugene Robinson, "Mission Accomplished," *Washington Post*, May 28, 2013.

47. Quoted in Mair and Kaldor, *Remote Control War.*

48. Quoted in Mayer, "The Predator War."

49. This passage draws on Hugh Gusterson, "An American Suicide Bomber?" *Bulletin of the Atomic Scientists*, January 20, 2010, http://thebulletin.org/american-suicide-bomber.

50. Talal Asad, *On Suicide Bombing* (New York: Columbia University Press, 2007), p. 35.

51. Quoted in Mair and Kaldor, *Remote Control War.*

52. Mayer, "The Predator War."

53. Elaine Scarry, "War and the Social Contract: Nuclear Policy, Distribution, and the Right to Bear Arms," 139 *University of Pennsylvania Law Review* (May 1991): 1257–1316. See also Elaine Scarry, "Citizenship in Emergency," *Boston Review*, October/November 2002, http://new.bostonreview.net/BR27.5/scarry.html.

54. Tom Englehardt, "The Arrival of the Warrior Corporation," TomDispatch.com, February 23, 2012, http://www.tomdispatch.com/blog/175507/.

55. Hastings, "The Rise of the Killer Drones." See also Greenwald, "Bravery and Drone Pilots."

56. Quoted in Mair and Kaldor, *Remote Control War.*

57. Hugh Gusterson, "Death by Drone," *Bulletin of the Atomic Scientists*, October 13, 2011, http://www.thebulletin.org/death-drone.

58. Charlie Savage and Mark Landler, "War Powers Act Does Not Apply to Libya, Obama Argues," *New York Times*, June 16, 2011; Charlie Savage, "2 Top Lawyers Lost to Obama in Libya War Policy Debate," *New York Times*, June 18, 2011.

59. Bradley Jay Strawser, "Moral Predators: The Duty to Employ Uninhabited Aerial Vehicles," *Journal of Military Ethics* 9, no. 4 (2010): 343.

60. Elaine Scarry, *The Body in Pain: The Making and Unmaking of the World* (New York: Oxford University Press, 1987).

61. Mary Ellen O'Connell, this volume. The International Committee on Robot Arms Control (ICRAC) has taken the lead in organizing against autonomous lethal drones. See http://icrac.net/.

62. Nick Turse, "A Drone-Eat-Drone world," *The Nation*, May 31, 2012.

63. Natasha Lennard, "Which Police Departments Want Drones?" *Salon.com*, February 11, 2013, http://www.salon.com/2013/02/11/which_police_departments_want_drones/.

64. See also http://www.geek.com for a dragonfly drone.

65. See the Web site of the DC Area Drone User Group, http://www.meetup.com/DC-Area-Drone-User-Group/.

66. Rosa Brooks, "Flying Mission Creep," paper presented at the conference The Drone Next Door, New America Foundation, Washington, DC, May 7, 2013. See also http://newamerica.net/events/2013/the_drone_next_door.

67. Peter Bergen, "A Dangerous New World of Drones," October 2, 2012, http://www.cnn.com/2012/10/01/opinion/bergen-world-of-drones/.

68. Adam Vaughan, "WWF Plans to Use Drones to Protect Wildlife," *The Guardian*, February 7, 2013.

11. What's Wrong with Drones?

Thank you to Amanda Weyler, who kindly provided significant support and many insightful comments throughout the development of this chapter. Thank you also to Dr. Laurens van Apeldoorn and Prof. Henry Shue for many helpful discussions and comments on the arguments I advance in this chapter. Of course, any errors or omissions in the chapter are mine alone.

1. Glenn Greenwald, "MLK's Vehement Condemnations of US militarism Are More Relevant Than Ever," 21 January 2013, http://www.guardian.co.uk/commentisfree/2013/jan/21/king-obama-drones-militarism-sanctions-iran.

2. Hugh Gusterson, this volume.

3. General counsel of the U.S. Department of Defense Jeh Johnson, quoted in Glenn Greenwald, "The We-Are-At-War! Mentality," 3 December 2011, http://www.salon.com/2011/12/03/the_we_are_at_war_mentality/singleton.

4. Bradley Jay Strawser, "Moral Predators: The Duty to Employ Uninhabited Aerial Vehicles," *Journal of Military Ethics* (2010): 342–68.

5. For example, the killing by drone strikes of the U.S. citizens Anwar al-Awlaki in Yemen on 30 September 2011 and his son, Abdulrahman Anwar al-Awlaki, two weeks later.

6. Philip Alston, *Report of the Special Rapporteur on Extrajudicial, Summary or Arbitrary Executions* (Geneva: UNHRC, 2010).

7. Whether or not conventional warfare, as it was historically understood, is a thing of the past is very much a subject of contention among scholars of military strategy. My view accords with those who argue that conventional war is no longer viable in most foreseeable conflict scenarios (in large part due to developments such as drones); see Martin van Creveld, *The Changing Face of War* (New York: Ballantine Books, 2006). As retired general Rupert Smith observes: " War as cognitively known to most non-combatants, war as battle in a field between men and machinery, war as a massive deciding event in a dispute in international affairs: such war no longer exists," quoted in Rupert Smith, *The Utility of Force* (New York: Alfred A. Knopf, 2007), 1.

8. This is the account conveyed in Walzer's influential *Just and Unjust Wars*, for example; Michael Walzer, *Just and Unjust Wars*, 4th ed. (New York: Basic Books, 2006).

9. I am following here the scholarship of Peter Haggenmacher and Gregory Reichberg on the antecedents of IHL as outlined in the following articles: Peter Haggenmacher, "Just War and Regular War in Sixteenth Century Spanish Doctrine," *International Review of the Red Cross* 32, special issue 290 (1992): 434–45; Gregory M. Reichberg, "Jus ad Bellum," in *War: Essays in Political Philosophy*, ed. Larry May

(Cambridge: Cambridge University Press, 2008), 11–29; Gregory M. Reichberg, "Just War and Regular War: Competing Paradigms," in *Just and Unjust Warriors*, ed. David Rodin and Henry Shue (Oxford: Oxford University Press, 2008), 193–213.

10. Augustine, quoted in Reichberg, "Just War and Regular War," 195.

11. This position represents a deep divide among political theorists that I will not expand on here.

12. Reichberg, "Just War and Regular War," 1; Carl Schmitt, *The Nomos of the Earth* (New York: Telos Press, 2003).

13. Thomas Aquinas, "Summa Theologiae," in *International Relations in Political Thought*, ed. Chris Brown, Terry Nardin, and Nicholas Rengger (Cambridge: Cambridge University Press, 2002), 218.

14. Schmitt, *The Nomos of the Earth*, 321. Carl Schmitt wrote extensively on IHL, especially the influence of regular war and just war theory on the development of modern laws of war.

15. Yves Sandoz, Christophe Swinarski, and Bruno Zimmermann, *Commentary on the Additional Protocols of 8 June 1977 to the Geneva Conventions of 12 August 1949* (Geneva: M. Nijhoff, 1987), 585.

16. Emer de Vattel, "War in Due Form," in *The Ethics of War: Classic and Contemporary Readings*, ed. Gregory M. Reichberg, Henrik Syse, and Endre Begby (Oxford: Blackwell, 2006), 514–15.

17. As Peter Haggenmacher explains: "The concept of regular war obtained in ancient Rome, although it was not expressed in an elaborate theory. It also prevailed in classical international law, between the end of the seventeenth and the beginning of the twentieth centuries. As such, it forms the basis of the Hague Conventions and Regulations respecting war on land of 1899/1907, and of humanitarian law as embodied in the Geneva Conventions." Haggenmacher, "Just War and Regular War," 136.

18. Francis Lieber, *Instructions for the Government of Armies of the United States in the Field*, originally issued as General Orders No. 100 (Washington, DC: Government Printing Office, 1898), Article 68, http://avalon.law.yale.edu/19th_century/lieber.asp#sec3.

19. Quoted in Reichberg, "Just War and Regular War," 211.

20. Schmitt, *The Nomos of the Earth*, 74.

21. The same is also true of eighteenth- and nineteenth-century strategic theories of war. For example, Clausewitz's famous analysis in *On War* is heavily influenced by the physical limitations of the eighteenth- and nineteenth-century battlefield.

22. Sandoz, Swinarski, and Zimmermann, *Commentary on the Additional Protocols*, 586.

23. Jean S. Pictet, *Commentary on the IV Geneva Convention* (Geneva: ICRC, 1958), 3.

24. Sandoz, Swinarski, and Zimmermann, *Commentary on the Additional Protocols*, 263.

25. ICRC, Convention (III) relative to the Treatment of Prisoners of War, Geneva, 12 August 1949, http://www.icrc.org/ihl/INTRO/375.

26. Pictet, *Commentary on the IV Geneva Convention*, 120–21.

27. Sandoz, Swinarski, and Zimmermann, *Commentary on the Additional Protocols*, 192.

28. David Muñoz-Rojas and Jean-Jacques Frésard, *The Roots of Behaviour: Understanding and Preventing IHL Violations* (Geneva: International Committee of the Red Cross, 2004).

29. Ibid.

30. Susan T. Fiske, Lasana T. Harris, and Amy J. C. Cuddy, "Why Ordinary People Torture Enemy Prisoners," *Science* 306 (26 November 2004): 1482–83.

31. Ervin Staub, "The Roots of Evil: Social Conditions, Culture, Personality, and Basic Human Needs," *Personality and Social Psychology Review* 3, no. 3 (1999): 179–92.

32. Schmitt, *The Nomos of the Earth*, 317.

33. Carl Schmitt, *The Theory of the Partisan* (New York: Telos Press, 2007).

34. Schmitt, *The Nomos of the Earth*, 320.

35. As Thomas Schelling writes, "Traditions or conventions are not simply an analogy for limits in war, or a curious aspect of them; tradition or precedent or convention is the essence of the limits"; Thomas Schelling, *The Strategy of Conflict* (Cambridge: Harvard University Press, 1980), 260.

36. Carl von Clausewitz, *On War* [1832] (Princeton, NJ: Princeton University Press, 1976), 605.

37. Clausewitz, *On War*.

38. Jonathan S. Landay, "Obama's Drone War Kills 'Others,' Not Just al Qaida Leaders," 13 April 2013, http://www.mcclatchydc.com/2013/04/09/188062/obamas-drone-war-kills-others.html.

39. Justin Randle, "Low-Flying Drones," LRB blog, 20 March 2013, http://www.lrb.co.uk/blog/2013/03/20/justin-randle/low-flying-drones/.

40. Jo Becker and Scott Shane, "Secret 'Kill List' Proves a Test of Obama's Principles and Will," *New York Times*, 29 May 2012.

41. US military documents leaked in 2010 recount an incident in 2007 during which two Iraqi "insurgents" attempted to surrender to an Apache helicopter but were killed after a military lawyer advised the crew that "they [the Iraqis] can not surrender to aircraft and are still valid targets." As Adam Roberts observes, the advice provided by the lawyer is incorrect; the issue is not whether they can surrender to an aircraft, but "that ground forces in such circumstances need to surrender in ways that are clear and unequivocal." Roberts quoted in David Leigh, "Iraq War Logs: Apache Crew Killed Insurgents Who Tried to Surrender," *The Guardian*, 22 October 2010. Such clarity seems difficult if not impossible to obtain in these circumstances.

42. Becker and Shane, "Secret 'Kill List.'"

43. UK Ministry of Defence, *The UK Approach to Unmanned Aircraft Systems*, JDN 2/11 (London: Ministry of Defence, 2011), 55.

44. Sandoz, Swinarski, and Zimmermann, *Commentary on the Additional Protocols*, 612.

45. Alston, *Report of the Special Rapporteur*, 25.

46. Charles Dunlap, this volume.

47. Ibid.

48. Psychological "fitness" for submarine service remains a significant issue for naval forces as the attrition rate for submariner candidates due to psychological unsuitability remains high. Techniques for predicting fitness and competence have been developed to aid training outcomes. Mark N. Bing, Alison America, Jerry Lamb, and Rick Severinghaus, *The Prediction of Submarine Officer Advanced Course Ascendency from SUBSCREEN Test Scores*, Naval Technical Report no. 1238 (Groton, CT: Naval Submarine Medical Research Laboratory, 2005).

49. Jeff Schogol and Markeshia Ricks, "Demand Grows for UAV Pilots, Sensor Operators," 21 April 2012, http://www.airforcetimes.com/news/2012/04/airforce-demand-grows-uav-pilots-sensor-operators-042112w/.

50. Elisabeth Bumiller, "Air Force Drone Operators Report High Levels of Stress," *New York Times*, 18 December 2011, my italics.

51. Ibid.

52. Stanford Law School and NYU School of Law, *Living Under Drones* (Stanford: Stanford University, 2012).

53. Ibid., viii.

54. Ibid., 133.

55. Ibid., 134.

56. Elisabeth Bumiller and Thom Shanker, "War Evolves with Drones, Some Tiny as Bugs," *New York Times*, 19 June 2011.

57. UK Ministry of Defence, *The UK Approach to Unmanned Aircraft Systems*, 5.9.

58. Ibid., 5.9.

59. Ibid., my italics.

60. "War Powers," Library of Congress Web site, 4 April 2011, http://www.loc.gov/law/help/war-powers.php.

61. Charlie Savage, "2 Top Lawyers Lost to Obama in Libya War Policy Debate," *New York Times,* 18 June 2011.

62. Becker and Shane, "Secret 'Kill List.'"

63. Columbia Law School, Human Rights Clinic, *Counting Drone Strike Deaths* (New York: Columbia Law School, 2012), 34–35.

64. Bumiller and Shanker, "War Evolves with Drones."

65. Jeff McMahan's recent arguments in support of asymmetric combatant rights exemplify this trend. Jeff McMahan, *Killing in War* (Oxford: Clarendon Press, 2009).

66. Strawser, "Moral Predators," 343.

67. This quotation originally appeared in an article from 1955 by Max Huber, "Quelques considerations sur une revision eventuelle des Conventions de La Haye relatives a la guerre," translated and quoted in Sandoz, Swinarski, and Zimmermann, *Commentary on the Additional Protocols*, 404.

68. Tami Biddle, this volume.

12. Banning Autonomous Killing

With thanks for research assistance to Akmal Niyazmatov, Conor McGuiness, and Sean Parish.

1. As Yale University ethicist Wendell Wallach writes, "For thousands of years the machines used in warfare have been extensions of human will and intention. Bad design and flawed programming have been the primary dangers posed by much of the computerized weaponry deployed to date, but this is rapidly changing as computer systems with some degree of artificial intelligence become increasingly autonomous and complex." Wendell Wallach, "Terminating the Terminator: What to Do about Autonomous Weapons," *Science Progress*, January 29, 2013, http://scienceprogress.org/2013/01/terminating-the-terminator-what-to-do-about-autonomous-wea

pons. For one of the first discussions of the ethics of autonomous killing, see, Robert Sparrow, "Killer Robots," *Journal of Applied Philosophy* 24, no. 1 (2007): 62–77.

2. For a good history of weapons development and legal regulation, see Robert L. O'Connell, *Of Arms and Men: A History of War, Weapons, and Aggression* (Oxford: Oxford University Press, 1989).

3. The terminology may be developing and proliferating faster than the technology it seeks to identify. Unmanned aerial, land, and sea vehicles are being referred to collectively as "unmanned systems," or UMS. Land-based unmanned vehicles are often referred to as robots, aerial unmanned vehicles as drones. Some prefer the term "remotely piloted aircraft," or RPAs. See Christof Heyns, "Report of the Special Rapporteur on Extrajudicial, Summary or Arbitrary Executions," UN document A/HRC/23/47, 9 April 2013; Russell Buchan, "Truly Automated Weapons and International Humanitarian Law," Cambridge Journal of International and Comparative Law blog, November 19, 2012, http://www.cjicl.org.uk/index.php/cjicl-blog/truly-automated-weapons-and-international-humanitarian-law; and Noel Sharkey, "Automating Warfare: Lessons Learned from the Drones," *Journal of Law, Information & Science* 21, no. 2 (2011) and accompanying text for terminology related to autonomous systems.

4. For a history of lethal drone operations and a discussion of the law, see Mary Ellen O'Connell, "Unlawful Killing with Combat Drones: A Case History of Pakistan 2004–2009," in *Shooting to Kill: Socio-Legal Perspectives on the Use of Lethal Force*, ed. Simon Bronitt, Miriam Gani, and Saskia Hufnagel (Oxford: Hart Publishing, 2012), 263.

5. Peter Finn, "A Future for Drones: Automated Killing," *Washington Post*, September 19, 2011.

6. Peter W. Singer, "In the Loop? Armed Robots and the Future of War," Brookings Web site, January 28, 2009, http://www.brookings.edu/articles/2009/0128_robots_singer.aspx.

7. "Autonomous and semi-autonomous weapon systems shall be designed to allow commanders and operators to exercise appropriate levels of human judgment over the use of force"; United States Department of Defense, Autonomy in Weapon Systems, Department of Defense Directive 3000.09, November 21, 2012, 2, http://www.dtic.mil/whs/directives/corres/pdf/300009p.pdf.

8. "States should prohibit the creation of weapons that have full autonomy to decide when to apply lethal force"; Human Rights Watch, "Losing Humanity: The Case against Killer Robots," November 19, 2012, 46, http://www.hrw.org/sites/default/files/reports/arms1112ForUpload_0_0.pdf.

9. The Campaign to Stop Killer Robots has played a major role in getting agreement for an expert workshop in 2014 with the aim of starting a process that could result in Protocol VI to the Convention on Prohibitions or Restrictions on the Use of Certain Conventional Weapons Which May Be Deemed to Be Excessively Injurious or to Have Indiscriminate Effects, http://www.icrc.org/eng/assets/files/other/icrc_002_0811.pdf. E-mail to the author from Noel E. Sharkey, Nov. 18, 2013 (on file with the author.)

10. Heyns, "Report of the Special Rapporteur on Extrajudicial, Summary or Arbitrary Executions."

11. Wallach, "Terminating the Terminator." Other ethicists and computer scientists have called for such a prohibition prior to 2013, see, e.g., Sparrow, "Killer Robots" and Noel E. Sharkey, "The Evitability of Autonomous Robot Warfare," *International Review of the Red Cross* 94 (2012), 787.

12. Heyns, "Report of the Special Rapporteur on Extrajudicial, Summary or Arbitrary Executions."

13. These systems are also designated LARs (for "lethal autonomous robotics"). See Heyns, Report of the Special Rapporteur on Extrajudicial, Summary or Arbitrary Executions. They have also been called "truly autonomous weapons (TAWs) to distinguish them from semi-automated weapons (SAWs)." See Buchan, "Truly Automated Weapons and International Humanitarian Law." For more on terminology, see Sharkey, "Automating Warfare."

14. United States Department of Defense, Autonomy in Weapon Systems, 1, 13–14.

15. Heyns, "Report of the Special Rapporteur on Extrajudicial, Summary or Arbitrary Executions," 7.

16. Deonna D. Neal, "In Defense of Humanity: Why Lethal Decision-Making Should Not Be Delegated to Machines," presentation at the meeting of the International Society for Military Ethics, San Diego, CA, January 26, 2011.

17. Armin Krishnan, *Killer Robots: Legality and Ethicality of Autonomous Weapons* (Surrey, England: Ashgate Publishing, 2009), 4, cited in Human Rights Watch, "Losing Humanity," 6.

18. Elizabeth Quintana, "The Ethics and Legal Implications of Military Unmanned Vehicles," Royal United Services Institute for Defence and Security Studies, 2008 5, http://www.rusi.org/downloads/assets/RUSI_ethics.pdf.

19. Neal, "In Defense of Humanity."

20. See Werner J. A. Dahm, "Remarks," Arizona State University Symposium on Drones, Remote Targeting and the Promise of Law, Washington, DC, February 24, 2011. Dahm is the director of Security and Defense Systems Initiative.

21. Dahm, "Remarks."

22. Scott Shane, "Coming Soon: The Drone Arms Race," *New York Times*, October 9, 2011.

23. Committee on Autonomous Vehicles in Support of Naval Operations National Research Council, *Autonomous Vehicles in Support of Naval Operations* (Washington, DC: National Academies Press, 2001), quoted in Sharkey, "Automating Warfare," 144.

24. DoD Press Briefing with Mr. Weatherington from the Pentagon Briefing Room, Arlington, VA, December 18, 2007, http://www.defense.gov/transcripts/transcript.aspx?transcriptid=4108, quoted in Sharkey, "Automating Warfare," 144.

25. Human Rights Watch, "Losing Humanity"; see also Jeffrey S. Thurnher, "The Law That Applies to Autonomous Weapons Systems," *ASIL Insights* 17, no. 4 (2012). Peter W. Singer writes, "Many experts have predicted that autonomous weapon systems will become the norm on the battlefield but the expected timeline for that to happen is about twenty years"; Peter W. Singer, *Wired for War: The Robotics Revolution and Conflict in the 21st Century* (New York: Penguin Press, 2009), 128.

26. Heyns, "Report of the Special Rapporteur on Extrajudicial, Summary or Arbitrary Executions."

27. Gary E. Marchant, "International Governance of Autonomous Military Robots," *Columbia Science & Technology Law Review* 12 (2011), 276. The Israeli Iron Dome missile defense system is another autonomous system that does not require human intervention but allows for it. See Michael N. Schmitt and Jeffrey S. Thurnher, "Out of the Loop: Autonomous Weapon Systems and the Law of Armed Conflict," *Harvard National Security Journal* 4, no. 231 (2013).

28. United Kingdom Ministry of Defence, *The UK Approach to Unmanned Aircraft Systems*, JDN 2/11 (London: Ministry of Defence, 2011), 502, https://www.gov.uk/government/uploads/system/uploads/attachment_data/file/33711/20110505JDN_211_UAS_v2U.pdf, citing Tony Gillespie and Robin West, "Requirements for Autonomous Unmanned Air Systems Set by Legal Issues," *The International C2 Journal* 4, no. 2 (2010), http://www.dodccrp.org/html4/journal_v4n2.html.

29. See Ken Anderson and Matthew Waxman, "Law and Ethics for Robot Soldiers," *Policy Review* 176 (2012), http://papers.ssrn.com/sol3/papers.cfm?abstract_id=2046375; John Markhoff, "War Machines: Recruiting Robots for Combat," *New York Times*, November 27, 2010; and Finn, "A Future for Drones."

30. Schmitt and Thurnher conclude: "Humans are never really 'out of the loop.' While autonomous weapon systems will increasingly be capable of solving complex problems, absent dramatic improvements in artificial intelligence, humans will decide when and where to deploy the system and what parameters to embed within it. . . . Although the subjective decisions may sometimes have to be made earlier in the targeting cycle than has traditionally been the case, this neither precludes the lawfulness of the decisions, nor represents an impediment to the lawful deployment of the systems." "Out of the Loop," 33.

31. Ibid.

32. Finn, "A Future for Drones."

33. For a general overview of the law applicable to resort to lethal force, see O'Connell, "Unlawful Killing with Combat Drones."

34. Article 51 provides: "Nothing in the present Charter shall impair the inherent right of individual or collective self-defence if an armed attack occurs against a Member of the United Nations, until the Security Council has taken measures necessary to maintain international peace and security. Measures taken by members in the exercise of this right of self-defence shall be immediately reported to the Security Council and shall not in any way affect the authority and responsibility of the Security Council under the present Charter to take at any time such action as it deems necessary in order to maintain or restore international peace and security." Charter of the United Nations, Chapter VII, http://www.un.org/en/documents/charter/chapter7.shtml.

35. *Corfu Channel (United Kingdom of Great Britain and Northern Ireland v. Albania)*, Judgment, 1949 I.C.J. 4 (July 9), http://www.icj-cij.org/docket/index.php?p1=3&p2=3&k=cd&case=1&code=cc&p3=4.

36. Judith Gardam, *Necessity, Proportionality and the Use of Force by States* (Cambridge: Cambridge University Press, 2004); see also Henry Shue, "Proportionality in War," in *The Encyclopedia of War*, ed. Gordon Martel (Malden, MA: Wiley-Blackwell, 2012).

37. 2005 World Summit Outcome, UN General Assembly document A/60/L.1, 15 September 2005, 22–23. Today all fully sovereign states are members of the United Nations.

38. Jean-Marie Henckaerts and Louise Doswald-Beck, eds., *Customary International Humanitarian Law* (Cambridge: ICRC, 2005), 13. See also two other initiatives sponsored by or associated with the ICRC: International Committee of the Red Cross, "Interpretative Guidance on the Notion of Direct Participation in Hostilities under International Humanitarian Law," *International Review of the Red Cross* 90, no. 872 (2009); and Program on Humanitarian Policy and Conflict Research at Harvard University, *HPCR Manual on International Law Applicable to Air and Missile Warfare* (May 2009), http://www.ihlresearch.org/amw/manual/.

39. "The Court observes that the protection of the International Covenant of Civil and Political Rights does not cease in times of war, excerpt by operation of Article 4 of the Covenant whereby certain provisions may be derogated from in a time of national emergency." Legality of the Threat or Use of Nuclear Weapons, 1996 I.C.J. 226, para. 25 (Advisory Opinion of July 8), http://www.icj-cij.org/docket/index.php?p1=3&p2=4&k=e1&p3=4&case=95. See also Louise Doswald-Beck, *Human Rights in Times of Conflict and Terrorism* (Oxford: Oxford University Press, 2011).

40. See *Isayeva, Yusopova and Bazayeva v. Russia*, nos. 57947/00, 57948/00, and 57949/00, European Court of Human Rights, 24 February 2005, http://www.humanrights.is/the-human-rights-project/humanrightscasesandmaterials/cases/regionalcases/europeancourtofhumanrights/nr/2615; and *Juan Carlos Abella v. Argentina*, Case 11.137, Report No. 55/97, Inter-American Court of Human Rights, OEA/Ser.L/V/II.98, Doc. 6. rev. 18 November 1997, paras. 149–51, http://www1.umn.edu/humanrts/cases/1997/argentina55-97a.html (distinguishing "internal disturbances" from armed conflict on the basis of the nature and level of violence).

41. See Germain G. Grisez, "Toward a Consistent Natural Law Ethics of Killing," *American Journal of Jurisprudence* 15 (1970): 76, cited in David Hollenbach, *Nuclear Ethics: A Christian Moral Argument* (Mahwah, NJ: Paulist Press, 1983), 18–19. Hollenbach describes how the just war tradition evolved from Aquinas's position presuming that war is sinful to one presuming war is just so long as it is waged by legitimate authorities. Hollenbach argues in favor of returning to the presumption that violent warfare is presumed to be morally wrong and that resort to war is justifiable only in exceptional situations. Hollenbach, *Nuclear Ethics*, 14–16. Hollenbach's position is consistent with current international law on the use of force, as reviewed above. See Jeremy Waldron, "Can Targeted Killing Work as a Neutral Principle?" Public Law Research Paper No. 11–20, NYU School of Law, March 24, 2011, http://papers.ssrn.com/sol3/papers.cfm?abstract_id=1788226.

42. International Committee of the Red Cross, "Interpretative Guidance on the Notion of Direct Participation in Hostilities under International Humanitarian Law," 80–81.

43. Markhoff, "War Machines."

44. Finn, "A Future for Drones."

45. Anderson and Waxman, "Law and Ethics for Robot Soldiers."

46. See Richard Jackson, Assistant to the Army Judge Advocate, "Remarks," Symposium on Drones, Remote Targeting and the Promise of Law, Washington DC, February 24, 2011, http://www.ustream.tv/recorded/12911842.

47. Ibid.

48. The problem of accountability was a critical issue that drove the UN Special Rapporteur to call for a moratorium on development of fully autonomous weapons.

Heyns, "Report of the Special Rapporteur on Extrajudicial, Summary or Arbitrary Executions."

49. For the ease of killing with drones, see Mary Ellen O'Connell, "Seductive Drones: Learning from a Decade of Lethal Operations," *Journal of Law, Information, & Science* 21, no. 2 (2011): 116–39.

50. Justin McClelland, "The Review of Weapons in Accordance with Article 36 of Additional Protocol I," *International Review of the Red Cross* 85, no. 850 (June 2003): 408, http://www.icrc.org/eng/assets/files/other/irrc_850_mcclelland.pdf; Patrick Worsnip, "U.N. Official Calls for Study of Ethics, Legality of Unmanned Weapons," *Washington Post*, October 24, 2010. See "The Statement of the 2010 Expert Workshop on Limiting Armed Tele-Operated and Autonomous Systems," International Committee for Robot Arms Control (ICRAC) statement, Berlin, 2010, at http://icrac.net/statements/.

51. Jackson, "Remarks."

52. See Convention on Certain Conventional Weapons, Protocol IV, Blinding Laser Weapons (May 3, 1996).

53. "Member States are called upon to prohibit all forms of human cloning inasmuch as they are incompatible with human dignity and the protection of human life"; United Nations Declaration on Human Cloning, UN General Assembly Resolution 59/280, March 23, 2005.

54. Ann Florini, "The Evolution of International Norms," *International Studies Quarterly* 40 (1996): 363–89. For another view on why some norms enter the agenda of norm-promoting groups, see Charli Carpenter, Sirin Duygulu, and Anna Tomaskovic-Devey, "Explaining the Advocacy Agenda: Insights from the Human Security Network," presentation at the University of California-Irvine, Irvine, CA, May 2011, http://www.socsci.uci.edu/files/internationalstudies/docs/carpenter2011.pdf.

55. Florini, "The Evolution of International Norms," 374.

56. Special Rapporteur on Extrajudicial, Summary or Arbitrary Executions, "Report of the Special Rapporteur on Extrajudicial, Summary or Arbitrary Executions, Philip Alston, Addendum, Study on Targeted Killings," UN Doc. A/HRC/14/24/Add.6, at 24 (May 28, 2010).

57. A 2012 poll of Americans on their attitudes toward killing with drones found that 76 percent of likely U.S. voters "approve of the use of the unmanned aircraft to kill terrorists." "Voters Are Gung-Ho for Use of Drones But Not over the United States," Rasmussen Reports Web site (survey conducted 10 February 2012), http://www.rasmussenreports.com/public_content/politics/current_events/afghanistan/voters_are_gung_ho_for_use_of_drones_but_not_over_the_united_states. This figure has declined following Senator Rand Paul's thirteen-hour filibuster against the nomination of John Brennan as director of the CIA and other consciousness-raising events, such as protests outside drone bases and media reports of young children and rescue workers being killed.

58. A recommitment to the ban on torture in the United States gained an essential boost when the National Religious Campaign Against Torture built a coalition of religious organizations and persons to protest U.S. use of torture. See http://www.nrcat.org/.

59. Florini, "The Evolution of International Norms," 376.
60. Ibid., 377.
61. For details on the treaty, see Convention on the Prohibition of the Use, Stockpiling, Production and Transfer of Anti-Personnel Mines and Their Destruction (known as the Ottawa Convention), http://legal.un.org/avl/ha/cpusptam/cpusptam.html.

List of Contributors

Tami Davis Biddle is Hoyt S. Vandenberg Chair of Aerospace Studies, U.S. Army War College. She is the author of *Rhetoric and Reality in Air Warfare: The Evolution of British and American Thinking on Strategic Bombing, 1914–1945* and many articles on air warfare, strategy, and the wars of the twentieth century.

Sahr Conway-Lanz is the Senior Archivist for American Diplomacy at the Yale University Library and the author of *Collateral Damage: Americans, Noncombatant Immunity, and Atrocity after World War II.*

Neta C. Crawford is professor of political science at Boston University and author of several books on international relations.

Janina Dill is Hedley Bull Fellow at the Department of Politics and International Relations of the University of Oxford and a Research Fellow at Merton College.

Charles J. Dunlap Jr., Major General (Ret.) and former deputy judge advocate general, U.S. Air Force, joined the Duke University law faculty in July 2010. His teaching and scholarly writing focus on national security, international law, civil-military relations, cyberwar, and military justice.

Matthew Evangelista is President White Professor of History and Political Science in the Department of Government at Cornell University.

Charles Garraway is a fellow of the Human Rights Centre, University of Essex.

Hugh Gusterson, professor of cultural studies and anthropology at George Mason University, has written or edited five books and writes for both peer-reviewed journals and the popular press.

Richard W. Miller is Hutchinson Professor in Ethics and Public Life and director of the Program on Ethics and Public Life in the Department of Philosophy at Cornell University.

Mary Ellen O'Connell holds the Robert and Marion Short Chair in Law and is Research Professor of International Dispute Resolution—Kroc Institute for International Peace Studies, University of Notre Dame.

Margarita H. Petrova is an assistant professor at Institut Barcelona d'Estudis Internacionals (IBEI).

CONTRIBUTORS

Klem Ryan received his doctorate in political theory from the University of Oxford, where his research focused on the philosophy of international humanitarian law. He currently works in South Sudan on disarmament and small arms regulation.

Henry Shue is a senior research fellow at the Centre for International Studies, Department of Politics and International Relations, University of Oxford, and a senior research fellow emeritus, Merton College, Oxford. He has edited three previous volumes on moral issues about war, including *Preemption*.

Index

Page numbers followed by n and nn indicate notes.

Accountability, and autonomous weapons, 85, 228, 236, 297n48
Acheson, Dean, 52
Action on Armed Violence, 189
"Advantageous environment," norm development and, 235–36
Aerial bombing, in early 20th century, 27–46
   foundational premises and civilian population expectations, 27–30
   interwar years, 33–39
   World War I and, 30–33
   World War II and increased violence to civilians, 39–46
Afghanistan, 2, 4, 11, 62
   drones used in, 193, 215, 224
   England's use of poison gas in, 8
   military target definitions and, 61
   post-9/11 invasion and changes in normative beliefs, 65, 80–85, 86
   United Nations Assistance Mission in, 19, 168, 280nn36, 44
   unjustified civilians deaths due to U.S. power strategy, 161, 166–70
AFP 110-31, *International Law—The Conduct of Armed Conflict and Air Operations* (U.S. Air Force), 75, 76
African National Congress (ANC), 168, 169
Agents, dissociation from violent acts, 213–14
   drones and, 217–18
Ahtisaari, Martii, 146, 153
Air Force Times, 218

Airpower: Myths and Facts (Meilinger), 111
Air Warfare (Sherman), 34, 38
Albania, 229
Albright, Madeleine, 126, 163
Alston, Philip, 23, 126–27, 208, 217
Anderson, Ken, 232
Anti-personnel mines, 118–19, 180. *See also* Land mines
Aquilla, John, 231–32
Aquinas, Thomas, 210, 230
Arab Spring, 109, 117
Arkin, Ronald, 232
Arnold, Henry "Hap," 42–43, 45
Artificial intelligence. *See* Autonomous weapons
Art of War, The (Sun Tzu), 116
Asad, Talal, 202
Atlantic magazine, 124
Augustine, Saint, 230
Autonomous weapons, 3, 224–36
   current lawful and ethical killing context, 229–32
   current terminology and technology, 226–27, 295n13
   norm development and, 224–25, 227–29
   proposal for and approaches to treaty ban on, 232–36
   see also Drones
Aziz, Tariq, 195

Baker, Deane-Peter, 221
Baker, James, 115

INDEX

Baker, Newton, 34, 39
Ball, George, 160
Balloons, in early aerial warfare, 90
Barbarism, U.S. power and
 interpretations of Taliban, 167–70
Battlefields
 drone use and, 196–99
 regular war and role of, 211–13
Behnke, Paul, 31
Behram, Noor, 194
Belgium, 182–83
Benford, Robert D., 178
Benjamin, Mark, 113
Bergen, Peter, 205
Best, Geoffrey, 120
Biddle, Tami Davis, 8, 23, 27–46, 67, 68, 223
Blair, Dennis, 221
Blum, Gabriella, 118
Body in Pain, The (Scarry), 204
Boorstin, Daniel, 117
Boothby, William H., 178
Bosnia, 78, 81
Brennan, John, 84
Brooks, Rosa, 205
Brown, Dr. Felix, 41
Brown, Harold, 13, 69
Brzezinski, Zbigniew, 267n101
Burridge, Brian, 200
Bush, George H. W., 75–76
Bush, George W., 84–85, 199, 215, 235

Calley, William, 159
Cambodia, 71, 74
Central Intelligence Agency (CIA),
 drones and, 193–94, 199
Chamberlain, Neville
 interwar years, 38, 39
 World War II proposed bombardment rules, 39, 92
Chanute, Octave, 29
"Chaotic State of the International Law Governing Bombardment, The" (Spaight), 36

Chechnya, 18, 98, 100, 102, 237n4
Chemical and biological weapons, 118
Christian Science Monitor, 124
Churchill, Winston
 poison gas use in 1919, 8–9
 World War II, 9, 10, 40, 42, 44, 246n65
Clark, Wesley, 79–80
Clausewitz, Carl von, 116, 130, 137, 199, 214, 220, 291n21
Cluster Munition Coalition (CMC), 186
Cluster munitions, 3, 233
 in Iraq, 77, 179
 Israel and, 183, 184
 in Kosovo, 78, 79–80, 256n70
 in Laos, 71
 nongovernmental organizations and, 21, 176, 177–88
 unintended consequences of restrictions on, 119
 in Vietnam, 74
Coercive air power, Warden's theory of, 76
Cohen, Elliot, 140–41
Cohen, William, 80
"Coherence," norm development and, 234, 235
Coll, Steven, 121
Columbia University Law School, Human Rights Clinic, 221
Combined air operations centers (CAOCs), of U.S., 113–14
Command of the Air, The (Douhet), 8, 35–36
Commentary on the Additional Protocols of 8 June 1977 (ICRC), 184
Committee on Autonomous Vehicles in Support of Naval Operations, 227
Conference on the Limitation of Armament (1921–1922), 91
Conventional war, questions of obsolescence of, 208, 290n1

Convention for the Protection of Human Rights and Fundamental Freedoms (1950), 93
Convention on Certain Conventional Weapons (CCW), 176, 179, 181, 182, 183, 188, 225, 235
Convention on Cluster Munitions, 186, 188, 190
Conway-Lanz, Sahr, 2, 11–13, 14–15, 23, 67, 47–63, 68, 82–83
Coppock, Kelvin, 82
Corfu Channel case (1948), 229
Counterinsurgency (COIN) operations
  changes in normative beliefs and, 66, 83
  counterinsurgency doctrine and public support, 62
  in 21st century, 19–20, 120–25, 166–70, 265n71
Crawford, J. W., 16, 154–55
Crawford, Neta C., 5, 11–13, 15, 16, 17, 20, 23, 64–86, 276n25
Cyprus, 98

Davis, Richard G., 242n3
Department of Defense, directive on autonomous weapons, 226, 227
Deptula, David, 127–28
Diagnostic framing, defined, 178
Dill, Janina, 16–18, 20, 23–24, 75, 131–44
Dissociation, and laws of war, 213
  of agents from violent acts, 22–23, 213–14
  of agents from violent acts, drones and, 217–18
  of public from violence committed on behalf of, 23, 214–15
  of public from violence committed on behalf of, drones and, 220–22
  of targets from source of violence, 23, 214

of targets from source of violence, drones and, 219
Distinction (discrimination), principle of, 7, 27
  defined, 6
  military objectives and, 13–14, 16–18, 131–33
Douhet, Giulio, 8, 13, 35–36
Dresden, bombing of, 10, 44, 45, 49, 93, 111–12, 248n85
Drones, 2, 3, 4, 113, 191–206, 235
  ambiguities and asymmetries regarding honor and courage, 200–203
  democratic process and, 203–5
  developing terminology, 294n3
  future scenarios, 205–6
  history and reasons for use of, 192–96
  military's serious treatment of moral issues and, 126–28
  Obama administration's use of, 65, 84–85, 200–201, 203–4, 207, 215, 219, 220–21, 235, 270n5
  respatialization of war and, 196–99, 207
  speed of war and, 199–200
  unintended consequences of restrictions on, 119–20
  see also Autonomous weapons; Drones, and international humanitarian law; *specific countries*
Drones, and international humanitarian law, 207–23
  dissociation and drones, 217–22
  dissociation's forms, 22–23, 213–15
  dissolution of battlefield and, 215–16, 292n41
  just war and "regular war" theories, 22, 209–11
  role of battlefield, 211–13
Dunant, Henry, 89
Dunham, Helen, 245n45

INDEX

Dunlap, Charles J. Jr., 13, 18, 20, 23, 83–84, 109–30, 141, 150–51, 198, 202, 217, 218, 241n54, 276n25
Dunn, Mike, 122
Dyer, Joe, 196
Dyson, Tim, 274n6

Effective contributions, targets and, 133–35, 270n8
Effects-based operations (EBOs), targets and, 138–39, 143, 272n31
Efficiency, logic of, 20, 132, 139–41
 logic of sufficiency and, 143–44
Eisenhower, Dwight, 43, 117
Electric grid, destroyed in Iraq, 16–17, 21, 76–78, 146–47, 152–57, 262–63, 276n33
England
 Albania and, 229
 cluster munitions and, 186
 drones and, 193
 interwar years, 34–39
 Ministry of Defense discussion paper, 216, 220
 public views on warfare in 1920s and 1930s, 35
 two standards for aerial bombing of civilians, 9–10
 use of poison gas in 1919–1922, 8–9
 World War I and, 31–33, 35
 World War II and, 39–44, 246nn65, 66, 69
Englehardt, Tom, 203
European Convention for the Protection of Human Rights and Fundamental Freedoms, 100
European Court of Human Rights, 18, 93, 97–99, 100, 103, 230
Evangelista, Matthew, 1–24
Exum, Andrew, 194, 201

Fair, Christine, 124
Falk, Richard, 120

Far East Air Forces (FEAF), 51–52, 53, 55–56, 57
Fast Assessment Strike Tool—Collateral Damage (FAST-CD), 82
Field Manual (FM) 3–24, of U.S. Army, 121–22, 125, 265n71
"Field Manual on the Joint Targeting Process" (1997), 135–36
Florini, Ann, 234–36
Food-supplying industry, effective contributions and military advantage, 133–34
Force protection
 changes in normative beliefs and, 66
 Iraq and triangular balance with military advantage and civilian care, 16, 145–57
Framing of issues
 collective action frames, 177–78
 counterframe deflation, 178, 282n12
 diagnostic framing, defined, 178, 282n11
 frame amplification, 178, 282n11
 motivational framing, defined, 178
 prognostic framing, defined, 178
France, 31, 33
Friedman, George and Meredith, 130
Fulgosius, Raphaël, 210
Fullerton, J. D., 29

Gaddafi, Muammar, 109
Garraway, Charles, 4, 14, 18–19, 87–105, 179
Gates, Robert, 167
GATOR mine system, 118
Gavotti, Giulio, 30
Generation Kill (Wright), 159–60, 163–64
Geneva Convention (1864), 89
Geneva Conventions (1949), 3, 46, 59–60, 93, 212, 230
 Article 14, 212
 Article 15, 98, 100, 212

Article 19, 212
  Common Article 3, 94–95, 100, 103–4
Geneva Conventions, Additional Protocols (1977), 3
  Protocol I, 24, 98, 115, 123, 132, 135, 148–49, 157
  Protocol I, Article 1(2), 105
  Protocol I, Article 48, 132
  Protocol I, Article 49(3), 95–96
  Protocol I, Article 50, 216
  Protocol I, Article 51(5)(b), 175
  Protocol I, Article 52(2), 13–14, 17–18, 132, 135–37, 141–42, 143–44
  Protocol I, Article 57, 148
  Protocol I, Article 57(3), 16, 153–54, 156
  Protocol II, 94–96, 104
  Protocol III, 104
Genocide Convention (1949), 65
Georgia, 99, 101
Germany
  air campaign of 1939, 9
  League of Nations and, 92
  World War I and, 31–33
  World War II and, 40–44, 47, 49–50
Gordon, Joy, 146–47, 163, 274nn6,7, 276n33
Grant, Rebecca, 111
Greenwald, Glenn, 200
Greenwood, Christopher, 150, 178, 275n20
Guardian, The, 164, 194
Gulf War. *See* Iraq, 1991 bombing of and sanctions against
Gusterson, Hugh, 22, 23, 24, 191–206, 207, 219

Hagel, Chuck, 200
Haggenmacher, Peter, 291n17
Hague Conventions, 9, 30, 61, 89–90, 212, 230

Regulations Respecting the Laws and Customs of War on Land (1907), 90, 93, 95
Hague Rules of Aerial Warfare (1923), 14, 36–37, 41, 91–96, 100, 103–5
Haig, Douglas, 31, 32
Handicap International, 182
Hansell, Haywood, 44–45
Harris, Arthur, 41–42, 44
Harvard University. *See Manual on International Law Applicable to Air and Missile Warfare*
Hastings, Michael, 203
Hayden, Mike, 85
Hazelton, Jill, 125
Heintzelman, Harry, 76
Heyns, Christof, 225, 226
Hiroshima, bombing of, 10–11, 12, 45–46, 161, 162
Hitler, Adolf, 9, 33, 37, 38, 40, 117
Hollenbach, David, 297n41
Hoover, Herbert, 37
Horner, John, 77
Huber, Max, 222
Hughes, Daniel J., 112
Hugo, Victor, 28–29
Hu Jintao, 129
Humanitarian intervention, paradox of, 7
Human rights laws. *See* International humanitarian law (IHL)
Human Rights Watch
  Afghanistan casualties and, 114
  autonomous weapons and, 226, 227
  cluster munitions and, 80, 181–82, 183, 256n70
  incendiary weapons and, 188
  Iraq combat operation and, 122
Hussein, Saddam, 15, 17, 20, 76, 117, 140, 162, 163. *See also Iraq entries*

Iliad (Homer), 191–93, 199, 201, 202
Incendiary weapons, NGOS and, 188–89

INDEX

India, 8
Infrastructure, effective contributions and military advantage, 134. *See also* Electric grid, destroyed in Iraq
"Initial prominence," norm development and, 234–35
Institute of International Law, 90
Insurgents. *See* Counterinsurgency (COIN) operations
Intelligence, reconnaissance, and surveillance (ISR), 112, 113
Intention, 11–12
    Korean War and increased emphasis on, 48, 58–60, 62–63, 68
Inter-American Commission, 230
Inter-American Court of Human Rights, 230
Intercontinental ballistic missiles (ICBMs), 234
International Committee of the Red Cross (ICRC)
    cluster munitions and, 179, 182, 187
    Commentary on the Additional Protocols of 8 June 1977, 184
    draft rules for protection of civilians, 60
    founding of, 89
    Interpretative Guidance on the Notion of Direct Participation in Hostilities under International Humanitarian Law, 230, 231, 233
    Korean safety zones and, 60
    military advantage and, 137
    non-international armed conflict and, 94
    report on violations of international humanitarian law outlines, 213
    rules governing warfare, 5–6, 7, 93
    Study into Customary International Humanitarian Law, 97
International Court of Justice, 101, 103, 229
International Covenant on Civil and Political Rights, 93, 100
International Covenants on Economic, Social and Cultural Rights, 93
International Criminal Court, 96–97
International humanitarian law (IHL), 6, 87–105, 245n45
    drones and, 198–99
    ethics and, 87–88, 104–5
    explosive weapons and, 189–90
    human rights laws and, 93–94
    laws of armed conflict and balance between military necessity and humanity, 88–93
    nongovernmental organizations and, 175, 176, 178, 181–84, 188–90
    non-international armed conflict and, 94–99
    targeting issues, 100–102
    unintended consequences of, 114–20
    see also Drones, and international humanitarian law
Interwar years, norms and, 33–39, 90–92
Invisible War (Gordon), 276n33
Iran, drones used in, 193
Iraq, 1919 and England's use of poison gas, 8–9
Iraq, 1991 bombing of and sanctions against, 4, 15–16, 20
    changes in normative beliefs and, 64, 75–78, 82
    cluster munitions and, 77, 179
    electricity grid as military target and effects on civilians, 16–17, 21, 76–78, 146–47, 152–57, 262–63, 276n33
    incorrect information on infrastructure destruction, 127
    law of armed conflict's interlocked rules and, 147–52

triangular balance of military advantage, force protection, and care of civilians, 145–57
unjustified civilians deaths due to U.S. power strategy, 11, 161–66
Iraq, 2003 invasion of
changes in normative beliefs, 65, 80–85, 86
COIN operations, 120–25
drones used in, 193, 215
logic of efficiency and, 140
unjustified civilians deaths due to U.S. power strategy, 159–60
Israel, 138, 237n4
cluster munitions and, 183, 184
drones and, 193, 194
Iron Dome defense system, 234, 293n27
Italo-Turkish War (1911–1912), 30, 90
Italy, 8, 92

Jackson, Richard, 233
Japan, 9, 44–46, 47, 49–50, 92
Jasanoff, Sheila, 196
Johnson, Lyndon, 69–70, 74, 160
Joint Doctrine for Targeting (2002), 136
Joint Fires and Targeting Handbook (2007), 156
Judge advocates general (JAGs), 3, 4, 77, 136
Just war theory, 27, 143–44
development of, 230
drones and, 221–22
guidelines for conduct, 6
Hague Conventions and, 36
international humanitarian law and, 216
"regular war" theory and, 22, 209–11
rejection of, 126

Kaempf, Sebastian, 73
Kaldor, Mary, 198

Kaltenthaler, Karl, 124
Kellenberger, Jakob, 228–29
Kennedy, David, 190
Kilcullen, David, 194
King, Colin, 184
Korean War, 11–13, 47–63, 67, 68
conduct of, 50–52
elastic definition of military targets, 47–48, 52–58, 61, 62
increased emphasis on intention, 48, 58–60, 62–63
legacies of, 60–63
World War II background and early influence, 47, 49–50
Kosovo campaign, 7, 15, 20, 98, 101, 111, 179
changes in normative beliefs and, 64, 78–80, 81, 82, 256n70
Kuwait. *See* Iraq, 1991 bombing of and sanctions against

Lancet, 164
Landmine Action, 180, 183, 186, 188–89
Land mines, 3, 21–22, 168, 227, 233
development of ban on, 233–34, 236
see also Anti-personnel mines; Cluster munitions
Laos, bombing of, 71, 74
Laser weapons, banning of blinding, 233, 234
Lauterpacht, Hersch, 87, 211
Law of armed conflict (LOAC), 6
civilian population and "innocence," 116–18
interlocked rules regarding military advantage, force protection, and civilian care, 147–52
see also International humanitarian law (IHL)
Law of War Manual (Department of Defense), 115
League of Nations, 37, 92

INDEX

Lebanon, 183, 184, 237n4
LeMay, Curtis, 44, 45
Lewy, Guenter, 72, 254n29
Lex specialis, 102–3
Libya, 7, 8, 15, 102, 109
   drones used in, 193, 201, 204, 215, 220–21, 235
   Italo-Turkish War and, 8, 30
Lieber, Francis, 211
Living Under Drones (Stanford University and New York University), 219
Lloyd George, David, 10, 34
Long War Journal, 84
Ludlow-Hewitt, Edgar, 38
Luftwaffe, 33, 38, 40, 42, 246n66

MacArthur, Douglas A., 51–55, 62
Maguire, T. Miller, 29
Malik, A. Rehman, 192
Manchuria, 9, 35, 52–53, 56
Manual of Military Ballooning (British War Office), 29
Manual on International Law Applicable to Air and Missile Warfare (Harvard University), 4, 102–3, 105, 115, 120
Martens Clause, 105
Mayer, Jane, 192, 193, 200–201, 202
McCaffrey, Barry, 122
McChrystal, Stanley, 4, 21, 83, 86, 123, 124
McCormack, Timothy L. H., 245n45
McDonald, Kent, 218
McMaster, H. R., 70
McNamara, Robert, 69–70, 72
McNaughton, John T., 72
Media installations, effective contributions and military advantage, 134
Mehsud, Baitullah, 192, 193, 199
Meilinger, Phillip, 111
Military advantage
   cluster munitions and, 179–80
   Iraq and triangular balance with force protection and civilian care, 145–47
   nongovernmental organizations and, 185–88
   targets and, 133–35, 137, 270n8
Military necessity, changes in normative beliefs about civilian casualties and, 66
Military supplies, Korea War and evolution of attitudes toward, 60–61
Military targets, 131–44
   drones and dissociation from source of violence, 214, 219
   effective contributions and military advantage, 133–35, 270n8
   effects-based operations doctrine and, 138–39, 143
   in humanitarian and human rights laws, 100–102
   interwar years, 36
   Korean War and elasticity of definition of, 47–48, 52–58, 61, 62, 67
   logic of efficiency, 132, 139–41, 143–44
   logic of sufficiency, 132, 141–44
   objectives and, 4
   principle of distinction between military and civilian targets, 13–14, 16–18, 131–33
   progress during operations defined, 137
   war-sustaining capability and, 135–36
Mill, John Stuart, 129, 170
Miller, Richard, 11, 16, 19–20, 24, 158–71
Miller, William, 124
Milošević, Slobodan, 79, 138, 140, 240n46
Mitchell, Billy, 8, 13, 27, 32, 33–34, 36
Mitchell, Jonathan, 37

Momyer, William, 70, 72–74
"Morale effect" of bombing, 2
  England and Germany in World War I and, 32
  England in World War II and, 41
  experience and, 116–18
"Moral Predators: The Duty to Employ Uninhabited Aerial Vehicles" (*Journal of Military Ethics*), 221–22
Munich Crisis, 38
My Lai massacre, 159

Nagasaki, bombing of, 10–11, 12, 45–46
NATO, 2, 78, 123
Neal, Deonna, 226
Necessity, autonomous weapons and principle of, 19, 229–32
New America Foundation, 128, 194
New England Journal of Medicine, 122
New Yorker, 192, 193, 195, 199, 200–201
New York Times, 109, 114, 127, 165, 194, 220, 221
Nexus, degree of, 17–18, 133, 135–36, 139
  logic of sufficiency and, 141–43
Nixon administration, 71, 74
"No-fly zones," over Iraq, 78, 256n2
Nomos of the Earth, The (Schmitt), 214
Nongovernmental organizations (NGOs), 3, 21, 175–90
  anti-cluster weapon campaigns, 188–90
  framing of issues and norm development, 177–78, 282nn11,12
  military necessity and humanitarian frame amplification, 180–84
  military necessity presumptions, 179–80
  military practice and legalities, 178–79
  military utility and counterframe deflation, 185–88

Non-international conflicts, laws and, 94–99, 102, 205, 230
Normative beliefs, defined, 5
Normative beliefs, evolution of changes in, 64–86
  Bosnia and Kosovo campaigns and, 64, 78–80, 81, 82
  drone strikes in Pakistan and, 81, 84–85, 86
  explanations for, 65–68, 85
  Gulf War and, 64, 75–78, 82
  Iraq and Afghanistan, post-9/11, 65, 80–85, 86
  nuclear weapons and, 253n1
  Vietnam War and, 64, 68–75, 86
Norm entrepreneurs, 5
Norms, governing aerial bombing
  focus on U.S. as standard-setter, 1–5, 6–8
  historical and theoretical perspectives, 8–13, 23
  legal restrictions interpreted and criticized, 13–20, 23–24
  new norm construction, 20 23, 24
  use of terms, 5–6
Norms, observations about debates regarding, 109–30
  counterinsurgencies and, 120–25, 265n71
  history of blending of laws and innovative technologies, 111–14
  laws, treaties, and unintended consequences, 114–20
  military's serious treatment of moral issues, 126–30
  utility of force in achieving peace, 129
Norstad, Lauris, 49
Northern Ireland, 97–98
Norway, 176, 183–84, 187
Norwegian Defense Research Establishment, 184
Norwegian People's Aid, 184

Nuclear deterrence
  during Cold War, 11–12
  current attitudes toward, 253n1
Nuremberg tribunals, 50, 93

Obama, Barack, drone use and, 65, 84–85, 200–201, 203–4, 207, 215, 219, 220–21, 235, 270n5
O'Connell, Mary Ellen, 19, 22–23, 24, 205, 224–36
O'Donnell, Emmett "Rosy," 51, 55
Off Target (Human Rights report), 164
Operational Law Handbooks, 136, 156
Operation Linebacker bombing campaign, in Vietnam, 71–74, 254nn27,29,30
Opinio iuris concept, 5–6
Oslo Declaration, 183–84, 186
Ottawa Convention, 118, 119, 236
Overy, Richard, 32, 45–46, 112

Pakistan, drones used in, 81, 84–85, 86, 124, 128, 192, 193, 194, 201, 203–4, 215, 220, 235
Pakistan Body Count, 84
Panetta, Leon, 200
Parks, W. Hayes, 13–14, 73–74
Partridge, Earle E., 53
Patton, George, 120
Pax Christi Netherlands, 180
Pershing, John, 33
Petraeus, David, 4, 121, 122, 123–24
Petrova, Margarita, 21–22, 24, 85, 175–90
Political-utility, changes in normative beliefs about civilian casualties and, 66–67
Polk, William R., 124–25
Portal, Charles, 41, 44, 246n65
Powell, Colin, 77, 126
Precision guided munitions (PGMs)
  drones and, 84
  in Gulf War, 76–77
  in Kosovo, 78–80, 256n70
  technological advances in, 112–13
  in Vietnam, 73, 74
  in World War II, 42
  see also Autonomous weapons; Drones
Prescriptive norms, defined, 5
Proportionality, principle of, 7, 124, 270nn5,7
  autonomous weapons and, 229, 232
  defined, 6
  Iraq and triangular balance of military advantage, force protection and civilian care, 16, 148, 149, 154–56, 275n20
  nongovernmental organizations and, 21, 22, 175–78, 181–82, 184, 188–90
  non-international armed conflict and, 95–96, 98–99
Public opinion
  drones and dissociation from violence committed on behalf of, 214–15, 220–22
  political-utility and civilian casualties, 66–67

RAND Corporation, 82, 121
"Regular war" (*guerre réglée*) theory, 209–11, 291n17
Regulations Respecting the Laws and Customs of War on Land (1907), 90, 93, 95
Remotely piloted aerial vehicles. *See* Drones
Reprieve, 194
Respatialization of war, drones and, 196–99, 207
Roberts, Adam, 292n41
Robinson, Eugene, 201
Rolling Thunder bombing campaign, in Vietnam, 13, 69–70, 73–74, 254n29
Roosevelt, Franklin D., 9, 10, 39, 49

INDEX

Royal Air Force (RAF), 34–35, 38, 40–41, 44, 49
Royal Flying Corps (RFC), 30
Royal Naval Air Service (RNAS), 30
Rumsfeld, Donald, 81
Russia, 7, 18, 99, 100, 101, 102, 237n4. *See also* Soviet Union
Rwanda, 96–97
Ryan, Klem, 22–23, 24, 198, 207–23

Scahill, Jeremy, 256n62
Scarry, Elaine, 203, 204
Schaffer, Ronald, 10
Schell, Jonathan, 201
Schelling, Thomas, 292n35
Schmitt, Carl, 211, 214, 219, 291n14
Schmitt, Michael N., 155, 156, 228, 296n30
Schools, effective contributions and military advantage, 135
Schwarzkopf, Norman, 77
"Sequencing," of military targets and goals, 142
Serbia, 2, 15, 78–80, 98, 138
Sewall, Susan, 170
Shapiro, Jeremy, 124
Shelton, Henry, 80
Sherman, William C., 34, 38
Sherry, Michael, 36
Short, Michael, 79
Shue, Henry, 15–16, 21, 24, 145–57
Signature strikes, 215
Singer, Peter, 127, 204, 217, 218, 295n25
Sino-Japanese War, 38
Sinuiju, Korea, bombing of, 52–55
Smith, Rupert, 290n7
Smuts, Jan Christian, 31
Snow, David A., 178
Social constructivism, 21
Somalia, 193, 194, 215
Soviet Union, 11–12, 50, 59, 117. *See also* Russia

Spaatz, Carl, 44
Spaight, J. M., 36
Spanish Civil War, 35, 38, 92
Speer, Albert, 42
Stalin, Josef, 117
Stars and Stripes, 197–98
Stimson, Henry L., 10, 49
St. Petersburg Declaration (1868), 89, 91, 115–16, 175
Strasser, Peter, 31
Stratemeyer, George E., 51–54
Strawser, Bradley Jay, 204, 221–22
Submariners, psychology and, 217–18, 292n48
Sufficiency, logic of, 17–18, 132, 141–43
  logic of efficiency and, 143–44
Suicide bombing, drones and ethics of, 202
On Suicide Bombing (Asad), 202
Summa Theologiae (Aquinas), 210
Sun Tzu, 116
Supreme Court, of U.S., 129
Swift, Christopher, 124
Sykes, Frederick, 244n26
Syria, 7, 129, 193

Tadić, Dusko, 97
Taliban, U.S. power and "barbarism," 167–70
Targets. *See* Military targets
Taylor, Frederick, 111–12
Technological determinism, of Douhet, 35–36
Thermobaric weapons, 119
Thomas, Ward, 79
Thompson, Hugh, 159
Thurnher, Jeffrey S., 228, 296n30
Time-sensitive targeting, 269n2
Tiverton, Lord, 32, 34
Tokyo, bombing of, 10, 44, 49
Trenchard, Hugh, 8–9, 32, 33, 34–35, 36, 38

Truman, Harry
  bombing of Hiroshima and Nagasaki and, 10–11, 49, 161, 162
  Korean War and, 51
Turkey, 98
Turse, Nick, 205

UNICEF, report on Iraq, 146–47, 153
United Nations
  Article 51 and self-defense, 229, 296n34
  Assistance Mission in Afghanistan (UNAMA), 19, 168, 280nn36,44
  human rights laws and, 93
  International Criminal Tribunals for the Former Yugoslavia and Rwanda, 96–97
  UNHRC report on targeted killings, 217
United States
  interwar years, 33–34
  nuclear deterrence and, 11–12
  as standard setter for norms, 1–5, 6–8
  World War I and, 32–33
  World War II and, 42–46
United States, power and strategy and
  changes in bombing practices, 158–71
  bias against insurgents, 167–70
  citizens' moral repugnance at previous wars, 159–60
  foreign policy effects, 158–59
  unjustified civilians deaths due to, 161–67
Universal Declaration of Human Rights, 65, 93
Unmanned aerial vehicles (UAVs). *See* Drones
U.S. Air Corps, 34, 38–39
U.S. Air Force, 2011 psychology report of, 218
U.S. Air Service, 34

U.S Army, field manuals of land warfare, 59, 61
U.S. Army/Marine Corps Counterinsurgency Field Manual, 19–20, 83, 161, 166
USA Today, 113

Vance, Cyrus, 160
Vattel, Emerich de, 22, 210–11, 213
Vietnam War, 12, 13, 179
  changes in normative beliefs and, 64, 68–75, 86
  limited civilian casualties and, 62, 63
  moral repugnance at U.S. violence in, 11, 159–60, 161
Virilio, Paul, 199
Vitoria Institute, 170

Wallach, Wendell, 225, 293n1
Walzer, Michael, 275n20
On War (Clausewitz), 291n21
Warden, John, 76
War Powers Act (1973), 220–21
War-sustaining capability, targets and, 135–36
Washington Naval Conference (1922), 36
Waters, Roger, 196
Waxman, Matthew, 232
"Weapons That Kill Civilians" (*New England Journal of Medicine*), 122
Weatherington, Dyke, 227
West, Francis J. "Bing," 125
Wever, Walther, 33
World Health Organization (WHO), report on Iraq, 146–47
Williams, Neil, 195
Williams, Jody, 201
Woodward, Bob, 85
World Disarmament Conference (1932), 9–10, 37

World Summit (2005), 230
World War I
  early laws of armed conflict and, 90
  evolution of role of aircraft and of bombing doctrine, 30–33
  morale and, 14
World War II, 2, 8, 9–11, 67, 68
  civilian deaths and casualties during, 28, 242n3
  early laws of armed conflict and, 92–93
  evolution of role of aircraft and of bombing doctrine, 39–46
  expanded zone of conflict and, 208–9
  morale and, 14–15
World Wildlife Federation, drones and, 205
Wright, Evan, 159–60, 163–64

Yemen, drones used in, 1, 84, 85, 193, 194, 203–4, 215, 220, 235
Yugoslavia
  Final Report to the Prosecutor by the Committee Established to Review the NATO Bombing Campaign against, 175–76
  International Criminal Tribunal for, 96
  NATO mission in Kosovo and Bosnia and, 78, 79, 80, 138, 140
  non-international armed conflict and, 96–97

Zardari, Asif Ali, 85
Zarqawi, Abu Musab al-, 113